COMPREHENSIVE BIOCHEMISTRY

ELSEVIER PUBLISHING COMPANY
335 Jan van Galenstraat, P.O. Box 211, Amsterdam, The Netherlands

ELSEVIER PUBLISHING COMPANY LIMITED
Barking, Essex, England

AMERICAN ELSEVIER PUBLISHING COMPANY, INC.
52 Vanderbilt Avenue, New York, N.Y. 10017

Library of Congress Card Number 62–10359
Standard Book Number 444–40695–6

With 48 illustrations and 34 tables

COPYRIGHT © 1970 BY ELSEVIER PUBLISHING COMPANY, AMSTERDAM

ALL RIGHTS RESERVED. NO PART OF THIS PUBLICATION MAY BE REPRODUCED, STORED IN A RETRIEVAL SYSTEM, OR TRANSMITTED IN ANY FORM OR BY ANY MEANS, ELECTRONIC, MECHANICAL, PHOTOCOPYING, RECORDING, OR OTHERWISE, WITHOUT THE PRIOR WRITTEN PERMISSION OF THE PUBLISHER,
ELSEVIER PUBLISHING COMPANY, JAN VAN GALENSTRAAT 335, AMSTERDAM

PRINTED IN THE NETHERLANDS

COMPREHENSIVE BIOCHEMISTRY

ADVISORY BOARD

Sir RUDOLF A. PETERS, M.C., M.D., D.Sc., F.R.S.
Emeritus Professor of Biochemistry, Oxford; Department of Biochemistry, Cambridge
Chairman

C.F. CORI, M.D., D.Sc.
Professor of Biochemistry, Massachusetts General Hospital, Boston, Mass.

J.N. DAVIDSON, D.Sc., M.D., F.R.S.
Professor of Biochemistry, The University of Glasgow, Scotland

E.F. GALE, D.Sc., F.R.S.
Professor of Chemical Microbiology, University of Cambridge

A. BAIRD HASTINGS, B.Sc., Ph.D., D.Sc.
Director of Biochemistry Division,
Scripps Clinic and Research Foundation, La Jolla, Calif.

E. LEDERER, Ph.D., D.Sc.
Professor of Biochemistry, Faculty of Science, University of Paris

F. LYNEN
Max Planck Institute for Cell Chemistry, Munich

R. NICOLAYSEN, M.D.
Professor of Nutrition Research, University of Oslo

S. OCHOA, B.A., M.D., Hon. LL.D., Hon. D.Sc.
Professor of Biochemistry, New York University School of Medicine, New York, N.Y.

J. ROCHE, D.Sc.
Professor of General and Comparative Biochemistry, Collège de France, Paris

KENNETH V. THIMANN
Professor of Biology, University of California, Santa Cruz, Calif.

A. W. K. TISELIUS, D.Sc., For.F.R.S.
Professor of Biochemistry, Institute of Biochemistry, Uppsala, Sweden

F. G. YOUNG, M.A., D.Sc., F.R.S.
Professor of Biochemistry, University of Cambridge

COMPREHENSIVE BIOCHEMISTRY

SECTION I (VOLUMES 1–4)
PHYSICO-CHEMICAL AND ORGANIC ASPECTS
OF BIOCHEMISTRY

SECTION II (VOLUMES 5–11)
CHEMISTRY OF BIOLOGICAL COMPOUNDS

SECTION III (VOLUMES 12–16)
BIOCHEMICAL REACTION MECHANISMS

SECTION IV (VOLUMES 17–21)
METABOLISM

SECTION V (VOLUMES 22–29)
CHEMICAL BIOLOGY

HISTORY OF BIOCHEMISTRY (VOLUME 30)
GENERAL INDEX (VOLUME 31)

COMPREHENSIVE BIOCHEMISTRY

EDITED BY

MARCEL FLORKIN

Professor of Biochemistry, University of Liège (Belgium)

AND

ELMER H. STOTZ

Professor of Biochemistry, University of Rochester, School of Medicine and Dentistry, Rochester, N.Y. (U.S.A.)

VOLUME 18

LIPID METABOLISM

ELSEVIER PUBLISHING COMPANY
AMSTERDAM · LONDON · NEW YORK
1970

CONTRIBUTORS TO THIS VOLUME

RUBIN BRESSLER, M.D.
Professor of Medicine and Pharmacology, Duke University Medical Center,
Durham, N.C. 27706 (U.S.A.)

SIDNEY S. CHERNICK, A.B., M.A., Ph.D.
Scientist Director, USPHS, Section on Endocrinology, Laboratory of Nutrition and Endocrinology, National Institute of Arthritis and Metabolic Diseases, National Institutes of Health, Bethesda, Md. 20014 (U.S.A.)

A. N. DAVISON, B. Pharm., B.Sc., Ph.D., D.Sc., F.P.S.
Professor of Biochemistry, Biochemistry Department, Charing Cross Hospital Medical School, 62–65, Chandos Place, London, W.C.2 (Great Britain)

JOHN M. JOHNSTON, Ph.D.
Professor of Biochemistry, Biochemistry Department, University of Texas, Southwestern Medical School at Dallas, 5323 Harry Hines Boulevard, Dallas, Texas 75235 (U.S.A.)

G. V. MARINETTI, B.Sc., Ph.D.
Professor of Biochemistry, University of Rochester School of Medicine and Dentistry, 260 Crittenden Boulevard, Rochester, N.Y. 14620 (U.S.A.)

WILLIAM M. O'LEARY, Ph.D.
Associate Professor of Microbiology, Department of Microbiology, Cornell University Medical College, 1300 York Avenue, New York, N.Y. 10021 (U.S.A.)

D. S. ROBINSON, Ph.D.
External Staff of the Medical Research Council, Department of Biochemistry, University of Oxford, South Parks Road, Oxford OX1 3QU (Great Britain)

ROBERT O. SCOW, A.B., M.A., M.D.
Chief, Section on Endocrinology, Laboratory of Nutrition and Endocrinology, National Institute of Arthritis and Metabolic Diseases, National Institutes of Health, Bethesda, Md. 20014 (U.S.A.)

PAUL K. STUMPF, B.A., Ph.D.
Professor of Biochemistry, Department of Biochemistry and Biophysics, University of California, Davis, Calif. 95616 (U.S.A.)

LARS SVENNERHOLM
Associate Professor of Neurochemistry, Head of the Department for Neurochemistry, Psychiatric Research Centre; Chief Physician, Neurochemical Service, Lillhagen and St. Jörgen Hospitals, Fack, 400 33 Göteborg (Sweden)

GUY A. THOMPSON Jr., Ph.D.
Associate Professor, Department of Botany, University of Texas, Austin, Texas 78712 (U.S.A.)

GENERAL PREFACE

The Editors are keenly aware that the literature of Biochemistry is already very large, in fact so widespread that it is increasingly difficult to assemble the most pertinent material in a given area. Beyond the ordinary textbook the subject matter of the rapidly expanding knowledge of biochemistry is spread among innumerable journals, monographs, and series of reviews. The Editors believe that there is a real place for an advanced treatise in biochemistry which assembles the principal areas of the subject in a single set of books.

It would be ideal if an individual or small group of biochemists could produce such an advanced treatise, and within the time to keep reasonably abreast of rapid advances, but this is at least difficult if not impossible. Instead, the Editors with the advice of the Advisory Board, have assembled what they consider the best possible sequence of chapters written by competent authors; they must take the responsibility for inevitable gaps of subject matter and duplication which may result from this procedure.

Most evident to the modern biochemist, apart from the body of knowledge of the chemistry and metabolism of biological substances, is the extent to which he must draw from recent concepts of physical and organic chemistry, and in turn project into the vast field of biology. Thus in the organization of Comprehensive Biochemistry, the middle three sections, Chemistry of Biological Compounds, Biochemical Reaction Mechanisms, and Metabolism may be considered classical biochemistry, while the first and last sections provide selected material on the origins and projections of the subject.

It is hoped that sub-division of the sections into bound volumes will not only be convenient, but will find favour among students concerned with specialized areas, and will permit easier future revisions of the individual volumes. Toward the latter end particularly, the Editors will welcome all comments in their effort to produce a useful and efficient source of biochemical knowledge.

Liège/Rochester

M. FLORKIN
E. H. STOTZ

PREFACE TO SECTION IV

(VOLUMES 17–21)

Metabolism in its broadest context may be regarded as the most dynamic aspect of biochemistry, yet depends entirely for its advances on progress in the knowledge of the structure of natural compounds, structure–function relationships in enzymes, bioenergetics, and cytochemistry, Approaches to the study of metabolism range from whole organism studies, with a limited possibility of revealing mechanisms, to cytochemical or even purified enzyme systems, sometimes with little attention to physiological conditions. Yet all approaches broaden our understanding of metabolism, and all of them may be recognized in the volumes assembled in Section IV on *Metabolism*. It is not unexpected, then, that previous sections of *Comprehensive Biochemistry* actually deal with some aspects under the broad heading of *Metabolism* and that the succeeding Section V on *Chemical Biology* will certainly draw heavily on a basic understanding of metabolism. Nevertheless Section IV attempts to bring together the broad outlines of the metabolism of amino acids, proteins, carbohydrates, lipids, and their derived products. The currently rapid advances in feed-back, hormonal, and genetic control of metabolism make it particularly difficult that these volumes be current, but the authors, editors, and publishers have made all possible efforts to include the most recent advances.

M. FLORKIN
Liège/Rochester
E. H. STOTZ

The chapters by JOHN M. LOWENSTEIN on *Pyruvate Oxidation* and *The Citric Acid Cycle*, earlier delayed for appearance in this volume were still not available as Volume 18 went to press. Also, the chapter by SALIH WAKIL on *Fatty Acid Oxidation and Synthesis* scheduled for this volume was not available. In the interests of early publication of those manuscripts available, the chapters referred to above will be published later as a supplementary volume in the Metabolism Section.

CONTENTS

VOLUME 18

LIPID METABOLISM

General Preface vii
Preface to Section IV viii

Chapter I. Assimilation, Distribution and Storage

Section A. Intestinal Absorption of Fats

by John M. Johnston

1. Introduction 1
2. Lumen phase 2
3. Penetration phase 5
4. Intracellular metabolism 9
   a. The monoglyceride pathway 11
   b. Protein synthesis and chylomicron formation 13
5. Absorption of cholesterol and phospholipids 14
   a. Cholesterol absorption 14
   b. Phospholipid absorption 15
6. Summary 16

Acknowledgement 16

References 17

Chapter I. Assimilation, Distribution and Storage

Section B. Mobilization, Transport and Utilization of Free Fatty Acids

by Robert O. Scow and Sidney S. Chernick, p. 19

1. Mobilization of FFA 20
   a. General aspects 20
   b. Effect of fasting 21
   c. Effect of diabetes and insulin 24
   d. Effect of sympathetic nervous system 29
   e. Effect of lipolytic hormones 30
      (i) Fast-acting lipolytic hormones, 31 – (ii) Slow-acting lipolytic hormones, 32 –
   f. Lipolytic enzymes in adipose tissue 33
2. Transport of FFA 34

3. Utilization of FFA 37
   a. Liver 38
   b. Kidney 40
   c. Skeletal muscle 40
   d. Heart 41
   e. Other tissues 42
      (i) Lung, 42 – (ii) Brain, 43 – (iii) Cornea, 43 – (iv) Adrenal, 43 – (v) Blood vessels, 43 –
4. Conclusions 43

Acknowledgement 44

References 45

Chapter I. Assimilation, Distribution and Storage

Section C. The Function of the Plasma Triglycerides in Fatty Acid Transport

by D. S. ROBINSON

1. Introduction 51
   a. The plasma triglycerides 51
   b. Triglyceride fatty acid transport 53
      (i) In the fasting state, 53 – (ii) Under conditions of enhanced fatty acid mobilization, 53 – (iii) In the fed state, 54 –
   c. The directive function of clearing factor lipase 55
2. The plasma lipoproteins 56
   a. The high- and low-density lipoproteins (HDL and LDL) 56
   b. The very-low-density lipoproteins (VLDL) 57
   c. The chylomicrons 59
      (i) In the thoracic duct, 59 – (ii) In the plasma, 60 –
   d. Formation and release of very-low-density lipoproteins (VLDL) . . 62
   e. Formation and release of chylomicrons 68
3. Entry of triglycerides into the blood 69
   a. From the intestine 69
   b. From the liver 71
      (i) In the fasting state, 75 – (ii) Under conditions of enhanced fatty acid mobilization, 76 – (iii) In the fed state, 78 – (iv) In various special conditions, 80 –
4. Removal of triglyceride fatty acids from the blood 81
   a. The mechanism of removal 82
      (i) By the extrahepatic tissues, 82 – (ii) By the liver, 84 –
   b. The role of particular extrahepatic tissues in triglyceride fatty acid removal . . 85
   c. The role of clearing factor lipase in triglyceride fatty acid removal 87
   d. Localization and states of clearing factor lipase in the tissues 93
   e. The regulation of tissue clearing factor lipase activity 95
   f. Tissue clearing factor lipase activity as a determinant of the plasma triglyceride concentration 97

(*i*) In different physiological situations, 98 – (*ii*) In different pathological conditions, 99 –
5. The fate of plasma triglyceride fatty acids in the extrahepatic tissues 102
   a. Oxidation 102
   b. Replenishment of adipose-tissue stores and formation of milk triglycerides . . 103
6. Concluding remarks 104

References 105

Chapter II. Biosynthesis of Triglycerides

by G. V. Marinetti

1. Introduction 117
2. Digestion, absorption and transport of triglycerides 121
3. Activation of fatty acids 122
4. Biosynthesis of triglycerides in liver 123
5. Biosynthesis of triglycerides in adipose tissue 134
6. Biosynthesis of triglycerides in intestine 137
7. Biosynthesis of mixed ester alkenyl or ester alkanyl glycerides 140
8. Control of triglyceride synthesis 140
9. Fatty livers 148
10. Phytanic acid-containing triglycerides 148
11. Hyperlipogenesis — Obesity 149
12. Problems in the study of lipid metabolism 150

Addendum 151

References 153

Chapter III. Phospholipid Metabolism

by Guy A. Thompson Jr.

1. Introduction 157
2. Biosynthesis 158
   a. Formation of phosphatidic acid 158
   b. Biosynthetic pathways utilizing phosphatidic acid as a precursor 161
      (*i*) Formation of phosphatidylcholine, 161 – (*ii*) Formation of phosphatidylethanolamine, 163 – (*iii*) Formation of phosphoinositides, 164 – (*iv*) Formation of phosphatidylserine, 166 – (*v*) Formation of phosphatidylglycerol, 166 – (*vi*) Formation of cardiolipin, 167 –
   c. Interconversions of phospholipids 168
   d. Formation of plasmalogens and glyceryl ether phospholipids 169
   e. Formation of phosphorus-containing sphingolipids 171
      (*i*) Formation of sphingosine, 171 – (*ii*) Formation of sphingomyelin, 173 – (*iii*) Analogs of sphingomyelin, 174 –
   f. Formation of lipids containing the carbon–phosphorus bond 174
   g. Summary 175

3. Catabolism 176
   a. Introduction 176
   b. Enzymes removing fatty acyl groups 176
      (*i*) Phospholipase A, 176 – (*ii*) Lysophospholipase, 178 – (*iii*) Enzymes catalyzing the deacylation of sphingolipids, 179 –
   c. Enzymes cleaving the phosphate–glycerol ester linkage 180
      (*i*) Phospholipase C, 180 – (*ii*) Phosphatidate phosphohydrolase, 181 –
   d. Enzymes cleaving the phosphate–base bond 182
   e. Phosphoinositide phosphomonoesterases 182
   f. Enzymes catalyzing the cleavage of ether bonds 183
      (*i*) Degradation of the vinylic ether linkage of plasmalogens, 183 – (*ii*) Degradation of the ether linkage of glyceryl ether phospholipids, 183 –
   g. Enzymes catalyzing sphingosine degradation 185
   h. Summary 186
4. Exchange processes 187
   a. Mechanisms for the turnover of specific phospholipid components 187
      (*i*) Acyl groups, 187 – (*ii*) Nitrogen bases, 191 –
   b. Exchange of intact molecules 191
   c. Summary and conclusions 192
5. Rates of lipid metabolism *in vivo* 193
   a. Introduction 193
   b. Rates of lipid metabolism in non-growing tissues 193
   c. Rates of lipid metabolism in growing tissues 194
   d. Possible physiological significance of phospholipid turnover 194
6. The role of lipid metabolism in the process of membrane fabrication 196

References 197

Chapter IV. Ganglioside Metabolism

by LARS SVENNERHOLM

1. Introduction 201
   a. Chemical structure and nomenclature 201
   b. Topographical distribution of gangliosides 204
2. Biosynthesis of gangliosides 207
   a. Biosynthesis of ceramide 207
   b. Biosynthesis of glucosylceramide and lactosylceramide 208
   c. CMP–NAN: ganglioside sialosyltransferases 210
   d. Elongation of the carbohydrate chain of ganglioside 213
      (*i*) UDP–*N*-acetylgalactosamine:monosialosyllactosylceramide *N*-acetylgalactosaminosyltransferase, 213 – (*ii*) UDP–galactose:monosialosyl–*N*-triglycosylceramide galactosyltransferase, 213 –
   e. Incorporation of labeled precursors *in vivo* 214
   f. Metabolic pathways for the biosynthesis 215
3. Biodegradation 218
   a. Neuraminidases 218
   b. Glycosidases 221
      (*i*) β-Galactosidases, 222 – (*ii*) β-Glucosidase, 223 – (*iii*) β-*N*-Acetylhexosaminidase, 223 –

4. Concluding remarks	224
References	226

Chapter V. Bacterial Lipid Metabolism
by WILLIAM M. O'LEARY

1. Introduction	229
2. Unique aspects of bacterial lipids	231
3. Biosynthesis of fatty acids	233
a. Saturated straight-chain fatty acids	233
b. Unsaturated fatty acids	237
c. Cyclopropane fatty acids	241
d. Branched-chain fatty acids	244
e. Hydroxy fatty acids	246
f. Fatty acids peculiar to mycobacteria and corynebacteria	246
4. Biosynthesis of complex lipids	248
a. Phospholipids	250
b. Glycolipids	250
c. Lipopolysaccharides	252
5. Cellular distribution of lipids	255
6. The functions of bacterial lipids	256
7. The effects of exogenous lipids on bacterial growth	259
8. Coda	261
References	263

Chapter VI. Fatty Acid Metabolism in Plant Tissues
by P. K. STUMPF

1. Introduction	265
2. Oxidative systems	266
a. Lipoxidase	266
b. α-Oxidation	269
c. β-Oxidation	271
d. Modified β-oxidation	273
3. Biosynthesis	275
a. General considerations and properties	275
b. Photobiosynthesis	282
c. Developmental aspects	285
4. Unsaturation	286
a. Mono- and dienoic acids	286
b. Hydroxylation and epoxy acid synthesis	288
5. Waxes	289
References	291

Chapter VII. Lipid Metabolism in Nervous Tissue
by A. N. Davison

1. Introduction	293
2. Neural lipid biochemistry	293
3. Cerebrosides and sulphatides	296
a. General properties	296
b. Estimation and separation from nervous tissue	297
c. Galactolipid metabolism within the nervous system	298
(*i*) Biosynthesis of cerebroside and sulphatide, 298 – (*ii*) Catabolism of cerebroside and sulphatide, 299 –	
4. Gangliosides	300
Metabolism of gangliosides	301
5. Cholesterol	302
a. Separation and analysis of sterols	302
b. Biosynthesis of cholesterol	302
(*i*) *In vitro*, 302 – (*ii*) *In vivo*, 305 –	
c. The catabolism of cholesterol in the nervous system	306
6. Neural phospholipids	308
a. Properties	309
b. Biosynthesis of phosphatides	309
c. Sphingomyelin	311
d. Inositol phospholipids and phosphatidic acids	311
7. Lipid metabolism in relation to anatomical structure	312
a. Lipid metabolism and the blood–brain barrier	315
b. Lipid metabolism at the cellular level	317
c. Lipid metabolism of subcellular structures	318
d. Lipid metabolism of the myelin sheath	320
8. Physiological significance of neural lipid metabolism	322
9. Conclusion	323
References	325

Chapter VIII. Fatty Acid Oxidation
by Rubin Bressler

1. Introduction	331
2. Fatty acid oxidation and carnitine	332
a. Mitochondrial compartmentalization and the role of carnitine in long-chain fatty acid oxidation	333
b. Long-chain acyl-CoA–carnitine acyltransferase (LCAT)	337
c. Acetyl-CoA: carnitine-acetyltransferase and short-chain fatty acid oxidation	339
d. The quantitative significance of carnitine in various other disease states	341
3. Fatty acid oxidation and gluconeogenesis	342
4. Ketogenesis	351
5. Glucose–fatty acid interactions	353
References	355
Subject Index	**361**

COMPREHENSIVE BIOCHEMISTRY

Section I—Physico-Chemical and Organic Aspects of Biochemistry
Volume 1.   Atomic and molecular structure
Volume 2.   Organic and physical chemistry
Volume 3.   Methods for the study of molecules
Volume 4.   Separation methods

Section II—Chemistry of Biological Compounds
Volume 5.   Carbohydrates
Volume 6.   Lipids — Amino acids and related compounds
Volume 7.   Proteins (Part 1)
Volume 8.   Proteins (Part 2) and Nucleic acids
Volume 9.   Pyrrole pigments, isoprenoid compounds, phenolic plant constituents
Volume 10. Sterols, bile acids and steroids
Volume 11. Water-soluble vitamins, hormones, antibiotics

Section III—Biochemical Reaction Mechanisms
Volume 12. Enzymes — general considerations
Volume 13. (second revised edition). Enzyme nomenclature
Volume 14. Biological oxidations
Volume 15. Group-transfer reactions
Volume 16. Hydrolytic reactions; cobamide and biotin coenzymes

Section IV—Metabolism
Volume 17. Carbohydrate metabolism
Volume 18. Lipid metabolism
Volume 19. Metabolism of amino acids, proteins, purines, and pyrimidines
Volume 20. Metabolism of porphyrins, steroids, isoprenoids, flavonoids and fungal substances
Volume 21. Vitamin and inorganic metabolism

Section V—Chemical Biology
Volume 22. Bioenergetics
Volume 23. Cytochemistry
Volume 24. Biological information transfer. Viruses. Chemical immunology
Volume 25. Regulatory functions, membrane phenomena
Volume 26. Part A. Extracellular and supporting structures
Volume 26. Part B. Extracellular and supporting structures (continued)
Volume 26. Part C. Extracellular and supporting structures (continued)
Volume 27. Photobiology, ionizing radiations
Volume 28. Morphogenesis, differentiation and development
Volume 29. Comparative biochemistry, molecular evolution

Volume 30. History of biochemistry

Volume 31. General index

Chapter I

Assimilation, Distribution and Storage

Section A

Intestinal Absorption of Fats*

JOHN M. JOHNSTON

Department of Biochemistry, University of Texas, Southwestern Medical School, Dallas, Texas (U.S.A.)

1. Introduction

The mechanism of the intestinal absorption of lipids has been the subject of intense controversy for almost 100 years. The center of this controversy has focused on the chemical moiety which penetrates the intestinal mucosa cell. In the late 19th century two investigators, Munk and Pflüger set the stage for two conflicting theories of fat absorption. These two theories were referred to as the "particulate" and "lipolytic" theories, respectively. Briefly this disagreement focuses on the question as to the quantitative amount of triglycerides that are hydrolyzed prior to absorption. It has been recognized for many years that the major dietary lipids are triglycerides. Munk's hypothesis was one in which only a limited hydrolysis of ingested triglycerides occurred and primarily triglycerides in the form of fine emulsified particles penetrated the intestinal mucosa. These views were challenged by Pflüger who subscribed to the theory that ingested triglycerides must be completely hydrolyzed to fatty

---

* Some of the investigations reported have been supported by grants from the National Institutes of Health, U. S. Public Health Service AM-3108 and the Robert A. Welch Foundation, Houston, Texas (U.S.A.).

References p. 17

acids and glycerol prior to absorption. The forceful arguments put forth by Pflüger almost completely dominated the literature for the next 30 or 40 years*. During this period the contributions by Professor Verzár and colleagues strongly supported and extended the "lipolytic" concept[5]. Verzár and colleagues stressed the importance of the formation of a soluble fatty acid–bile complex in fat absorption and also suggested that α-glycerophosphate may be of importance in resynthesis of triglycerides. In 1938, Frazer[6] published a paper that reopened the question of particulate absorption. In summary, Frazer suggested that a part of the ingested triglycerides were hydrolyzed forming fatty acids, monoglycerides and diglycerides, which in combination with bile acids provided the necessary components needed to emulsify the remaining triglycerides. This complex, containing primarily triglycerides, was negatively charged and had a diameter which was less than 0.5 μ. The glycerides passed into the lymphatic circulation, and the fatty acids were partitioned into the portal system. Thus, this concept was referred to as the "partition theory", and was viewed as a modification of the "particulate theory"[7,8].

Parallel to the earlier biochemical investigation, were a series of morphological studies on the problem of fat absorption which in general supported the particulate theory of fat absorption. In the last decade, numerous biochemical and morphological investigations have been published. As an outgrowth of these studies, a fairly comprehensive understanding of the mechanism of fat absorption has evolved in which there is general agreement. The presently accepted mechanism represents a merger of the "lipolytic" and "particulate" theories.

Although the major area to be discussed will be the digestion and absorption of triglycerides, the subject of cholesterol and phospholipid absorption will be briefly referred to.

2. Lumen phase

Dietary triglycerides are not appreciably affected by any of the enzymatic processes involved in the gastrointestinal tract until the fat reaches the small intestine. At this point, the triglycerides are exposed to two secretions which are important in their digestion and absorption. The first of these is the bile of which the bile salts play a major role. Because of their surface-active prop-

* For a more complete discussion of the historical development of this subject, see the reviews of refs. 1–4.

erties, the bile salts emulsify the dietary triglyceride into particles with a diameter of approximately 1 μ. The second secretion is the pancreatic juice. The predominate component of this fluid which is involved in the digestion and absorption of triglycerides is the enzyme, pancreatic lipase (EC 3.1.1.3). The formation of lower glycerides and fatty acids by the action of pancreatic lipase on triglycerides was demonstrated a number of years ago[9]. Frazer and his colleagues[10,11] clearly pointed out the importance of the hydrolytic products in the suspension of the fats in the intestinal lumen. The specificity of the hydrolytic cleavage of triglycerides by pancreatic lipase was first demonstrated by Mattson and Beck[12] who showed that the action of this enzyme was to hydrolytically cleave the fatty acids present in the 1 and 3 positions of the triglyceride molecule resulting in the formation of fatty acids derived from the 1 and 3 positions and a 2-monoglyceride. Although the importance of pancreatic lipase in the digestion and absorption of fats has been recognized for many years, only recently has this enzyme been purified and its properties thoroughly studied, primarily by Desnuelle and his colleagues[13]. This laboratory demonstrated that pancreatic lipase acts at the interface of emulsion particles. Therefore, its activity can be increased by increasing the surface area. Although the purified enzyme demonstrates a marked specificity for the hydrolysis of the 1 and 3 positions of the triglyceride molecule, different rates of hydrolysis at these two positions can occur, depending on the fatty acid present[14]. As has been mentioned, in addition to pancreatic lipase, the bile salts are of primary importance in the emulsification of triglycerides. The first report as to their specific function at the enzymatic level was that of Borgström[15] who demonstrated that the addition of taurocholate to a preparation containing emulsified triglycerides and pancreatic lipase, displaced the pH optimum of this enzyme from approximately pH 6 to 8. The latter pH is the approximate pH of the intraluminary contents of the upper jejunum. In addition to the above functions, bile salts have recently been shown to play a fundamental role in the formation of the intraluminary micellar phase.

As an outgrowth of his studies on the clearing effect of pancreatic lipase on emulsions of triglycerides, Borgström suggested that the products of this enzymatic action, namely fatty acids and 2-monoglycerides, in combination with bile salts, form a micellar phase[16].

The importance of micelles in the absorptive process was clearly documented several years later in a series of important papers by Hofmann and Borgström[17-20], encompassing a spectrum from the physical properties of micellar solutions to discussions of the relations of micelles to the absorption

References p. 17

of fats in man. The following general concepts have emerged regarding the importance of micelles in the absorption of fats. Pancreatic lipase acting at the interface of triglyceride emulsions hydrolyzes the fatty acids present in the 1 and 3 positions. The resulting 2-monoglyceride and fatty acids in combination with bile salts form negatively charged polymolecular aggregates termed "micelles", with a diameter in the range of 50 Å. It has been established that only a very small quantity of di- and triglycerides are present in the micellar phase[19]. It is evident that in the conversion of triglycerides from the emulsion phase into its hydrolytic products, mainly monoglycerides and fatty acids in the micellar phase, the diameter of the particles has been reduced. In terms of volume, it can be estimated that one emulsion particle can form in the order of $1 \cdot 10^6$ micelles. Borgström[20] established that in the presence of pancreatic lipase and bile salts a similar equilibrium mixture of fatty acid, monoglyceride, diglyceride and triglyceride was obtained when the initial reaction mixture contained either triglyceride emulsions or fatty acids and 2-monoglycerides added to the system in a micellar suspension. These results provide a mechanism by which a continuous supply of micelles containing predominately monoglycerides and fatty acids is being presented to the intestine for absorption. As will be discussed in a later section, the absorbed monoglycerides and fatty acids from micellar dispersion are resynthesized into triglycerides in the intestinal mucosa.

Although the morphological studies concerned with the absorption of fats are presented in a separate section, certain spatial considerations with regard to the mucosal cell and micellar solutions should be considered. First of all, the size of the micelle (40–50 Å) is small enough to allow free access to the intermicrovillous spaces, which are approximately 500–1 000 Å. The development of the microvilli of the mucosal cell no doubt occurs because of the specialized function of this cell, which results in a tremendous increase in the absorptive surface. The formation of a micellar phase with particles of the dimension mentioned above takes full advantage of the increased surface area since these particles have free access to the entire absorptive surface. Secondly, the specificity of the absorptive sequence may in part reside in the ability of a substance to be incorporated into a micellar phase. An example is the very small quantity of di- and triglycerides absorbed by the mucosal cell and similarly, the very small quantity of these two components that are actually present in the micellar phase. The physico-chemical properties of micelles have been recently discussed by Bangham[21]. The relationships of the emulsion and micellar phases are depicted in Fig. 1 as taken from Borgström[22].

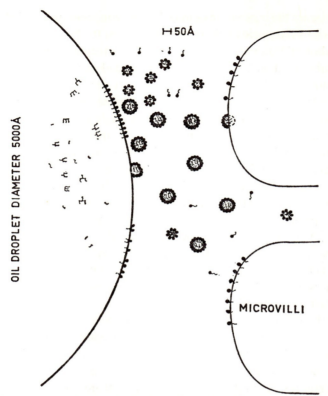

Fig. 1. Scheme representing the physical transformation of fats from emulsion to micellar phases and transportation into the brush border (•~ and ⊢ represent bile salts and triglycerides respectively). From Borgström[22].

3. Penetration phase

As has been discussed, one of the more controversial subjects with regard to the absorption of fats by the intestinal mucosa has been the chemical nature of the material which penetrates the intestinal mucosa. In consideration of this question several years ago, an investigation was carried out in which intestinal slices were incubated with micellar solutions containing labeled monoglycerides and fatty acids[23]. It was found that monoglycerides and fatty acids were rapidly taken up by the slices and converted into triglycerides *via* the reaction sequence to be discussed. The penetration phase was suggested to be energetically (in terms of ATP requirements) and enzymatically in-

References p. 17

dependent since the incubation of micellar solutions containing labeled monoglycerides and fatty acids with intestinal slices at 0°, heat-inactivated slices, and metabolically inhibited systems, resulted in uptake similar to that of slices incubated at 37°. In addition to intestinal slices, it was shown that labeled monoglycerides and fatty acids which were added in the form of micellar solutions were taken up to a high degree by isolated brush-border preparations. These studies suggest rather strongly that monoglycerides and fatty acids dispersed in a micellar physical state, are taken up by the intestinal mucosa and transformed into triglycerides. No conclusions were drawn from these studies as to whether or not the intact micelle, including the bile acids, was absorbed under these conditions, or simply the monoglycerides and fatty acids were released from the micellar physical state into a monomolecular dispersion prior to their uptake. Recent studies by Gordon and Kern[24] and Dietschy[25] would suggest that the intact micelle is being taken up by this tissue. Recently, Simmons et al.[26] have questioned the necessity of micellarization of derived fats as a prerequisite for absorption and suggested that the fatty acids were transferred to the absorptive cell in a monomolecular form. Earlier in vitro investigations employing everted sacs had established that the fatty acids bound to the albumin were capable of penetrating the limiting membrane of the intestinal mucosa and were enzymatically incorporated into triglycerides[27]. Although the question as to whether or not monoglycerides and fatty acids penetrate the mucosal cell in the form of micelles or as monomolecular dispersions or both is still to be answered; there can be no question as to the existence and importance of the micellar phase in the overall absorptive process.

For some time, both light- and electron-microscopic investigations on the digestion and absorption of fats have been at variance with the biochemical observations. Recently, a series of investigations have been reported in which the morphological and biochemical events are incorporated into a mechanism in which there is general agreement. Strauss[28] incubated micellar solutions containing monoglycerides, and ^{3}H-labeled fatty acids at 0° with intestinal slices and demonstrated by the use of autoradiographic and electron-microscopic techniques that the labeled fatty acids penetrated the mucosal cell wall below the terminal webb (Fig. 2). Although no dense osmiophilic granules were observed and the radioactivity was still present in the form of fatty acids, when the slices were removed from the micellar solution, washed and reincubated at 37° for short periods, large intracellular lipid droplets were evident which were similar to those observed in the mucosal cell during fat absorption. Parallel to the formation of the lipid particles was the incorporation of the free fatty acids into triglycerides. Recently, Dermer[29,30] has reported a similar uptake from micellar solutions by intestinal slices; however, it was concluded that a predominance of the absorbed lipids was present in the microvillus membrane when slices were incubated with micelles at 0°. Porter

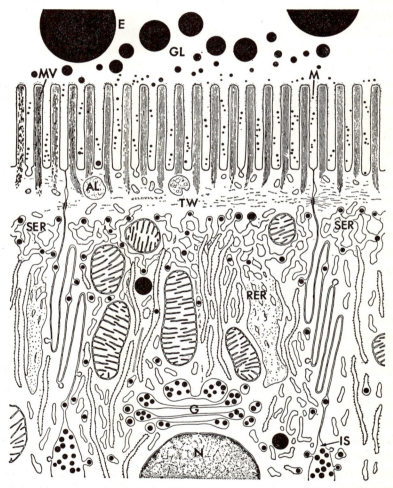

Fig. 2. A diagrammatical summary of the apical portion of an intestinal epithelial cell during fat absorption based on electron micrograph. GL, gut lumen; AL, apical lysosomes; TW = terminal web; SER = smooth endoplasmic reticulum; RER = rough endoplasmic reticulum; G, Golgi complex; N, nucleus; IS, intercellular space; E, emulsion particles; MV, microvilli; M, micelles.

References p. 17

and colleagues[31] have recently published a most comprehensive electromicrographic study on the absorption of fat. These authors, by the use of electron-opaque particles suspended in corn oil, were unable to demonstrate the uptake of the marker under conditions in which the absorption of fat was clearly evident. In addition, it was suggested that the smooth endoplasmic reticulum was the major site for the resynthesis of triglycerides and during the fat-fed state there was a conversion of rough to smooth endoplasmic reticulum. These results lend additional support to the hypothesis that fat is absorbed by selective diffusion of monoglycerides and fatty acids from micelles rather than by pinocytosis[32] of unhydrolyzed triglycerides. The results reported by Porter et al.[31] as they are incorporated into our present understanding of fat absorption, have been very nicely illustrated by this laboratory as is shown in Fig. 2.

Certain objections to the concept of pinocytosis being an important consideration in the absorption of fats had previously been expressed by Sjöstrand[33] on the basis that the membranes which bound the globular intracellular fat were not of the plasma-membrane type.

In addition, a recent investigation of Sjöstrand and Borgström[34] has further substantiated these concepts by fractionating the intracellular organelles into several constitutive components related to fat absorption. They obtained a vesicule fraction which was rich in triglycerides and fatty acids and suggested that this fraction was the site of triglycerides synthesized.

The fundamental question still unsolved regarding the penetration of fat to the intestinal mucosa is whether the intact micelle or the constitutive monoglycerides and fatty acids penetrate the brush-border membrane. Furthermore, it is unclear how the absorbed monoglycerides and fatty acids reach the site of triglyceride biosynthesis in the endoplasmic reticulum. Although claims have been made to having seen particles whose size is the same as micelles attached and associated with the intercellular microvilli spaces[35], Dermer[29] has questioned these results. He suggests that the observed particles are lead precipitates which appear quite abundantly in lead-stained material. In addition, it was also reported that these artifacts were seen in the electron micrograph of non-fat-absorbing intestinal cells. Although numerous questions regarding the penetration of fats by the intestine are still to be answered, there is almost unanimous agreement that pinocytosis is not involved to any appreciable extent in this process.

The site of absorption of fats has been demonstrated in a number of species to be predominately in the upper jejunum. In addition, the enzymes responsible for the resynthesis of triglycerides from monoglycerides and fatty acids are present in a much higher concentration in this area of the gut. In contrast, bile salts have been shown to be actively absorbed in the lower ileum[36]. However, it has been suggested that bile salts do penetrate the cells of the upper jejunum by passive diffusion[37]. Since at present it is not clear whether or not the intact micelle penetrates the jejunal cell, and there is conflicting evidence with regard to the intracellular effects of bile salts in relation to fat absorption[38-40], further investigations will be necessary in order to clarify these questions.

4. Intracellular metabolism

Until the last 10 years, very little information was available concerning the types of lipids found in the intestinal mucosal cell, or more important, the enzymatic activity of this tissue. In the last few years, a sizeable number of papers have appeared describing the function of the mucosal cell in terms of enzymatic activity. These results have been of great help in the elucidation of the basic mechanisms involved in the digestion and absorption of fats. Most of the metabolic reactions of the intestinal mucosa with regard to fat absorption have centered around the enzymatic steps involved in the resynthesis of triglycerides. The occurrence of the α-glycerophosphate pathway for the synthesis of triglycerides in the intestinal mucosa was suggested in the late fifties[41]. This reaction sequence had been demonstrated by Weiss and coworkers[42,43] and Stein et al.[44] in other tissues. The enzymatic reactions which are related to fat absorption are shown in Fig. 3.

As can be seen from the figure, α-glycerophosphate is acylated forming the intermediate, phosphatidic acid, which in turn is dephosphorylated giving rise to the 1,2-diglycerides which are further acylated to form triglycerides (reactions 16, 18, 19). It was originally reported that free glycerol could not be utilized by the intestinal mucosal cell for the formation of α-glycerophosphate[45] since the enzyme, glycerol kinase (EC 2.7.1.30) (reaction 14), could not be demonstrated. This question was re-opened by Buchs and Favarger[46] who suggested that the intestinal mucosa was capable of utilizing free glycerol for glyceride biosynthesis. These observations were confirmed and further extended by the report of Saunders and Dawson[47] who demonstrated the oxidation and incorporation into glyceride-glycerol of free glycerol by segments of the small intestine. Furthermore, when the presence of the enzyme glycerol kinase was re-examined by two laboratories employing more refined

References p. 17

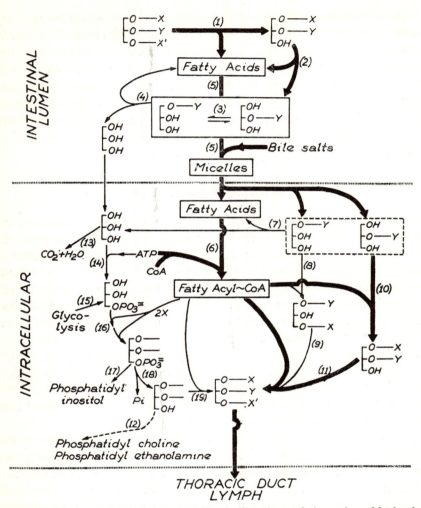

Fig. 3. The biochemical reactions involved in the digestion and absorption of fat by the intestine. Heavy arrows indicate the major pathway of triglyceride resynthesis (monoglyceride pathway). Note: *Via* this reaction, only the fatty acids present in the 1 and 3 positions exchange and the 1,2-diglycerides synthesized by the α-glycerophosphate pathway, and monoglycerides do not equilibrate. (-Y, -X, -X' and — denote fatty acids present in ester linkages.)

methods, this enzyme was shown to be present in the intestinal mucosa[48,49]. The quantitative significance of the incorporation of free glycerol into glyceride-glycerol is still not fully evaluated.

The activation of fatty acids to their CoA derivatives in this tissue by the enzyme referred to as either fatty acid: CoA ligase (AMP) (EC 6.2.1.3), acyl-CoA synthetase or fatty acid thiokinase (reaction 6) was shown by several laboratories[50-52]. Recent evidence would suggest that a functional portion of CoA, possibly of an acyl-carrier protein type, may function in the activation of fatty acids in this tissue[53]. Although it is generally conceded that most of the enzymes which are responsible for the synthesis of triglycerides *via* the α-glycerophosphate pathway are localized in the microsomal fraction, some peculiar properties of the enzyme L-α-phosphatidate phosphohydrolase (EC 3.1.3.4) (reaction 18) do exist. It has been reported that the supernatant fraction markedly stimulates incorporation of both glycerophosphate and fatty acids into the triglyceride moiety[54]. The major function of the supernatant fraction can be attributed to the presence of the enzyme L-α-phosphatidate phosphohydrolase[55,56]. The difference between the microsomal and supernatant phosphohydrolase resides in the substrate specificity. If exogenous phosphatidic acid is added, the microsomal enzyme shows the greatest enzymatic activity[57,58]. However, if biosynthetic phosphatidic acid bound to microsomes is employed as the substrate, the supernatant enzyme is the most active towards this substrate[55].

(a) The monoglyceride pathway (Reactions 6, 10, 11)

In 1898, Frank first related monoglycerides to the problem of fat absorption[59]. As mentioned previously, Artom and Reale[9] were the first to report that monoglycerides were present in the intestinal lumen. This was further amplified by the reports by Frazer and Sammons[10], and in the early 1950's, Reiser and Williams[60] and Skipski *et al.*[61], provided the first evidence for the absorption of monoglycerides. In 1960, Clark and Hübscher[50,62] provided the first direct evidence, at the enzyme level, for the biosynthesis of triglycerides from monoglycerides in the intestinal mucosa. These authors demonstrated that the addition of monoglycerides markedly stimulated the incorporation of [^{14}C]palmitic acid into higher glycerides under conditions in which the α-glycerophosphate pathway was inhibited. Senior and Isselbacher[63] further showed that [^{14}C]glycerol labeled 1-monopalmitin was incorporated into di- and triglycerides. The final confirmation for the existence of the monoglyceride pathway was provided by a double labeled experiment in which ^{3}H in the glycerol and ^{14}C in the fatty acid portion of the molecule were shown to be incorporated intact into di- and triglycerides[64]. Most of the earlier studies

References p. 17

were carried out employing the 1-monoglyceride isomer. Of more importance from a physiological standpoint, is the question as to the relative degree to which 2-monoglycerides were acylated, since this isomer is the product absorbed in the intestinal cell. It was found that 2-monoglycerides were preferred as a substrate[65-67]. Furthermore, it was clearly established that one enzyme, monoglyceride transacylase, accepts both the 1- and 2-monoglycerides. When the 1-isomer was employed as the substrate, it was acylated to give the 1,3-diglyceride intermediate[65-67].

Recently the enzymes involved in the monoglyceride pathway have been purified 70–100 fold from hamster intestinal mucosa[68]. During the process of purification, it was apparent that the enzymes acyl-CoA synthetase (EC 6.2.1.2, 3) (reaction 6), monoglyceride transacylase (reaction 10), and diglyceride transacylase (reaction 11) were purified simultaneously. Based on these findings, the complex has been termed "triglyceride synthetase"[68]. Recent evidence has suggested that these enzymes are confined to the membrane portion of the endoplasmic reticulum (C. M. Schiller and J. M. Johnston, unpublished observation). Sucrose-density-gradient studies suggested that the substrates and products remain enzyme-bound[68].

The intestinal mucosa offers the unique opportunity to study the α-glycerophosphate and monoglyceride pathways simultaneously. Both *in vivo*[69-71] and *in vitro*[55,72] studies suggested that the monoglyceride pathway is the major pathway for the synthesis of triglycerides.

As can be seen from Fig. 3, 1,2-diglycerides are intermediates in the synthesis of triglycerides by either the α-glycerophosphate or monoglyceride pathways. Since both reaction sequences occur primarily in the microsomal fraction, the question as to whether or not the 1,2-diglycerides synthesized by each of the pathways equilibrate was recently examined[73]. It was concluded from these investigations that the diglycerides formed by each of the two pathways do not equilibrate in the microsomes. These results were further substantiated by the observation that only the 1,2-diglycerides synthesized *via* the α-glycerophosphate pathway could serve as a precursor of phosphatidyl choline when CDP–choline was added to the incubation mixture (J. M. Johnston, unpublished results). This observation would, in part, explain the reports by numerous laboratories of the fatty acid composition of the phospholipids and triglycerides found in the thoracic duct lymph following the feeding of triglycerides. The lymphatic triglycerides are similar to the dietary triglycerides in the distribution of fatty acids between the 1, 3 and 2 positions since they are synthesized primarily *via* the monoglyceride pathway. On the other hand, phosphatidyl choline has been shown to be synthesized *via* the separate intermediates of the α-glycerophosphate pathway and, therefore, the fatty acid distribution between the 1 and 2 positions do not parallel the dietary fat.

(b) *Protein synthesis and chylomicron formation*

As was discussed previously, in the digestion and absorption of lipids, the components are absorbed *via* the thoracic duct lymph. Furthermore, the components are incorporated into a distinct particle termed the chylomicron. Although some minor disagreements have occurred as to the composition of these particles, in general, the chemical composition is approximately 86% triglycerides, 8.5% phospholipids, 3% cholesterol and cholesterol esters and 2% protein. In the previous sections, the microsomal synthesis of triglycerides and phospholipids has been discussed. The absorption of phospholipids and cholesterol will be briefly discussed in a later section. Therefore, of the chemical components of the chylomicron, only the protein has not been taken into consideration. Until recently, very little information was available with regard to the synthesis or characterization of protein associated with the chylomicron. Hatch *et al.*[74] have recently investigated the proteins associated with the chylomicron and attempted to relate these results to the solubility of intracellular proteins of the mucosal cell. A very interesting property of the chylomicron protein was reported. These proteins contained a predominate amount of amino acids with polar side-chains and demonstrated considerable solubility in organic solvents. This property would make them ideally suited for the formation of a lipoprotein complex necessary for coating the chylomicron. Isselbacher's laboratory has recently studied protein synthesis as it relates to fat absorption by the use of certain inhibitors of protein synthesis[75,76]. The pretreatment of rats with inhibitors of protein synthesis, such as puromycin, resulted in a massive accumulation of chemically identified triglycerides within the intestinal cell following the oral administration of triglycerides. The treated animals failed to develop a postprandial hyperlipemia which normally occurs in animals, and showed a decrease in the incorporation of a labeled amino acid into protein, and a decrease in the plasma β-lipoproteins. These results would strongly suggest the absolute requirement of protein synthesis for the formation of the chylomicrons. This same laboratory[76] has also related their findings to a rare hereditary disorder, acanthocytosis[77], which is characterized by decreased ability to synthesize

References p. 17

β-lipoproteins and a similar accumulation of large droplets of lipids within the intestinal mucosal cell. Definitive evidence regarding the mechanism of the formation of chylomicrons awaits future investigations. However, based on the intracellular locations for the biosynthesis of protein, triglycerides, phospholipids and other constituents required for the formation of chylomicrons, it would appear that these processes are very closely associated. The mechanism by which chylomicrons leave the mucosal cell is still not clear. It has been suggested[30] that chylomicrons are discharged into the extracellular space by means of the reversal of the phenomenon of pinocytosis. In addition, it has also been proposed that certain modifications of the released components in these intercellular spaces may occur[31]. From the recent observations of Salpeter and Zilversmit[78,79] concerning the formation of chylomicrons, the data would suggest that the surface coat of the chylomicron is acquired intracellularly. Further investigation is necessary in order to establish the mechanism of transfer of the constituents of the chylomicron from the intracellular to extracellular compartments.

5. Absorption of cholesterol and phospholipids

It has been previously mentioned that the final products of the absorptive process are the chylomicrons. Of the constituents present in the chylomicrons, only cholesterol and phospholipids have not been discussed. For the sake of completeness, the absorption of these two substancess will be briefly discussed in the following section.

For a complete and comprehensive discussion with regard to our present understanding of cholesterol absorption, the reader is referred to an excellent chapter by Treadwell and Vahouny[80]. The major points regarding this process are briefly discussed in the following paragraphs.

(a) Cholesterol absorption

The dietary and endogenous cholesterol, which includes the biliary cholesterol as well as the cholesterol derived from the desquamated cells, are completely mixed in the intestinal lumen. The cholesterol esters are hydrolyzed to free cholesterol and fatty acids by the pancreatic enzyme, cholesterol esterase, (EC 3.1.1.13), which is activated by trihydroxy bile acids. It has been convincingly established that only free cholesterol is absorbed. The cholesterol dissolves in the emulsion phase of the intestinal contents containing primarily triglycerides. As has been discussed in the absorption of fats, the triglycerides are hydrolyzed at the 1,3 positions to form fatty acids and 2-monoglycerides by the action of pancreatic lipase which in combination with bile salts form a micellar phase. The cholesterol distributes itself between the emulsion and micellar phases.

Present evidence would suggest that the cholesterol in the micelles is absorbed by passive diffusion. Upon entrance into the intracellular compartment of the mucosal cell, the cholesterol is mixed with the intracellular pool of cholesterol which has been synthesized by this cell. Between 80–90% of the cholesterol present in the lymph is in the esterified form. Therefore, prior to the inclusion of cholesterol into the chylomicron, a significant quantity of the intracellular cholesterol must be esterified. The enzyme responsible for the esterification reaction has also been referred to as cholesterol esterase. This enzyme utilizes fatty acid-CoA for the transacylation and is also activated by certain bile salts. Following the esterification, the cholesterol esters are incorporated into the chylomicron structure and leave the intestinal mucosa in that form. It has been suggested that the esterification of the cholesterol occurs at a latter stage of the formation of the chylomicron since cholesterol esters do not accumulate in the mucosal cell to any appreciable degree.

The rate of cholesterol penetration in the mucosal cell can be affected by a number of conditions such as the presence of triglycerides in the diet, or the amount of cholesterol being fed. Recently, Sylven and Borgström[81] have attempted to define some of the kinetic parameters of cholesterol absorption. In the rat, it was suggested that the rate-limiting step in the cholesterol-transport system was the transfer from the intracellular compartment of the mucosal cell to the lymph. A similar conclusion was reached with regard to the absorption of the triglycerides[23]. Although the general pattern of cholesterol absorption parallels to some degree the route of triglyceride absorption as outlined previously, in general the absorption of cholesterol occurs over a longer time period than the absorption of triglycerides.

(b) Phospholipid absorption

It has been recognized since the early work of Artom and Swanson[82] that at least some of the phospholipids may be absorbed without the necessity of complete hydrolytic breakdown. These findings were further corroborated by the investigations by Bloom et al.[83] and Blomstrand[84]. The problem with the early investigations was the fact that it was difficult to differentiate whether a portion of the phospholipids was absorbed intact or the phospholipid was partially hydrolyzed and the resulting hydrolytic products absorbed

References p. 17

without the necessity of complete hydrolysis. In addition, it is difficult to assess the absorptive patterns of the phospholipid fractions when this entire class of compounds is employed. Recently Scow et al.[85] have re-examined the question of phospholipid absorption employing phosphatidylcholine which has been labeled specifically with ^3H in the fatty acid at the one position, ^{14}C in the fatty acid in the two position, and ^{32}P in the phosphate of this molecule. The authors concluded from this investigation that the 1-lysophosphatidylcholine moiety could be absorbed as an intact unit. In general, similar conclusions have been reported by Nilsson[86]. The enzyme responsible for the cleavage of the fatty acid at the two position of phosphatidylcholine has been demonstrated in the pancreatic secretion. Although the lysophosphatidylcholine can be absorbed as an intact unit, it should be pointed out that a large proportion of the phosphatidylcholine is completely hydrolyzed during the digestive process. Although specific evidence regarding the absorptive sequences has been reported only for phosphatidylcholine, a similar mechanism may be involved in the absorption of other glycerophospholipids.

6. Summary

The biochemical mechanisms involved in the absorption from the intestine of dietary constituents have been the subject of numerous investigations over the past one hundred years. In the case of fat absorption, we now have a more comprehensive understanding of this process at the cellular and subcellular level than of any of the other major foodstuffs. This can probably be attributed to the fact that during the absorptive process, a chemical change occurs, which enabled investigators to study this process employing an additional tool. Based on the subcellular distribution of the enzymes responsible for triglyceride biosynthesis and the additional enzymes necessary for the formation of the other components of the chylomicron, the endoplasmic reticulum contains the necessary enzymes for the synthesis of the chylomicron, and the evidence would suggest that the synthesis occurs in an organized complex of the endoplasmic membrane. Furthermore, the mechanism of absorption, as outlined in this section, is consistent with both the recent morphological and biochemical investigations and concepts in which there is now general agreement.

ACKNOWLEDGEMENT

The assistance of Addie Uranga in the preparation of this manuscript is sincerely appreciated.

REFERENCES

1. J. M. JOHNSTON, *Handbook of Physiology*, Vol. 3, American Physiological Society, Washington, D. C., 1967, p. 1353.
2. J. M. JOHNSTON, *Advan. Lipid Res.*, 1 (1963) 105.
3. J. R. SENIOR, *J. Lipid Res.*, 5 (1964) 495.
4. G. CLÉMENT, *J. Physiol. (Paris)*, 56 (1964) 111.
5. F. VERZÁR AND E. J. MCDOUGALL, *Absorption from the Intestine*, Longmans, Green, London, 1936, p. 150.
6. A. C. FRAZER, *Analyst*, 63 (1938) 308.
7. A. C. FRAZER, *Physiol. Rev.*, 26 (1946) 103.
8. A. C. FRAZER, *Brit. Med. Bull.*, 14 (1958) 212.
9. C. ARTOM AND L. REALE, *Arch. Sci. Biol. (Bologna)*, 21 (1935) 368.
10. A. C. FRAZER AND H. G. SAMMONS, *Biochem. J.*, 36 (1945) 122.
11. A. C. FRAZER, J. H. SCHULMAN AND H. C. STEWART, *J. Physiol. (London)*, 103 (1944) 306.
12. F. H. MATTSON AND L. W. BECK, *J. Biol. Chem.*, 219 (1956) 129.
13. P. DESNUELLE, *Advan. Enzymol.*, 23 (1961) 129.
14. B. L. ENTRESSANGLES, L. PASERO, P. SAVARY, L. SARDA AND P. DESNUELLE, *Bull. Soc. Chim. Biol.*, 43 (1961) 581.
15. B. BORGSTRÖM, *Biochim. Biophys. Acta*, 13 (1954) 491.
16. B. BORGSTRÖM, *Biochemical Problems of Lipids*, Butterworth, London, 1955, p. 179.
17. A. F. HOFMANN, *The Enzymes of Lipid Metabolism*, Pergamon, Oxford, 1961, p. 158.
18. A. F. HOFMANN AND B. BORGSTRÖM, *Gastroenterology*, 50 (1966) 56.
19. A. F. HOFMANN AND B. BORGSTRÖM, *Federation Proc.*, 21 (1962) 43.
20. B. BORGSTRÖM, *J. Lipid Res.*, 5 (1964) 522.
21. A. D. BANGHAM, *Advan. Lipid Res.*, 1 (1963) 65.
22. B. BORGSTRÖM, *Metabolism and Physiological Significance of Lipids*, Wiley, New York, 1964, p. 221.
23. J. M. JOHNSTON AND B. BORGSTRÖM, *Biochim. Biophys. Acta*, 84 (1964) 412.
24. S. G. GORDON AND F. KERN JR., *Biochim. Biophys. Acta*, 52 (1968) 372.
25. N. C. SMALL AND J. M. DIETSCHY, *Gastroenterology*, 54 (1968) 168.
26. W. J. SIMMONDS, T. G. REDGRAVE AND R. L. S. WILLIX, *J. Clin. Invest.*, 47 (1968) 1015.
27. J. M. JOHNSTON, *J. Biol. Chem.*, 234 (1959) 1065.
28. E. W. STRAUSS, *Handbook of Physiology*, Vol. 3, American Physiological Society, Washington D. C., 1967, p. 1377.
29. G. B. DERMER, *J. Ultrastruct. Res.*, 20 (1967) 51.
30. G. B. DERMER, *J. Ultrastruct. Res.*, 22 (1968) 312.
31. R. R. CARDELL JR., S. BADENHAUSEN AND K. PORTER, *J. Cell Biol.*, 34 (1967) 123.
32. S. L. PALAY AND L. J. KARLIN, *J. Biophys. Biochem. Cytol.*, 5 (1959) 373.
33. F. S. SJÖSTRAND, in A. C. FRAZER (Ed.), *Biochemical Problems of Lipids*, Elsevier, Amsterdam, 1963, p. 91.
34. F. S. SJÖSTRAND AND B. BORGSTRÖM, *J. Ultrastruct. Res.*, 20 (1967) 140.
35. J. ROSTGAARD AND R. J. BARNETT, *Anat. Record*, 152 (1965) 325.
36. L. LACK AND I. M. WEINER, *Federation Proc.*, 22 (1963) 1334.
37. J. M. DIETSCHY, *J. Lipid Res.*, 9 (1968) 297.
38. A. M. DAWSON AND K. J. ISSELBACHER, *J. Clin. Invest.*, 39 (1960) 730.
39. P. R. HOLT, H. A. HAESSLER AND K. J. ISSELBACHER, *J. Clin. Invest.*, 42 (1963) 777.
40. J. L. POPE, T. M. PARKINSON AND J. A. OLSON, *Biochim. Biophys. Acta*, 130 (1966) 218.
41. J. M. JOHNSTON, *Proc. Soc. Exptl. Biol. Med.*, 100 (1959) 669.
42. S. B. WEISS AND E. P. KENNEDY, *J. Am. Chem. Soc.*, 78 (1956) 3550.
43. S. B. WEISS, E. P. KENNEDY AND J. Y. KIYASU, *J. Biol. Chem.*, 235 (1960) 40.

44 Y. STEIN, A. TIETZ AND B. SHAPIRO, *Biochim. Biophys. Acta*, 26 (1957) 286.
45 E. P. KENNEDY, *Proc. 5th Intern. Congr. Biochem. Moscow*, Vol. 7, Pergamon, Oxford, 1961, p. 113.
46 A. BUCHS AND R. REISER, *J. Biol. Chem.*, 234 (1959) 217.
47 D. R. SAUNDERS AND A. M. DAWSON, *Biochem. J.*, 82 (1962) 477.
48 B. CLARK AND G. HÜBSCHER, *Nature*, 195 (1962) 599.
49 H. A. HAESSLER AND K. J. ISSELBACHER, *Biochim. Biophys. Acta*, 73 (1963) 427.
50 B. CLARK AND G. HÜBSCHER, *Nature*, 185 (1960) 35.
51 B. CLARK AND G. HÜBSCHER, *Biochim. Biophys. Acta*, 46 (1961) 479.
52 A. M. DAWSON AND K. J. ISSELBACHER, *J. Clin. Invest.*, 39 (1960) 150.
53 G. A. RAO AND J. M. JOHNSTON, *Biochim. Biophys. Acta*, 144 (1967) 25.
54 D. N. BRINDLEY AND G. HÜBSCHER, *Biochim. Biophys. Acta*, 106 (1965) 495.
55 J. M. JOHNSTON, G. A. RAO, P. A. LOWE AND B. SWARTZ, *Lipids*, 2 (1967) 1.
56 G. HÜBSCHER, D. N. BRINDLEY, M. E. SMITH AND B. SEDQWICK, *Nature*, 216 (1967) 449.
57 J. M. JOHNSTON AND J. H. BEARDEN, *Biochim. Biophys. Acta*, 56 (1962) 365.
58 R. COLEMAN AND G. HÜBSCHER, *Biochim. Biophys. Acta*, 56 (1962) 479.
59 O. FRANK, *Z. Biol.*, 36 (1898) 568.
60 R. REISER AND M. C. WILLIAMS, *J. Biol. Chem.*, 202 (1953) 815.
61 V. P. SKIPSKI, M. G. MOREHOUSE AND H. J. DEUEL JR., *Arch. Biochem. Biophys.*, 81 (1959) 93.
62 G. HÜBSCHER, *Proc. 5th Intern. Congr. Biochem., Moscow, 1961*, Vol. 8. Pergamon, Oxford, 1962, pp. 139–148.
63 J. R. SENIOR AND K. J. ISSELBACHER, *J. Biol. Chem.*, 237 (1962) 1454.
64 J. M. JOHNSTON AND J. L. BROWN, *Biochim. Biophys. Acta*, 59 (1962) 500.
65 G. D. AILHAUD, D. SAMUEL, M. LAZDUNSKI AND P. DESNUELLE, *Biochim. Biophys. Acta*, 84 (1964) 643.
66 J. L. BROWN AND J. M. JOHNSTON, *Biochim. Biophys. Acta*, 84 (1964) 448.
67 J. M. JOHNSTON AND J. L. BROWN, in A. C. FRAZER (Ed.), *Biochemical Problems of Lipids*, Elsevier, Amsterdam, 1963, p. 210.
68 G. A. RAO AND J. M. JOHNSTON, *Biochim. Biophys. Acta*, 125 (1966) 465.
69 F. H. MATTSON AND R. A. VOLPENHEIN, *J. Biol. Chem.*, 237 (1962) 53.
70 F. H. MATTSON AND R. A. VOLPENHEIN, *J. Biol. Chem.*, 239 (1964) 2772.
71 P. SAVARY, M. J. CONSTANTIN AND P. DESNUELLE, *Biochim. Biophys. Acta*, 48 (1961) 562.
72 F. KERN AND B. BORGSTRÖM, *Biochim. Biophys. Acta*, 98 (1965) 520.
73 J. M. JOHNSTON, G. A. RAO AND P. A. LOWE, *Biochim. Biophys. Acta*, 137 (1967) 578.
74 F. T. HATCH, A. YOSHIRO, L. M. HAGOPIAN AND J. J. RUBENSTEIN, *J. Biol. Chem.*, 241 (1966) 1655.
75 K. J. ISSELBACHER AND D. BUDZ, *Nature*, 200 (1963) 364.
76 S. M. SABESIN AND K. J. ISSELBACHER, *Science*, 147 (1965) 1149.
77 H. B. SALT, O. H. WOLFF, J. K. LLOYD, A. S. FOSBROOKE, A. H. CAMERON AND D. V. HUBBLE, *Lancet*, (1960) 325.
78 D. B. ZILVERSMIT, *J. Lipid Res.*, 9 (1968) 180.
79 M. M. SALPETER AND D. B. ZILVERSMIT, *J. Lipid Res.*, 9 (1968) 187.
80 C. R. TREADWELL AND G. V. VAHOUNY, *Handbook of Physiology*, Vol. 3, American Physiological Society, Washington, D. C., 1967, p. 1407.
81 C. SYLVEN AND B. BORGSTRÖM, *J. Lipid Res.*, 9 (1968) 596.
82 C. ARTOM AND M. A. SWANSON, *J. Biol. Chem.*, 175 (1948) 871.
83 B. BLOOM, J. Y. KIYASU, W. O. REINHARDT AND I. H. CHAIKOFF, *Am. J. Physiol.*, 177 (1954) 84.
84 R. BLOMSTRAND, *Acta Physiol. Scand.*, 34 (1955) 147.
85 R. O. SCOW, Y. STEIN AND O. STEIN, *J. Biol. Chem.*, 242 (1967) 4919.
86 A. NILSSON, *Biochim. Biophys. Acta*, 152 (1968) 379.

Chapter I

Assimilation, Distribution and Storage

Section B

Mobilization, Transport and Utilization of Free Fatty Acids

ROBERT O. SCOW AND SIDNEY S. CHERNICK

Laboratory of Nutrition and Endocrinology, National Institute of Arthritis and Metabolic Diseases, National Institutes of Health, Bethesda, Md. (U.S.A.)

Long-chain fatty acids are important in mammals because they are a form of energy that can be either readily used by many different cells in the body or stored in adipose tissue until needed[45,47,181]. They are also constituents of complex lipids necessary for the structural and functional integrity of cells. The most commonly found fatty acids in mammalian tissues are palmitic (16:0), oleic (18:1), linoleic (18:2), palmitoleic (16:1), stearic (18:0), myristic (14:0) and arachidonic (20:4)[78]. The relative amounts of each present depend on the tissue, the species and the composition of the diet.

Adipose tissue stores fatty acids in the form of triglyceride[118]. When these fatty acids are mobilized, they are deesterified and released as free fatty acids (FFA) to the blood plasma[77,79,171]. The FFA are then transported, in the blood stream, to the tissues that utilize them for energy and other purposes[45,47]. Some of the physiological and biochemical processes involved in the mobilization, transport and utilization of FFA are described in this paper. Nutritional, hormonal and neural factors that regulate these processes are also discussed.

References p. 45

1. Mobilization of FFA

(a) General aspects

White adipose tissue is the primary storage site for fatty acids in mammals[118]. Brown adipose tissue also stores fatty acids, but it is too small to be an important source of mobilizable fatty acids except, perhaps, in animals that hibernate[80]. Although there are other tissues that store triglyceride, such as liver, kidney, and muscle, they do not release appreciable amounts of FFA to the blood stream[61]. White adipose tissue is located throughout the body —under the skin, within the abdomen and thorax, in skeletal muscle, and in bone marrow. The tissue in some areas actually forms discrete fat bodies, such as the epididymal and parametrial fat pads[181]. The primary cell type of adipose tissue is the fat cell, or adipocyte, which esterifies and stores fatty acids as triglyceride[127]. In addition, adipose tissue contains mast cells, macrophages, fibroblasts and other connective tissue elements[99,150,184]. Some of the adipose tissue, such as that in mesentery and epididymal fat bodies, is enveloped by a layer of mesothelial cells. Adipose tissue has a rich blood vascular supply and is innervated by the autonomic nervous system[23,98,99,141,184,187].

More than 95% of the fatty acids in adipose tissue are present in the form of triglyceride, 0.5% in phospholipids, less than 0.2% in diglyceride and FFA, and traces in monoglyceride and other lipids[70,78,141,154]. Palmitic, oleic and linoleic acids usually account for more than 2/3 of the fatty acids in adipose tissue. Although fatty acids are synthesized in fat cells, most of the fatty acids stored are derived from dietary triglyceride[70]. Fatty acids are carried from the intestines to adipose tissue in blood as triglyceride, in chylomicrons[45] and in very-low-density lipoprotein[186]. When they are taken up, most of the triglyceride is hydrolyzed to FFA by lipropotein lipase near or within the capillary endothelium[123,129]. The FFA then cross the capillary wall and interstitial space to the fat cells where they are esterified and deposited as triglyceride in intracellular lipid droplets[158]. Plasma FFA can also be taken up by adipose tissue[12].

The pathway of triglyceride synthesis in adipose tissue is similar to that found in liver[149,175]. The process requires an ATP-dependent system for activating fatty acids to fatty acyl-CoA and a source of glycerol 3-phosphate. The esterification process occurs within the microsomes and mitochondria[131], and the newly formed triglyceride is transferred directly to the nearest lipid droplet for storage[158].

When fatty acids are mobilized, triglyceride in fat cells is hydrolyzed to FFA, glycerol and diglyceride[141,165], and FFA are released to the blood stream. The glycerol produced is also released, whereas the diglyceride is retained in the tissue. Some of the FFA produced is reesterified and retained in the adipose tissue without entering the blood stream[141,149,165]. The plasma concentrations of both FFA and glycerol are increased during the mobilization of FFA[21,26,64,148].

In order to study clearly the factors that regulate FFA mobilization from adipose tissue, it is essential to distinguish between *lipolysis*, the breakdown of triglyceride to FFA, and *FFA release*, the delivery of FFA to the bloodstream. The amount of triglyceride broken down, or hydrolyzed, can be determined by measuring the amounts of glycerol and diglyceride formed[141,165,175]. The amount of glycerol released by the tissue can be used as a measure of that formed by hydrolysis since glycerol utilization by adipose tissue is negligible[165]. Diglyceride formation is measured by the amount of diglyceride that accumulates in the tissue[141]. Since partial hydrolysis seldom occurs in adipose tissue (see section 1f, p. 33), *lipolysis* is usually measured by the amount of glycerol released by the tissue. *FFA release* is measured as the amount of FFA delivered to the blood or to the medium.

The effect of nutritional, hormonal and neural factors on FFA release have been studied *in vivo* and *in vitro*[14,64]. Most studies *in vivo* have employed indirect measurements[25] such as changes in plasma FFA concentration[10,32,57], turnover of FFA in the plasma[61], venous–arterial differences in FFA concentration across adipose tissue[157] or parts of the body rich in adipose tissue[58], changes in triglyceride content of the liver[90], and changes in weight of the adipose tissue[169]. Direct measurements of FFA release *in vivo* have been made by determining the venous–arterial difference in plasma FFA concentration and the rate of blood plasma flow in adipose tissue[83,110,151]. Release of FFA, and also glycerol, has been measured *in vitro* with incubation techniques using pieces of adipose tissue[165] and isolated fat cells[127] and with perfusion techniques using rat parametrial[122,141] and epididymal fat bodies[72].

(b) Effect of fasting

FFA mobilization is increased when the energy needs of the organism exceed that which can be met by food in the gastrointestinal tract or by carbohydrate stored in the liver and muscle[21,45]. This occurs in normal individuals whenever food is withheld for more than 15 h. It is associated with decreased glucose and insulin and increased FFA and glycerol concentrations in the plasma[21]. The effect of fasting is exacerbated in late pregnancy[144] and phloridzin poisoning[180] (Scow and Chernick, unpublished observations), both of which lower markedly the blood glucose in fasting animals. Part of the effect of glucose lack may be mediated through the pancreas by a decrease in insulin secretion since insulin injection alone will decrease FFA mobilization[144] (Scow and Chernick, unpublished observations).

Studies with incubated[24,172] and perfused[141] adipose tissue and with incubated fat cells[127] have shown that there is a continuous breakdown of triglyceride (as measured by glycerol release) but practically no FFA release in adipose tissue of fed rats when glucose is present (Table I). Fasting the rats for one day increases FFA release without changing the rate of lipolysis, whereas fasting two days increases both lipolysis and FFA release (Tables I and II). The effect of fasting on lipolysis persists for several hours after the tissue is removed from the animal, even in tissue perfused extensively[122] and in isolated fat cells washed with many volumes of incubation medium[41].

Glucose added *in vitro* markedly suppresses FFA release in tissues of fed and fasted rats but does not decrease lipolysis (Table I); under some conditions, it may actually increase lipolysis. Insulin added *in vitro* decreases FFA release if there is glucose either in the tissue or in the medium (Table III). Insulin may also suppress lipolysis but only in tissues of fed animals

TABLE I

EFFECT OF FASTING AND GLUCOSE ON FFA AND GLYCEROL RELEASE IN INCUBATED EPIDIDYMAL FAT PADS OF RATS[24]

Days fasted	Number of incubations	FFA release		Glycerol release	
		Without glucose	With glucose	Without glucose	With glucose
		(μmoles per g tissue per h)			
0	8	4.1 ± 0.3	0.5 ± 0.1	1.1 ± 0.1	2.1 ± 0.1
1	8	2.9 ± 0.2	1.0 ± 0.1	1.1 ± 0.8	1.7 ± 0.1
2	8	5.2 ± 0.2	2.3 ± 0.3	2.5 ± 0.1	2.7 ± 0.1

Pieces of adipose tissue were incubated for 1 h in Krebs–Ringer bicarbonate-buffered solution containing albumin, 20 mg/ml. Glucose was added at a concentration of 11.1 mM. Values given are means ± standard error.

and animals fasted one day[41,82]; insulin has no anti-lipolytic effect in tissues of rats fasted for two or more days (Tables III and IV). These observations suggest that decreased reesterification of fatty acids alone accounts for FFA mobilized during the early part of fasting and that lipolysis is not increased until late in fasting. The early inhibitory effect of glucose and

TABLE II

GLYCEROL PRODUCTION BY INCUBATED EPIDIDYMAL FAT PADS OF RATS[51,a]

Treatment of rats[b]	Glycerol produced (μmoles per g tissue per h)	
	Without insulin	With insulin[c]
Normal fed	0.66 ± 0.6	0.87 ± 0.05
Normal fasted 2 days	1.30 ± 0.19	1.27 ± 0.15
Alloxan diabetic	1.15 ± 0.10	1.26 ± 0.03
Hypophysectomized	—	0.45 ± 0.04
Hypophysectomized alloxan diabetic	—	0.58 ± 0.03
Hypophysectomized alloxan diabetic + growth hormone + cortisol	—	1.67 ± 0.37
Adrenalectomized	—	0.52 ± 0.06
Adrenalectomized alloxan diabetic	—	0.70 ± 0.09

[a] Tissues from adult male rats were incubated in bicarbonate-buffered medium (Krebs and Henseleit) containing glucose 8.3 mM. Minimum number of observations per group was 5. Values given are means ± standard error.
[b] Alloxan was injected into fed rats 48 h before experiment. Hypophysectomy was done 2 weeks before, and adrenalectomy 48 h before the experiment. Bovine growth hormone, 0.1 mg/100 g, was injected at 24, 12 and 4 h and cortisol, 2.5 mg/100 g at 24 h and 1.25 mg/100 g at 4 h before the experiment.
[c] Insulin concentration in medium was 100 mU/ml.

TABLE III

EFFECT OF INSULIN AND GLUCOSE ON FFA AND GLYCEROL RELEASE IN FAT CELLS ISOLATED FROM PARAMETRIAL ADIPOSE TISSUE OF RATS FASTED 18 h[41]

Glucose (mmolar)	FFA release		Glycerol release	
	Without insulin	With insulin	Without insulin	With insulin
	(μmoles per g cells per h)			
0	64 ± 10	40 ± 12	25 ± 4	9 ± 2
2.4	15 ± 7	−9 ± 7	22 ± 4	10 ± 5

About 7 mg of fat cells were incubated for 4 h in 2 ml of Krebs–Ringer bicarbonate-buffered solution containing albumin, 40 mg/ml. Values given are means ± standard error.

References p. 45

TABLE IV

EFFECT OF GLUCOSE AND INSULIN ON FFA AND GLYCEROL RELEASE BY INCUBATED PARAMETRIAL ADIPOSE TISSUE OF FASTING PANCREATECTOMIZED AND NORMAL RATS[a]

(Fain and Scow, unpublished observations)

Group	Number of animals	Glucose in medium (mmolar)	FFA release[b]		Glycerol release	
			Without insulin	With insulin	Without insulin	With insulin
			(μmoles per g tissue per h)			
Normal	6	0	2.2	2.3	0.82	0.82
		2.8	−0.3	−0.3	0.85	1.15
Pancreatectomized	6	0	6.9	2.8[c]	2.38	1.72[c]
		2.8	2.1	+0.3[c]	2.05	2.35[c]

[a] All rats were fasted for 3 days before the experiment; pancreatectomy was performed 48 h before the experiment. Pieces of tissue weighing 80–100 mg were incubated for 4 h in 3 ml of bicarbonate buffered-solution containing albumin, 40 mg/ml.
[b] Measured as accumulation of FFA in the tissue and medium.
[c] Effect of insulin was statistically significant ($P = 0.05$).

insulin administration on FFA release in fasting animals is probably secondary to the stimulation of glycerol 3-phosphate formation and reesterification in the adipose tissue.

Although the pituitary and adrenal glands are not needed for FFA mobilization in fasting animals, removal of either gland decreases lipid mobilization[2,54,90,142]. Several studies have shown that growth hormone and glucocorticoid are, respectively, the pituitary and adrenal hormones involved in lipid mobilization in fasting animals[21,35,54,90,143]. There is evidence that the sympathetic nervous system may also be important in FFA mobilization during fasting[14,64] but its role has been difficult to evaluate[56,69]. Effects of hormones and the sympathetic nervous system on fatty acid mobilization are discussed in more detail below.

(c) *Effect of diabetes and insulin*

In well nourished animals diabetes causes severe ketosis, hypertriglyceridemia, and accumulation of triglyceride in liver and kidneys[142]. It also increases markedly the concentration of FFA in plasma and the release of FFA by adipose tissue[83,141]. The plasma FFA concentration and FFA release in diabetic animals are immediately decreased to normal by insulin administration whereas the other abnormalities are not fully corrected for many hours[42,83,142].

Incubation studies with isolated fat cells and pieces of adipose tissue have shown that lipolysis and FFA release are much greater in pancreatectomized fasted rats than in normal rats fasted 2 or 3 days (Tables IV and V). Addition of glucose decreases release of FFA but has no effect on glycerol release in tissue of the diabetic animals (Table IV). Insulin enhances the effect of glucose on FFA release but does not affect glycerol release.

TABLE V

EFFECT OF PANCREATECTOMY IN FASTING RATS ON BLOOD GLUCOSE, BLOOD KETONE BODIES AND PLASMA FFA, AND ON RELEASE OF FFA AND GLYCEROL BY FAT CELLS ISOLATED FROM PARAMETRIAL ADIPOSE TISSUE[a]

(Fain and Scow, unpublished experiments)

Group[b]	Number of rats	Blood glucose	Blood ketone bodies	Plasma FFA (μequiv./ml)	FFA release		Glycerol release	
		(mg/100 ml)			0–2 h	0–4 h	0–2 h	0–4 h
					(μmoles per g cells per h)			
Normal	7	50 ± 5	13 ± 2	0.75 ± 0.04	40	50	25	38
Pancreatectomized	6	350 ± 20	47 ± 14	1.12 ± 0.07	265	205[c]	100	85

[a] Approximately 7 mg of fat cells were incubated in 2 ml of Krebs–Ringer bicarbonate-buffered solution which contained albumin, 40 mg/ml, and glucose, 2.4 mM. The cells were incubated for either 2 or 4 h. Values given are means ± standard error.
[b] All rats were fasted 42 h before the experiment; pancreatectomy was performed 24 h after onset of fasting.
[c] The molar ratio of FFA to albumin in the medium at the end of the incubation was approximately 5.0.

Removal of the pituitary or adrenal glands prevents development of ketosis and accumulation of triglyceride in blood, liver and kidney in stressed insulin-deficient animals (Table VI)[142,170]. Studies made in pancreatectomized rats immediately after removal of the pituitary or adrenal glands show that ketosis, hyperlipemia and fatty liver develop in hypophysectomized–pancreatectomized rats when either cortisone or ACTH (adrenocorticotropin) is given and in adrenalectomized–pancreatectomized rats when cortisone is given; ACTH is ineffective in the absence of the

References p. 45

TABLE VI

EFFECTS OF HORMONES ON BLOOD KETONE BODIES, BLOOD LIPID AND LIVER TRIGLYCERIDE CONTENT IN HYPOPHYSECTOMIZED AND ADRENALECTOMIZED-PANCREATECTOMIZED RATS[145]

Group	Dose (mg/day)	Number of rats	Hours after start of experiment	Final blood ketone bodies (mg/100 ml)	Change in blood total lipid (mg/100 ml)	Liver Weight (g)	Triglyceride (mg)
$\bar{P}$	—	7	48	25 ± 7	33 ± 47	5.1	26 ± 13
$\bar{P}$-sham $\bar{A}$	—	6	48	167 ± 22	526 ± 180	5.6	328 ± 81
$\overline{AP}$	—	5	48	21 ± 4	-81 ± 69	4.7	24 ± 7
$\overline{HP}$	—	7	48	12 ± 4	-24 ± 90	4.7	23 ± 3
$\overline{HP}$ + GH	0.3	4	48	14 ± 1	-145 ± 38	4.5	31 ± 8
	2.0	3	48	11 ± 3	-100 ± 19	4.1	12 ± 3
	5.0	4	48	7 ± 1	-33 ± 61	4.4	38 ± 19
$\overline{HP}$ + cortisone	0.25	3	48	41 ± 32	64 ± 92	4.7	18 ± 28
	1.0	6	48	61 ± 30	188 ± 88	4.5	52 ± 24
	2.0	5	28	108 ± 10	180 ± 80	4.8	144 ± 25
	4.0	5	20	122 ± 28	302 ± 200	5.2	135 ± 37
$\bar{H}$ + cortisone	4.0	4	24	8 ± 1	-75 ± 30	4.9	130 ± 39
N + cortisone	4.0	3	24	11 ± 1	-25 ± 46	5.3	66 ± 8
$\overline{AP}$ + cortisone	2.0	5	48	155 ± 19	895 ± 365	5.8	434 ± 141
$\overline{AP}$ + DOCA	3.0	5	48	28 ± 4	1 ± 52	4.7	47 ± 9

H̄P + ACTH	0.3–0.4	5	48	120±26	−41±28	4.8	117±81
ĀP + ACTH	0.3–0.4	6	48	7±1	−95±35	4.2	6±3
H̄	—	3	48	13±3	−31±20	4.1	29±7
H̄ + ACTH	0.3–0.4	6	48	5±1	−42±21	4.3	69±19
H̄P + TSH	1.2	7	48	20±7	29±63	4.5	32±8
H̄P + prolactin	8.0	4	48	16±3	−148±17	4.2	29±3

[a] Rats were "totally" pancreatectomized 7–12 days before the experiment and maintained thereafter with insulin and food; the last insulin injection and feeding were given 17 h before the start of the experiment. All hypophysectomies and adrenalectomies were made at the start of the experiment. The hormones were given at the 1st and 20th h of the experiment, except ACTH which was given 3–4 times daily. The blood ketone body concentration was less than 10 mg/100 ml, blood lipids between 150 and 250 mg/100 ml and blood glucose between 200 and 350 mg/100 ml in pancreatectomized rats at the start of the experiment. Values are means ± standard error.

[b] Abbreviations:
P̄ = pancreatectomized; H̄ = hypophysectomized; Ā = adrenalectomized; N = normal; GH = bovine-growth hormone; DOCA = deoxycorticosterone acetate; ACTH = adrenocorticotrophic hormone; and TSH = thyroid-stimulating hormone.

References p. 45

adrenal glands and growth hormone is not needed (Table VI)[145]. However, when the pituitary is removed several days before the study, growth hormone is required in addition to glucocorticoid (dexamethasone) for the development of ketosis (Fig. 1)[140]. Explanation for this difference is not known. Growth hormone and glucocorticoid are both needed for development of diabetic ketosis and hyperlipemia in the dog[170].

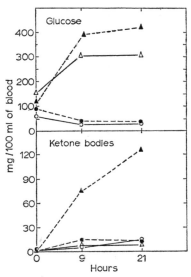

Fig. 1. Effect of growth hormone (5 mg) and dexamethasone (2.5 µg) in rats pancreatectomized for 2 weeks and hypophysectomized for 1 week. The hormones were given 17 h after the last insulin injection and feeding. (From ref. 140) o——o, control; •---•, growth hormone; △—△, dexamethasone; ▲---▲, GH + dexamethasone.

Extirpation of either the anterior pituitary or adrenal glands in insulin-deficient rats decreases the release of FFA and glycerol by adipose tissue when measured *in vitro* (Table II)[42,51]. Treatment of alloxanized–hypophysectomized rats with growth hormone and cortisol for 24 h increases glycerol release in their adipose tissue (*in vitro*), by two-fold, to the same level as that in alloxan-diabetic controls (Table II).

These findings suggest that glucocorticoid and growth hormone in the absence of insulin stimulate lipolysis in adipose tissue, and this results in increased FFA mobilization, hyperketonemia and accumulation of triglyceride in blood, liver and kidneys. These effects are not seen in normal animals unless there is a marked glucose deficiency, as in fasting pregnant[143] or fasting phlorizin-treated rats (Scow and Chernick, unpublished observations). This enhancement of lipolysis is slow in onset and may be prevented by insulin, but it is not readily reversed by insulin. Insulin, however, can quickly suppress FFA release in diabetic adipose tissue[42,83,141]. The effects of these hormones on lipolysis are discussed in more detail below.

There is suggestive evidence that the sympathetic nervous system is also involved in the mobilization of lipid in uncontrolled diabetes[64,69]. However, the nature of its role has not been determined, whether it is altering lipolysis or the rate of blood flow through the tissue.

(d) Effect of sympathetic nervous system

Study of the role of the sympathetic nervous system in fatty acid mobilization has been facilitated by the use of pharmacologic agents that either block transmission of impulses through the sympathetic nerves[106,176] or block the response of effector cells to the impulses[106]. The drugs used have been ganglion blocking (hexamethonium), adrenergic blocking (dichloroisoproterenol (DCI), pronethalol, propranolol), and antiadrenergic agents (guanethedine and reserpine)[14,37,38,64,69].

It has been found that the sympathetic nervous system is the chief stimulator of fatty acid mobilization when there is a sudden demand for energy, as in exercise or exposure to cold[24,64], or when the individual is subjected to a frightening or stressful situation[11]. Although the sympathetic nervous system may facilitate, it does not seem to initiate mobilization of fatty acids during fasting and diabetes[64,69].

The sympathetic nervous system increases fatty acid mobilization primarily by accelerating the breakdown of triglyceride in adipose tissue. The effect, fast in onset and short in duration, resembles the action of catecholamines in incubated adipose tissue[174] and fat cells[127]. Norepinephrine is undoubtedly the neurohumoral factor that stimulates lipolysis, since it is the chemical mediator of postganglionic sympathetic nerve fibers[75] and it is present in adipose tissue in relatively large amounts (12 µg/mg protein)[14]. Incubation studies with adipose tissue and fat cells have shown that the lipolytic effect of catecholamines is mediated through beta receptors[20,38,129], probably located in the cell membrane of the fat cells[124]. The sympathetic nerve fibers in adipose tissue are located primarily along arteries, arterioles and

References p. 45

venules, and are seen occasionally in the intercellular space, but are never closely associated with fat cells[99,187]. The means by which norepinephrine passes from nerve fibers to fat cells has not been determined.

The sympathetic nervous system may also affect fatty acid mobilization by altering the flow of blood through the adipose tissue. Experiments *in vivo* and in perfused isolated fat pads have shown that stimulation of the nerves[105,133] or intraarterial injection of either epinephrine or norephinephrine[141] (Kovacev and Scow, unpublished observations) causes an immediate slowing of blood flow through adipose tissue. In perfused adipose tissue the dosage of epinephrine needed to stimulate FFA release usually retards or stops blood flow through the tissue[141]. Transport of FFA from adipose tissue, of course, would be reduced with blood flow if there developed a shortage of available fatty acid binding sites in the blood flowing through the tissue (see below)[83,141]. The nature and action of adrenergic receptors in the vascular bed of adipose tissue, unfortunately, have not been adequately studied. Norepinephrine and epinephrine, which act on both alpha and beta receptors[75], retard blood flow in adipose tissue[141] whereas isoproterenol, which acts primarily on beta receptors[75], does not affect blood flow in perfused adipose tissue (Hamosh and Scow, unpublished data). The difference in vasomotor effects between the naturally occurring catecholamines and isoproterenol suggests that activation of alpha receptors within the vascular bed causes vasoconstriction in adipose tissue[105]. It is possible, of course, that norepinephrine might have very little effect on alpha receptors in blood vessels of adipose tissue when it is released by nerve endings under natural stimulation.

(e) *Effect of lipolytic hormones*

There are many hormones that stimulate lipolysis in adipose tissue. Most of them act immediately, such as norepinephrine, epinephrine, ACTH, TSH (thyroid-stimulating hormone), glucagon and various hypophyseal peptides[36,38,40,83,127,134,174]. However, growth hormone and glucocorticoids have a latent period of more than 1 h (Fig. 2)[40]. There is now considerable evidence that these two groups of hormones stimulate lipolysis by different mechanisms. The fast-acting lipolytic hormones act through cyclic AMP (adenosine 3',5'-monophosphate)[20] whereas the slow-acting hormones act primarily through a mechanism involving RNA and protein synthesis[40].

(i) *Fast-acting lipolytic hormones*

The fast-acting hormones have an effect on lipolysis that is rapid in onset (Fig. 2) and short in duration[174]. They differ from each other in origin, chemical structure, and ability to stimulate lipolysis in adipose tissue of different species. The catecholamines and TSH, for example, are very effective in the rat but not the rabbit, whereas some of the other pituitary

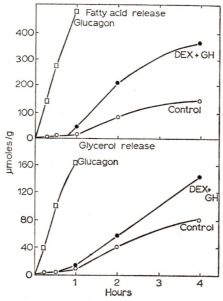

Fig. 2. Comparison of the effect of glucagon (1 μg/ml) with that of dexamethasone (0.016 μg/ml) and growth hormone (1 μg/ml) on isolated fat cells of rats fasted 18 h. The incubation medium contained 40 mg of bovine serum albumin and 2.4 μmoles of glucose per ml. (From ref. 36)

lipolytic peptides (α- and β-melanocyte-stimulating hormone, vasopressin and fraction H) are active in the rabbit but not the rat[134]. The responsiveness of adipose tissue to these hormones does not seem to be dependent on the nutritional state of the animal.

The lipolytic effect of the fast-acting hormones *in vitro* can be suppressed by insulin[41,82], and prostaglandin E_1 (refs. 37, 38, 166), but only if the lipolytic stimulus is small; the lipolytic action predominates when larger amounts of lipolytic hormones are used[36,37,41]. Interpretation of the anti-

References p. 45

lipolytic action of insulin *in vitro* is complicated by the finding that insulin under some conditions actually enhances the action of the lipolytic hormones[55] (Scow, unpublished observations).

The fast-acting lipolytic hormones increase the cyclic AMP content of adipose tissue at the same time they stimulate lipolysis[17,18,20]. They also activate adenyl cyclase, the enzyme that catalyzes cyclic AMP formation, in subcellular preparations of fat cells[128]. Additional evidence that the lipolytic action of the hormones may be mediated through cyclic AMP is the finding that N^6-2'-O-dibutyryl cyclic AMP, a derivative of cyclic AMP, stimulates lipolysis[18].

The action of these hormones on lipolysis is greatly enhanced by methyl xanthines (theophylline and caffeine)[20] and unaffected by substances that block RNA (actinomycin D) and protein synthesis (puromycin, cyclohexamide)[38]. Methyl xanthines enhance the hormone action presumably by decreasing the activity of cyclic nucleotide phosphodiesterase[17,18], the enzyme that degrades cyclic AMP[19]. Insulin and prostaglandin E_1, both antilipolytic agents under certain conditions, lower the cyclic AMP content in adipose tissue[17,18]. Their site of action, however, whether on formation or degradation of cyclic AMP, is not known.

The fact that many different lipolytic hormones—catecholamines, ACTH, glucagon and TSH—increase cyclic AMP in adipose tissue suggests that there may be several hormone-receptors for activating adenyl cyclase[20,38,124,128].

Although many of the fast-acting hormones are potent lipolytic agents *in vitro* and some may act *in vivo*, norepinephrine probably is the only one that has physiological significance in FFA mobilization[4,64,69,88]. Norepinephrine, as described above, is the chemical mediator between the sympathetic nervous system and the fat cells. The role of the other hormones in lipid mobilization *in vivo* requires further study.

(ii) Slow-acting lipolytic hormones

Growth hormone and glucocorticoid together increase lipid mobilization in animals deprived of either food or insulin (see above). The effect on FFA mobilization has a time lag of at least 1 h and lasts for several hours after the injection is given[83]. Mobilization of FFA by the slow-acting hormones is readily prevented by refeeding in animals with an intact pancreas or by insulin administration in normal and diabetic animals[83,142].

Growth hormone and glucocorticoid added together stimulate, after 1 h, lipolysis in incubated adipose tissue and fat cells of fasted rats (Fig. 2)[40]. This lipolytic effect is easily blocked by insulin and is not observed in tissues or cells of fed animals. The effect is much less than that of the fast-acting lipolytic hormones[36,40]. The lipolytic action of growth hormone and glucocorticoid is blocked by inhibitors of DNA-dependent RNA synthesis (actinomycin D) and protein synthesis (puromycin, cycloheximide), suggesting that their action involves newly synthesized RNA and protein[36,40]. Their action is enhanced by theophylline, suggesting that they may act through the formation of cyclic AMP, perhaps by stimulating synthesis of protein involved in cyclic AMP formation[39].

The role of the individual hormones, growth hormone and glucocorticoid, in the stimulation of lipolysis has not been clearly defined. Neither hormone alone is effective at low concentrations *in vitro*. Addition of the hormones separately shows that glucocorticoid is not needed during the lag period[36]. It is possible that glucocorticoid may be acting through or on the protein(s) synthesized under the influence of growth hormone. This is supported by the observation that glucocorticoid given alone is a potent lipid mobilizer in acutely hypophysectomized and acutely adrenalectomized-diabetic rats (Table VI)[145], but not in chronically hypophysectomized-diabetic rats (Fig. 1)[142].

(f) Lipolytic enzymes in adipose tissue

Adipose tissue has two enzyme systems that hydrolyze triglyceride to FFA and glycerol. One of them acts on tissue triglyceride and is involved in the production of FFA during mobilization of lipid from adipose tissue[60,119,120,167]. The other, lipoprotein or clearing factor lipase, acts on serum triglyceride in or near the capillary endothelium and is involved in the uptake of serum triglyceride fatty acids by adipose tissue[123,125,129,135].

The tissue-triglyceride lipase system consists of two, or perhaps three enzymes, that hydrolyze tissue triglyceride step-wise to FFA and glycerol[60,119,120,167]. The first enzyme hydrolyzes triglyceride to diglyceride and FFA. It has been called the hormone-sensitive triglyceride lipase because its activity in tissue homogenates is increased if the tissue is incubated with a fast-acting lipolytic hormone for a few minutes prior to homogenization[119,167]. This enzyme acts on either tissue triglyceride or triglyceride emulsions and does not require serum activation of the triglyceride. Although some studies suggest that tissue diglyceride is hydrolyzed

References p. 45

by a single enzyme to FFA and glycerol, the presence of two enzymes for this process has not been excluded, namely, that one enzyme hydrolyzes diglyceride to monoglyceride and FFA, and another hydrolyzes monoglyceride to glycerol and FFA[60,167].

The hormone-sensitive triglyceride lipase is generally considered to be the rate-limiting step in lipolysis in adipose tissue[167,175]. This is based on the finding in many different kinds of adipose-tissue preparations (*in vivo*, incubated fat pads, isolated fat cells) that diglyceride and monoglyceride do not accumulate when FFA are being mobilized under the influence of lipolytic hormones[165,175]. In addition, diglyceride and monoglyceride lipase activities are 3–20 times greater in homogenates than is triglyceride lipase activity. However, accumulation of diglyceride in adipose tissue may occur *in vivo*[177] and in perfused isolated tissue[141] when lipolysis is maximally stimulated by hormones. This would suggest that under certain conditions hydrolysis of diglyceride may be rate-limiting in the breakdown of triglyceride[146].

The maximal rate of hydrolysis of triglyceride in intact adipose tissue, whether incubated[74,115,174], perfused[141] or *in vivo*[83], is 10–20 μequiv. FFA/g/h. The rate in homogenates of adipose tissue is 35–50 μequiv./g/h[167,173]. In isolated fat cells, however, the maximal rate is many times greater, between 300 and 600 μequiv. FFA/g/h[127]. This very rapid rate, of course, is contingent on an excess of albumin in the incubation medium (see below). Since lipolysis is readily inhibited by FFA accumulation in fat cells, it is possible that the faster lipolytic rate in isolated fat cells reflects a faster diffusion of FFA to the medium and a lower intracellular FFA concentration than in intact tissue[126].

The fast-acting hormones probably activate the hormone-sensitive triglyceride lipase through the formation of cyclic AMP[119]. The short duration of action suggests that the lipase may return to an inactive form when the hormone and cyclic-AMP concentrations fall. The slow and prolonged lipolytic activity associated with fasting, diabetes, and growth hormone-glucocorticoid, however, probably involves mechanisms other than lipase activation by cyclic AMP[40].

2. Transport of FFA

FFA are transported in blood bound primarily to albumin, and, to a lesser extent, to α- and β-lipoprotein[45,53,147]. Distribution of FFA between albumin and lipoprotein depends on chain length and degree of saturation of the fatty acids. Albumin binds more than 90% of palmitic, oleic, linoleic and linolenic, 85% of stearic, and 77% of arachidonic acid in plasma, and α- and β-lipoprotein bind the remainder[147]. Less than 2% of FFA in plasma are bound to red blood cells and less than 0.01% are free in solution[53].

Albumin at normal serum concentrations increases more than 1000 times the solubility of FFA in aqueous solutions[153]. Human and bovine serum albumins have 6–7 different high-energy FFA-binding sites and a large number of weak sites[22,53,153], but only the high-energy binding sites are involved in physiological and biochemical processes. The highest molar ratio of FFA to albumin obtained seldom exceeds 7, and never 8, in studies with incubated adipose tissue[22] and fat cell[127], and with lipolytic enzymes[59]. Albumin is required in the medium for release of FFA by adipose tissue and by isolated fat cells and for lipolysis in fat cells[127], but not for lipolysis in adipose tissue[22,82]. Adipose tissue incubated in albumin-free medium can bind 20–40 μequiv. of FFA per g of tissue when lipolysis is stimulated with epinephrine or ACTH[22,82], whereas fat cells similarly treated bind less than 4 μequiv./g[127]. The FFA content in tissue is usually less than 5 μequiv./g when the tissue is incubated in medium containing an excess of albumin[174]. The above suggest that some cellular or extracellular component of adipose tissue can bind FFA but its affinity for FFA is probably less than that of the high-energy binding sites of albumin.

Perfusion studies of rat adipose tissue demonstrate that the rate of FFA release is dependent on the molar ratio of FFA to albumin in blood and on the rate of blood flow through the tissue[141]. FFA release, but not lipolysis, is reduced when the first two binding sites of plasma albumin are saturated, suggesting that the reesterification mechanism has less affinity for FFA than the first 2 binding sites and more affinity than the next 4–5 sites. Blood flow becomes an important determinant of FFA release when the FFA–albumin molar ratio in venous blood exceeds 2 and that in arterial blood is less than 2. The FFA–albumin ratio is 2 in plasma of the rat when the FFA concentration is 0.9 μequiv./ml (plasma albumin, 3.2%) and 2 in plasma of man when the concentration is 1.3 μequiv./ml (plasma albumin, 4.4%). It is evident that release of FFA by adipose tissue *in vivo* is dependent not only on the production of FFA but also on the number and kinds of FFA binding sites available in the blood stream[141].

Uptake of FFA from blood by most tissues is proportional to the plasma concentration of FFA (see below). It may also be influenced by the type or

References p. 45

number of binding sites on albumin being used to carry FFA to the tissue since the affinity of albumin for FFA decreases exponentially as the plasma FFA concentration, or molar ratio of FFA bound to albumin, increases (Fig. 3). FFA uptake by certain tissues may even be dependent on high FFA–albumin ratios in the plasma. The form in which FFA crosses the capillary

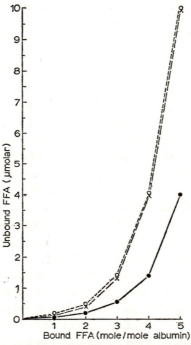

Fig. 3. Effect of FFA–albumin molar ratio on dissociation of FFA bound to bovine serum albumin in phosphate-buffered salt solution at pH 7.4 and 37°. (Adapted from ref. 153) ○---○, linoleic acid; ×—×, oleic acid; ●—●, palmitic acid.

endothelium and interstitial space is not known. It is unlikely that albumin-bound FFA crosses the capillary wall since fractional transfer of plasma FFA to tissue (30–50%) is many times greater than that of plasma albumin (less than 0.05%)[178]. Movement of FFA across the capillary endothelium and extracellular space probably involves a non-energy requiring process since it readily occurs in perfused heart[161] at 0°. The transfer process may involve different protein or lipid carriers in each of the various compartments and membranes traversed by the FFA.

3. Utilization of FFA

There is a very rapid turnover of FFA in the blood stream[12,46]. When [^{14}C]palmitate is injected intravenously into rats, fed and fasted, about 99% disappears from the plasma within 10 min. About 95% of that removed in fed rats is incorporated into tissue lipid and 5% is oxidized to CO_2, whereas in fasted rats, half is immediately oxidized and half is esterified[12,95]. The labeled palmitate is taken up by liver, muscle, adipose tissue, kidney, lung and heart. A small fraction (1%) of the injected palmitate remains in the plasma as FFA for at least 3 h, suggesting that plasma FFA may exchange with a very large pool of fatty acids in tissue.

Turnover of FFA in the circulation is proportional to the concentration of FFA in the plasma[3,46,76]. In a steady state the turnover rate is a measure of both the rate of FFA release from adipose tissue and other sources, and the rate of FFA uptake by all tissues. Differences between species in turnover of FFA are reduced when the rate is expressed per unit surface area[76]. The rate in man and in dog is about 0.5 mequiv./m^2 per minute at a plasma concentration of 1.0 µequiv. FFA/ml.

The plasma FFA pool has at least 10 different fatty acids, of which palmitic, oleic, linoleic, stearic and palmitoleic are the more common. Radioisotope studies have shown that palmitate, linoleate and oleate in plasma have similar initial disappearance rates in man[46] and that labeled palmitate is a representative tracer for plasma FFA in both man and dog[76,96]. Some of the FFA, however, are removed and utilized at different rates and have different metabolic fates in certain tissues[97]. For example, palmitate is preferentially utilized by kidney[52] and linoleate is preferentially incorporated by liver into plasma cholesterol ester and phospholipid[111].

Various hormonal and nutritional factors increase the concentration of FFA in the plasma and, thereby, accelerate the turnover of plasma FFA. The concentration in rats is increased by fasting (1–2 days), from 0.3–0.4 µequiv./ml to 0.6–0.8 µequiv./ml, and by diabetes, to 1.4–1.7 µequiv./ml[83]. In man, the plasma FFA concentration is 0.4–0.5 µequiv./ml after an overnight fast, 0.8–1.2 µequiv./ml after 2–6 days of fasting, and is increased to 1.8 µequiv./ml in diabetes[6,21]. The FFA concentration in plasma may be as high as 3.0 µequiv./ml when FFA mobilization is maximally stimulated, as in man with severe diabetic ketoacidosis[6], in diabetic dogs given growth hormone[16], in normal dogs given epinephrine[25], and in normal rabbits given either ACTH or Fraction H (anterior-pituitary adipokinetic substance)[71].

References p. 45

Both concentration and turnover of FFA in plasma are decreased by administration of glucose in fasting and by insulin in diabetic dogs[65,96]. These effects are secondary to the suppression of FFA release from adipose tissue (Tables I, III and IV).

(a) Liver

About one-third of the FFA removed from the blood stream in non-exercising animals are taken up by the liver[7,12,107]. Fractional clearance of plasma FFA by the liver is 25–40% and is unaffected by fasting, diabetes, and hormones[5,44,68]. Most of the FFA taken up are first incorporated into triglycerides, phospholipids and cholesterol esters and then either retained in the liver or secreted in lipoprotein; some of the FFA are directly oxidized to CO_2 and ketone bodies. FFA are diverted from oxidation to glyceride synthesis by the metabolism of carbohydrate[34,91].

The individual FFA are taken up at different rates and have different metabolic fates in the liver[47,97,102,162]. The means by which the liver discriminates between the different fatty acids is not known. Plasma FFA mix with hepatic fatty acids in intracellular FFA pools before they are utilized[7]; the FFA content of the liver is 0.6 μequiv./g. The fatty acid composition of the liver FFA pools resembles that of plasma FFA more than that of the hepatic neutral lipids and phospholipids[132]. Unsaturated fatty acids are incorporated mostly into the C-2 position, and saturated fatty acids into the C-1 position, of phospholipids and triglycerides in the liver. The discrimination may occur when glycerides are synthesized *de novo*[13], or when lysophosphatides are acylated[87]. The turnover of liver glyceride, which involves deacylation and reacylation, is unaffected by fasting and diabetes[29].

About 50% of the plasma FFA taken up by liver in normal rats are incorporated into triglyceride, 25% into phospholipid, and smaller amounts into cholesterol esters[7,107]. The esterified fatty acids are either incorporated into structural elements, secreted as lipoproteins, or stored in lipid droplets[81,160]. The liver secretes fatty acids mostly as triglyceride, about 10% as phospholipid and cholesterol ester, and practically none as FFA. The liver secretes about 0.1 μmole of triglyceride per g per min. About one-third of the fatty acids secreted by liver in rats fasted overnight are derived directly from plasma FFA and about two-thirds are from hepatic lipids[7]. Studies with perfused liver have shown, however, that lipoprotein secretion under certain conditions may not be related to the uptake of FFA[66].

Puromycin and orotic acid block the formation and secretion of lipoprotein without affecting uptake and esterification of FFA[81,185]. When the rate of formation of triglyceride exceeds the rate of secretion, the excess triglyceride is stored in lipid droplets in the liver.

The uptake of FFA by liver slices and the proportion incorporated into triglyceride increase with the FFA concentration in the medium[132]. Such preferential synthesis of triglyceride occurs also in livers perfused with plasma containing high concentrations of FFA[103].

Fatty acid oxidation in the liver accounts for about half of the FFA oxidized in the whole body at rest. About two-thirds of the FFA oxidized in the liver are oxidized completely to CO_2 and the remainder, to ketone bodies[47]. FFA oxidation in the liver is decreased by feeding and is greatly increased by starvation and by diabetes. The latter conditions increase also the proportion of FFA oxidized to ketone bodies.

Plasma FFA and ketone body concentrations increase together in fasting and diabetic animals[83,114]. The FFA concentration in plasma, however, decreases before the ketone bodies when glucose is given to normals and when insulin is given to diabetics[83,114]. Both FFA and ketone body concentrations in plasma are higher in untreated diabetic than in fasting normal animals (Table V). There is a relationship in rats between the plasma ketone body concentration *in vivo* and the rate of hepatic ketogenesis measured *in vitro*, but, the ketonemia associated with any given rate of ketogenesis is always higher in the diabetics than in the normals[142].

The rate of ketogenesis in perfused livers of fed and fasted rats is accelerated by increasing the FFA concentration in the perfusing fluid[43,113]. Ketogenesis in liver slices and in perfused livers is related also to the triglyceride content of the liver; both are greatly increased by diabetes[142]. Administration of insulin to diabetic rats lowers hepatic ketone body production 25–30% without changing the liver triglyceride content. Since insulin has no direct effect on ketogenesis in perfused isolated liver[113,142], it probably decreases ketone body production *in vivo* by suppressing FFA mobilization to the liver[83]. Ketosis and hepatic accumulation of triglyceride do not occur during diabetes or fasting in animals with depleted body fat stores[94,142]. It is evident, from the above, that ketone bodies can be derived from either plasma FFA or hepatic triglyceride fatty acids.

Addition of FFA to the perfusing medium increases gluconeogenesis and decreases glucose utilization in liver[168,183]. These effects have been attributed to increased concentrations of intermediates of fatty acid oxidation, such

References p. 45

as NADH and acetyl-CoA[183]. This interpretation, however, has been challenged[34,152].

(b) Kidney

Fatty acids, derived from plasma FFA and tissue lipid, are the major oxidative fuel of the kidney[31,84,89]. Plasma FFA uptake by the kidney is roughly proportional to the arterial FFA concentration; the fractional removal, however, is small, less than 10%[58,73]. The principal FFA taken up by the kidney in dogs is palmitic acid[52]. The kidney in fasted dogs removes 0.13 µequiv. of FFA per g per min when the plasma FFA is about 0.6 µequiv./ml[8]. When the plasma FFA concentration exceeds 1 µequiv./ml, the kidney extracts more FFA than it can oxidize[8]; the excess FFA is esterified and deposited as triglyceride in lipid droplets in the convoluted tubules of the renal cortex[142].

Slices of rat-kidney cortex take up 0.2 µequiv./g/min when the FFA content in the medium is 2 µequiv./ml. About 25% of the FFA taken up are oxidized to CO_2, 2–5% are oxidized to ketone bodies and the rest are esterified. About half of the CO_2 and ketone bodies produced are derived directly from added FFA and the remainder, from endogenous lipid[182]. Because of the rapid oxidation of ketone bodies by kidney-cortex slice, maximal ketogenesis is observed *in vitro* only when the tricarboxylic acid cycle is depressed by inhibitors[182]. Gluconeogenesis by kidney slices is readily stimulated by fructose, glycerol, and pyruvate; it is not augmented by supplementation of the medium with palmitate[89].

The cells of the proximal convoluted tubules of the kidney remove organic acids from the blood stream and secrete them into the lumen of the tubules. This transport mechanism, which is dependent on energy derived from cellular metabolism, is inhibited by α-ketoglutarate, probenecid and chlorothiazide. Since these substances also inhibit FFA uptake by the kidney, it has been suggested that the organic acid transport mechanism may be involved in the uptake of plasma FFA by the cells of the convoluted tubules[8].

(c) Skeletal muscle

Fatty acid oxidation provides more than 50% of the energy required by skeletal muscle at rest[47]. About one-fifth of the fatty acids oxidized come from plasma FFA and the remainder from muscle lipid[155]. Plasma palmitate, oleate and linoleate are utilized equally by skeletal muscle of the dog.

Extraction of FFA by resting muscle is proportional to FFA concentration in arterial plasma (40%) and is unaffected by either fasting (4 days) or diabetes[156]. In fasting, arterial plasma FFA concentration is increased 50% and blood flow is decreased, whereas in diabetes the FFA concentration is increased 200% and blood flow is unchanged. Thus the total amount of plasma FFA taken up by skeletal muscle is unaffected by fasting and increased two-fold by diabetes[156]. Plasma FFA concentration and FFA uptake by muscle in diabetic dogs are both lowered by insulin. Muscular contraction, in response to electrical stimulation, increases blood flow, FFA uptake and CO_2 production[156]. About 10% of the CO_2 produced is from oxidation of plasma FFA in normal and fasting dogs and about 20% in diabetics[156].

Uptake of FFA by rat diaphragm *in vitro* is a function of the molar ratio of FFA to albumin in the medium[137]. FFA, before they are utilized, mix in an intracellular pool with FFA produced by lipolysis of muscle lipid[48,49,100,101]. The rate of lipolysis in diaphragm muscle, as measured by release of glycerol, is unaffected by fasting for 40 h and is increased 50% by alloxan diabetes[51].

Insulin added *in vitro* does not reduce but, instead, augments lipolysis by 50% or more. Lipolysis in the diaphragm is decreased by hypophysectomy in alloxan-diabetic rats and increased by growth hormone and glucocorticoid treatment in hypophysectomized-diabetic rats[51]. In the diaphragm of the normal rat the turnover time of fatty acids is about 2–3 h in triglycerides and lecithin[30].

Although oxygen consumption by isolated diaphragm is slightly decreased by chronic diabetes and starvation (4 days), oxidation of FFA to CO_2 is increased two-fold[48]. Glucose decreases FFA oxidation in diaphragm of starved and diabetic rats, and insulin, which alone is ineffective, augments the effect of glucose[48,49,139]. Glucose also enhances incorporation of [^{14}C]palmitate into neutral lipid without altering FFA uptake[49,139]. The main effect of glucose on fatty acid metabolism in muscle is to divert FFA from oxidation to glyceride formation[49,100,139].

(*d*) *Heart*

The heart removes 30–50% of the plasma FFA in arterial blood[97]. Although uptake is usually proportional to arterial FFA concentration, the fractional removal rate may decrease when the FFA concentration is very high, as in diabetes[155,156]. The substrate oxidized by the heart shifts between carbo-

References p. 45

hydrate and fatty acids, depending on the nutritional and hormonal state of the animal[108]. Nearly all of the oxygen consumed by the heart is used for carbohydrate oxidation in fed dogs (cardiac RQ=1.0), about half is used for carbohydrate and half for fatty acid oxidation in dogs fasted overnight (RQ=0.86), and all is used for fatty acid oxidation in starved (4 days) and diabetic dogs (RQ=0.7). About two-thirds of the fatty acids oxidized by the heart, in all groups, are plasma FFA and the rest are derived from myocardial lipids[156]. Acetoacetate infusion decreases the extraction and oxidation of plasma FFA by the heart in dogs[9]. Part of this effect is due to the lowering of plasma FFA concentration by the ketone body, perhaps, mediated through its stimulatory effect on insulin secretion[93].

The uptake of FFA and their fate in the perfused heart are very similar to that in the heart *in vivo*; uptake is proportional to the circulating FFA concentration and about two-thirds of that taken up are oxidized to CO_2 and the rest are esterified[33,85]. Release of glycerol and FFA to the medium is increased in perfused hearts of diabetic rats, indicating an accelerated breakdown of myocardial lipids in these animals[85,117]. Epinephrine increases O_2 consumption, glucose uptake and glucose oxidation in the perfused heart of fed rats. It also stimulates lipolysis of tissue lipid in the heart. Epinephrine has no effect on the uptake of plasma FFA, but it changes their metabolic fate in the heart by diverting about 20% from oxidation to esterification[86]. Oxygen consumption is increased and glucose oxidation is decreased when the heart is perfused with medium containing a high concentration of palmitate (FFA–albumin molar ratio=3.4). These findings suggest that the calorigenic effect of epinephrine on the heart may be secondary to its lipolytic action, which would increase the FFA concentration within the tissue[27,28].

Electron-microscopic radioautographic studies of perfused heart show that FFA are routed from the cell surface through the cisternae of the sarcoplasmic (or endoplasmic) reticulum either to mitochondria, where they are immediately oxidized, or to the lateral sacs of the sarcoplasmic reticulum, where they are esterified[161]. The newly synthesized triglyceride is deposited in lipid droplets near mitochondria, to be oxidized at a later time.

(e) Other tissues

(i) Lung

The lung produces a unique phospholipid, dipalmitoyl phosphatidyl choline, that lowers the surface tension of the alveoli and is important for gaseous exchange[15,67]. The surfactant is formed from plasma palmitate by lung[50]. The rate of formation and secretion of the surfactant is markedly increased at birth when breathing ensues[50]. Radioautographic studies show that palmitate esterification occurs mainly in large alveolar cells of the lung and that radioactive lecithin is present in the lining of the alveoli[16]. Phospholipid synthesis in rabbit-lung slices is accelerated by the addition of palmitate to the medium[136]. Synthesis is also influenced, *in vivo*, by the nutritional state of the animal and by the gaseous environment and blood flow in the lung[67].

(ii) Brain

The brain extracts virtually no FFA from arterial blood[112]. Minced brain, however, oxidizes small amounts of palmitate but not enough to maintain a normal rate of oxygen consumption. Palmitate does not decrease glucose oxidation[109].

(iii) Cornea

The cornea of the eye *in vitro* takes up and incorporates oleic acid into triglyceride. It oxidizes only very small amounts of the oleic acid[1].

(iv) Adrenal

Adrenal slices incorporate palmitate into neutral lipid and phospholipid, and oxidize some to CO_2. Although ACTH stimulates FFA release by the adrenal gland, it does not affect the uptake and metabolism of palmitate added to the medium[92].

(v) Blood vessels

The aorta and vena cava, as shown by studies with tissue slices, can incorporate FFA into phospholipid and neutral lipid[159,163]. Uptake and esterification both increase with the FFA concentration in the medium, and esterification is facilitated by glucose.

4. Conclusions

Fatty acids are a readily mobilizable source of energy for virtually all tissues in the body except brain. They are stored in the adipose tissue as triglyceride and transported in the blood as free fatty acids (FFA). Although

References p. 45

there is a continuous breakdown of triglyceride to FFA and glycerol in adipose tissue, FFA are not released unless the glucose or insulin content of the blood is low. FFA mobilization in fasting and diabetes is augmented by growth hormone and glucocorticoid. These hormones, which act slowly, accelerate lipolysis through the formation of protein. The stimulation of FFA mobilization by growth hormone and glucocorticoid is suppressed by glucose and insulin. Sudden demands of the body for extra energy, as in exercise, are met by FFA mobilization through the action of the sympathetic nervous system. The chemical mediator between sympathetic nerves and fat cells is norepinephrine. It stimulates lipolysis through the formation of cyclic AMP. There are other lipolytic hormones that act through cyclic AMP, such as ACTH, glucagon, TSH and epinephrine, but their physiologic roles in FFA mobilization have not been established.

The concentration of FFA in blood plasma is related to the rate of FFA mobilization. FFA in plasma are bound primarily to albumin which has 6–7 high-energy FFA-binding sites. The affinity of the sites for FFA decreases exponentially as the molar ratio of FFA bound to albumin is increased. The reduced affinity associated with higher plasma FFA concentrations probably plays an important role in the delivery of FFA from plasma to individual tissues.

The uptake of FFA by each tissue is a function of the FFA concentration (or FFA–albumin molar ratio) in the plasma, the rate of blood flow through the tissue, and the avidity of the tissue for FFA. The fate of the FFA in the tissue is determined by the specific metabolic paths present and by the concomitant metabolism of carbohydrate. In most tissues FFA are either oxidized at once or stored as triglyceride, to be oxidized later. Some of the triglyceride formed in liver, however, is secreted as lipoprotein. In nearly all tissues, but not in kidney, FFA are diverted from oxidation to triglyceride formation when glucose is metabolized.

ACKNOWLEDGEMENT

The authors are very grateful to Mrs. Jeanne Heffernan for her expert assistance in the preparation of the manuscript.

REFERENCES

1 ANDREWS, J. S., AND T. KUWABARA, *Biochim. Biophys. Acta*, 54 (1961) 315.
2 ARMATRUDA JR., T. T., AND F. L. ENGEL, *Yale J. Biol. Med.*, 31 (1959) 303.
3 ARMSTRONG, D. T., R. STEELE, N. ALTSZULER, A. DUNN, J. S. BISHOP AND R. C. DEBODO, *Am. J. Physiol.*, 201 (1961) 9.
4 ASTWOOD, E. B., *Handbook of Physiology*, Section 5, *Adipose Tissue*, Am. Physiol. Soc., Washington, D. C., 1965, pp. 529–532.
5 AYDIN, A., AND J. E. SOKAL, *Am. J. Physiol.*, 205 (1963) 667.
6 BAGDADE, J. D., D. PORTE JR. AND E. L. BIERMAN, *Diabetes*, 17 (1968) 127.
7 BAKER, N., AND M. C. SCHOTZ, *J. Lipid Res.*, 8 (1967) 646.
8 BARAC-NIETO, M., AND J. J. COHEN, *Am. J. Physiol.*, 215 (1968) 98.
9 BASSENGE, E., V. E. WENDT, P. SCHOLLMEYER, G. BLÜMCHEN, S. GUDBJARNASON AND R. J. BING, *Am. J. Physiol.*, 208 (1965) 162.
10 BIERMAN, E. L., I. L. SCHWARTZ AND V. P. DOLE, *Am. J. Physiol.*, 191 (1957) 359.
11 BOGDONOFF, M. D., AND C. R. NICHOLS, *Handbook of Physiology*, Section 5, *Adipose Tissue*, Am. Physiol. Soc., Washington, D. C., 1965, pp. 613–616.
12 BRAGDON, J. H., AND R. S. GORDON JR., *J. Clin. Invest.*, 37 (1958) 574.
13 BROCKERHOFF, H., R. J. HOYLE AND P. C. HWANG, *Biochim. Biophys. Acta*, 144 (1967) 541.
14 BRODIE, B. B., R. P. MAICKEL AND D. N. STERN, *Handbook of Physiology*, Section 5, *Adipose Tissue*, Am. Physiol. Soc., Washington, D. C., 1965, pp. 583–600.
15 BROWN, E. S., *Am. J. Physiol.*, 207 (1964) 402.
16 BUCKINGHAM, S., H. O. HEINEMAN, S. C. SOMMERS AND W. F. MCNARY, *Am. J. Pathol.*, 48 (1966) 1027.
17 BUTCHER, R. W., C. E. BAIRD AND E. W. SUTHERLAND, *J. Biol. Chem.*, 243 (1968) 1705.
18 BUTCHER, R. W., R. J. HO, H. C. MENG AND E. W. SUTHERLAND, *J. Biol. Chem.*, 240 (1965) 4515.
19 BUTCHER, R. W., AND E. W. SUTHERLAND, *J. Biol. Chem.*, 237 (1962) 1244.
20 BUTCHER, R. W., AND E. W. SUTHERLAND, *Ann. N.Y. Acad. Sci.*, 139 (1967) 849.
21 CAHILL JR., G. F., M. G. HERRERA, A. P. MORGAN, J. S. SOELDNER, J. STEINKE, P. L. LEVY, G. A. REICHARD JR. AND D. M. KIPNIS, *J. Clin. Invest.*, 45 (1966) 1751.
22 CAMPBELL, J., A. D. MARTUCCI AND G. R. GREEN, *Biochem. J.*, 93 (1964) 183.
23 CANTU, R. C., AND H. M. GOODMAN, *Am. J. Physiol.*, 212 (1967) 207.
24 CARLSON, L. A., *Ann. N.Y. Acad. Sci.*, 131 (1965) 119.
25 CARLSON, L. A., AND P. R. BALLY, *Handbook of Physiology*, Section 5, *Adipose Tissue*, Am. Physiol. Soc., Washington, D. C., 1965, pp. 557–574.
26 CARLSON, L. A., AND L. ORO, *Metabolism*, 12 (1963) 132.
27 CHALLONER, D. R., AND D. STEINBERG, *Am. J. Physiol.*, 210 (1966) 280.
28 CHALLONER, D. R., AND D. STEINBERG, *Am. J. Physiol.*, 211 (1966) 897.
29 CHERNICK, S. S., AND R. O. SCOW, *J. Biol. Chem.*, 239 (1964) 2416.
30 COLODZIN, M., E. M. NEPTUNE JR. AND H. C. SUDDUTH, *J. Lipid Res.*, 3 (1962) 234.
31 DICKENS, F., AND F. SIMER, *Biochem. J.*, 24 (1930) 1301.
32 ENGEL, H. R., D. M. BERGENSTAL, W. E. NIXON AND J. A. PATTEN, *Proc. Soc. Exptl. Biol. Med.*, 100 (1959) 699.
33 EVANS, J. R., L. H. OPIE AND J. C. SHIPP, *Am. J. Physiol.*, 205 (1963) 766.
34 EXTON, J. H., AND C. R. PARK, *J. Biol. Chem.*, 242 (1967) 2622.
35 FAIN, J. N., *Endocrinology*, 71 (1962) 633.
36 FAIN, J. N., *Advances in Enzyme Regulation*, Vol. 5, Pergamon, Oxford, 1967, p. 39.
37 FAIN, J. N., in P. W. RAMWELL AND J. E. SHAW (Eds.), *Prostaglandin Symposium*

    at the Worcester Foundation for Experimental Biology, Interscience, New York, 1967, pp. 67–75.
38 FAIN, J. N., *Ann. N.Y. Acad. Sci.*, 139 (1967) 879.
39 FAIN, J. N., *Endocrinology*, 82 (1968) 825.
40 FAIN, J. N., V. P. KOVACEV AND R. O. SCOW, *J. Biol. Chem.*, 240 (1965) 3522.
41 FAIN, J. N., V. P. KOVACEV AND R. O. SCOW, *Endocrinology*, 78 (1966) 773.
42 FAIN, J. N., AND R. O. SCOW, *Endocrinology*, 77 (1965) 547.
43 FELTS, J. M., in E. COWGILL AND L. H. KINSELL (Eds.), *Proc. Deuel Conference on Lipids*, U. S. Govt. Print. Office, Washington, D. C., 1967, p. 211.
44 FINE, M. B., AND R. H. WILLIAMS, *Am. J. Physiol.*, 199 (1960) 403.
45 FREDRICKSON, D. S., AND R. S. GORDON JR., *Physiol. Rev.*, 38 (1958) 585.
46 FREDRICKSON, D. S., AND R. S. GORDON JR., *J. Clin. Invest.*, 37 (1958) 1504.
47 FRITZ, I. B., *Physiol. Rev.*, 41 (1961) 52.
48 FRITZ, I. B., AND E. KAPLAN, *Am. J. Physiol.*, 198 (1960) 39.
49 FRITZ, I. B., AND E. KAPLAN, *Am. J. Physiol.*, 200 (1961) 1047.
50 FUJIWARA, T., AND F. H. ADAMS, *Proc. Soc. Exptl. Biol. Med.*, 128 (1968) 88.
51 GARLAND, P. B., AND R. J. RANDLE, *Biochem. J.*, 93 (1964) 678.
52 GOLD, M., AND J. J. SPITZER, *Am. J. Physiol.*, 206 (1964) 153.
53 GOODMAN, D. S., *J. Am. Chem. Soc.*, 80 (1958) 3892.
54 GOODMAN, H. M., AND E. KNOBIL, *Am. J. Physiol.*, 201 (1961) 1.
55 GOODRIDGE, A. G., *Am. J. Physiol.*, 214 (1968) 902.
56 GOLDFIEN, A., K. S. GULLIXSON AND G. HARGROVE, *J. Lipid Res.*, 7 (1966) 357.
57 GORDON JR., R. S., AND A. CHERKES, *J. Clin. Invest.*, 35 (1956) 206.
58 GORDON JR., R. S., *J. Clin. Invest.*, 36 (1957) 810.
59 GORDON JR., R. S., E. BOYLE, R. K. BROWN, A. CHERKES AND C. B. ANFINSEN, *Proc. Soc. Exptl. Biol.*, 84 (1953) 168.
60 GORIN, E., AND E. SHAFRIR, *Biochim. Biophys. Acta*, 84 (1964) 24.
61 HAVEL, R. J., *Proc. of the First Intern. Pharm. Meeting*, Vol. 2, Macmillan, London, 1963, pp. 43–62.
62 HAVEL, R. J., *Handbook of Physiology*, Section 5, *Adipose Tissue*, Am. Physiol. Soc., Washington, D. C., 1965, pp. 575–582.
63 HAVEL, R. J., *Ann. N.Y. Acad. Sci.*, 131 (1965) 91.
64 HAVEL, R. J., AND L. A. CARLSON, *Life Sci.*, 9 (1963) 651.
65 HAVEL, R. J., A. NAIMARK AND C. F. BORCHGREVINK, *J. Clin. Invest.*, 42 (1963) 1054.
66 HEIMBERG, M., A. DUNKERLEY AND T. O. BROWN, *Biochim. Biophys. Acta*, 125 (1966) 252.
67 HEINEMANN, H. O., AND A. P. FISHMAN, *Physiol. Rev.*, 49 (1969) 1.
68 HILLYARD, L. A., C. E. CORNELIUS AND I. L. CHAIKOFF, *J. Biol. Chem.*, 234 (1959) 2240.
69 HIMMS-HAGEN, J., *Pharmacol. Rev.*, 19 (1967) 367.
70 HIRSCH, J., *Handbook of Physiology*, Section 5, *Adipose Tissue*, Am. Physiol. Soc., Washington, D. C. 1965, pp. 181–189.
71 HIRSCH, R. L., D. RUDMAN, R. IRELAND AND R. K. SKRALY, *J. Lipid Res.*, 4 (1963) 289.
72 HO, R. J., AND H. C. MENG, *J. Lipid Res.*, 5 (1964) 203.
73 HOHENLEITNER, F. J., AND J. J. SPITZER, *Am. J. Physiol.*, 200 (1961) 1095.
74 HOLLENBERG, C. H., M. S. RABEN AND E. B. ASTWOOD, *Endocrinology*, 68 (1961) 589.
75 INNES, I. R., AND M. NICKERSON, in L. S. GOODMAN AND A. GILMAN (Eds.), *The Pharmacological Basis of Therapeutics*, 3rd ed., Macmillan, London, 1967, pp. 477–520.
76 ISSEKUTZ JR., B., W. M. BORTZ, H. I. MILLER AND P. PAUL, *Metabolism*, 16 (1967) 1001.

REFERENCES

77 JEANRENAUD, B., *Metabolism*, 10 (1961) 535.
78 JEANRENAUD, B., *Handbook of Physiology*, Section 5, *Adipose Tissue*, Am. Physiol. Soc., Washington, D. C., 1965, pp. 169–177.
79 JEANRENAUD, B., *Ergeb. Physiol. Biol. Chem. Exptl. Pharmakol.*, 60 (1968) 57.
80 JOEL, C. D., *Handbook of Physiology*, Section 5, *Adipose Tissue*, Am. Physiol. Soc., Washington, D. C., 1965, pp. 59–85.
81 JONES, A. L., N. B. RUDERMAN AND M. G. HERRERA, *J. Lipid Res.*, 8 (1967) 429.
82 JUNGAS, R. L., AND E. G. BALL, *Biochemistry*, 2 (1963) 383.
83 KOVACEV, V. P., AND R. O. SCOW, *Am. J. Physiol.*, 210 (1966) 1199.
84 KREBS, H. A., *Proc. Intern. Union of Physical Sci.*, VI (1968) 227.
85 KREISBERG, R. A., *Am. J. Physiol.*, 210 (1966) 379.
86 KREISBERG, R. A., *Am. J. Physiol.*, 210 (1966) 385.
87 LANDS, W. E. M., *Ann. Rev. Biochem.*, 34 (1965) 313.
88 LEBOVITZ, H. E., AND F. L. ENGEL, *Handbook of Physiology*, Section 5, *Adipose Tissue*, Am. Physiol. Soc., Washington, D. C., 1965, pp. 541–548.
89 LEE, J. B., V. K. VANCE AND G. F. CAHILL JR., *Am. J. Physiol.*, 203 (1962) 27.
90 LEVIN, L., AND R. K. FARBER, *Recent Progr. Hormone Res.*, 7 (1952) 399.
91 LOSSOW, W. J., AND I. L. CHAIKOFF, *Arch. Biochem. Biophys.*, 57 (1955) 23.
92 MACHO, L., AND M. SAFFRAN, *Endocrinology*, 81 (1967) 179.
93 MADISON, L. L., D. MEBANE, R. H. UNGER AND A. LOCHNER, *J. Clin. Invest.*, 43 (1964) 408.
94 MAYES, P. A., *Metabolism*, 11 (1962) 781.
95 MCCALLA, C., H. S. GATES JR. AND R. S. GORDON JR., *Arch. Biochem. Biophys.*, 71 (1957) 346.
96 MILLER, H. I., *Metabolism*, 16 (1967) 1096.
97 MILLER, H. I., M. GOLD AND J. J. SPITZER, *Am. J. Physiol.*, 202 (1962) 370.
98 NAPOLITANO, L. M., *Ann. N.Y. Acad. Sci.*, 131 (1965) 34.
99 NAPOLITANO, L., *Handbook of Physiology*, Section 5, *Adipose Tissue*, Am. Physiol. Soc., Washington, D. C., 1965, pp. 109–123.
100 NEPTUNE JR., E. M., H. C. SUDDUTH, M. COLODZIN AND J. J. REISH JR., *J. Lipid Res.*, 3 (1962) 229.
101 NEPTUNE JR., E. M., H. C. SUDDUTH, D. R. FOREMAN AND F. J. FASH, *J. Lipid Res.*, 1 (1960) 229.
102 NESTEL, P. J., A. BEZMAN AND R. J. HAVEL, *Am. J. Physiol.*, 203 (1962) 914.
103 NESTEL, P. J., AND D. STEINBERG, *J. Lipid Res.*, 4 (1963) 461.
104 NEWMAN, D., AND A. NAIMARK, *Am. J. Physiol.*, 214 (1968) 305.
105 NGAI, S. H., S. ROSELL AND L. R. WALLENBERG, *Acta Physiol. Scand.*, 68 (1966) 397.
106 NICKERSON, M., in L. S. GOODMAN AND A. GILMAN (Eds.), *The Pharmocologic Basis of Therapeutics*, 3rd ed., Macmillan, London, 1967, pp. 546–577.
107 OLIVECRONA, T., *Acta Physiol. Scand.*, 54 (1962) 295.
108 OLSON, R. E., *Handbook of Physiology*, Section 2, *Circulation*, Vol. 1, Am. Physiol. Soc., Washington, D. C., 1962, pp. 220–223.
109 OPENSHAW, H., AND W. M. BORTZ, *Diabetes*, 17 (1968) 90.
110 ORO, L., L. WALLENBERG AND S. ROSELL, *Nature*, 205 (1965) 178.
111 ORTH, R. D., M. B. FINE AND R. H. WILLIAMS, *Proc. Soc. Exptl. Biol. Med.*, 106 (1961) 339.
112 OWEN, O. E., A. P. MORGAN, H. G. KEMP, J. P. SULLIVAN, M. G. HERRARA AND G. F. CAHILL JR., *J. Clin. Invest.*, 46 (1967) 1589.
113 PENHOS, J. C., C. H. WU, A. LEMBERG, J. DAUNAS, B. BRODOFF, A. SODERO AND R. LEVINE, *Metabolism*, 17 (1968) 246.
114 PERSSON, B., G. STERKY AND J. THORELL, *Metabolism*, 16 (1967) 714.

115 RABEN, M. S., *Handbook of Physiology*, Section 5, *Adipose Tissue*, Am. Physiol. Soc., Washington, D. C., 1965, pp. 331–334.
116 RABEN, M. S., AND C. H. HOLLENBERG, *J. Clin. Invest.*, 38 (1959) 484.
117 RANDLE, P. J., P. B. GARLAND, C. N. HALES, E. A. NEWSHOLME, R. M. DENTON AND G. J. POGSON, *Rec. Progr. Hormone Res.*, 22 (1966) 1.
118 RENOLD, A. E., AND G. F. CAHILL JR. (Eds.), *Handbook of Physiology*, Section 5, *Adipose tissue*, Am. Physiol. Soc., Washington, D. C., 1965.
119 RIZACK, M. A., *J. Biol. Chem.*, 236 (1961) 657.
120 RIZACK, M. A., *J. Biol. Chem.*, 239 (1964) 392.
121 RIZACK, M. A., *Handbook of Physiology*, Section 5, *Adipose Tissue*, Am. Physiol. Soc., Washington, D. C., 1965, pp. 309–311.
122 ROBERT, A., AND R. O. SCOW, *Am. J. Physiol.*, 205 (1963) 405.
123 ROBINSON, D. S., *Advan. Lipid Res.*, 1 (1963) 133–182.
124 ROBISON, G. A., R. W. BUTCHER AND E. W. SUTHERLAND, *Ann. N.Y. Acad. Sci.*, 139 (1967) 703.
125 RODBELL, M., *J. Biol. Chem.*, 239 (1964) 753.
126 RODBELL, M., *Ann. N.Y. Acad. Sci.*, 131 (1965) 302.
127 RODBELL, M., *Handbook of Physiology*, Section 5, *Adipose Tissue*, Am. Physiol. Soc., Washington, D. C., 1965, pp. 471–482.
128 RODBELL, M., A. B. JONES, G. E. CHIAPPE DE CINGOLANI AND L. BIRNBAUMER, *Rec. Progr. Hormone Res.*, 24 (1968) 215.
129 RODBELL, M., AND R. O. SCOW, *Handbook of Physiology*, Section 5, *Adipose Tissue*, Am. Physiol. Soc., Washington, D. C., 1965, pp. 491–498.
130 RODBELL, M., R. O. SCOW AND S. S. CHERNICK, *J. Biol. Chem.*, 239 (1964) 385.
131 ROSE, G., AND B. SHAPIRO, *Bull. Res. Council Israel*, 9-A (1960) 15.
132 ROSE, H., M. VAUGHAN AND D. STEINBERG, *Am. J. Physiol.*, 206 (1964) 345.
133 ROSELL, S., *Acta Physiol. Scand.*, 67 (1966) 343.
134 RUDMAN, D., M. DI GIROLAMO, M. R. MALKIN AND L. A. GARCIA, *Handbook of Physiology*, Section 5, *Adipose Tissue*, Am. Physiol. Soc., Washington, D. C., 1965, pp. 533–539.
135 SALAMAN, M. R., AND D. S. ROBINSON, *Biochem. J.*, 99 (1966) 640.
136 SALISBURY-MURPHY, S., D. RUBENSTEIN AND J. C. BECK, *Am. J. Physiol.*, 211 (1966) 988.
137 SCHONFELD, G., *J. Lipid Res.*, 9 (1968) 453.
138 SCHONFELD, G., AND D. M. KIPNIS, *Diabetes*, 17 (1968) 422.
139 SCHWARTZMAN, L. I., AND J. BROWN, *Am. J. Physiol.*, 199 (1960) 235.
140 SCOW, R. O., in C. F. CORI, V. G. FOGLIA, L. F. LELOIR AND S. OCHOA (Eds.), *Perspectives in Biology*, Elsevier, Amsterdam, 1963, pp. 150–157.
141 SCOW, R. O., *Handbook of Physiology*, Section 5, *Adipose Tissue*, Am. Physiol. Soc., Washington, D. C., 1965, pp. 437–453.
142 SCOW, R. O., AND S. S. CHERNICK, *Rec. Progr. Hormone Res.*, 16 (1960) 497.
143 SCOW, R. O., AND S. S. CHERNICK, *Am. J. Physiol.*, 210 (1966) 1.
144 SCOW, R. O., S. S. CHERNICK AND M. S. BRINLEY, *Am. J. Physiol.*, 206 (1964) 796.
145 SCOW, R. O., S. S. CHERNICK AND B. A. GUARCO, *Diabetes*, 8 (1959) 132.
146 SCOW, R. O., F. A. STRICKER, T. Y. PICK AND T. R. CLARY, *Ann. N.Y. Acad. Sci.*, 131 (1965) 288.
147 SHAFRIR, E., S. GATT AND S. KHASIS, *Biochim. Biophys. Acta*, 98 (1965) 365.
148 SHAFRIR, E., AND E. GORIN, *Metabolism*, 12 (1963) 580.
149 SHAPIRO, B., *Handbook of Physiology*, Section 5, *Adipose Tissue*, Am. Physiol. Soc., Washington, D. C., 1965, pp. 217–221.
150 SHELDON, H., *Handbook of Physiology*, Section 5, *Adipose Tissue*, Am. Physiol. Soc., Washington, D. C., 1965, pp. 125–139.

REFERENCES

151 SHOEMAKER, W. C., P. J. CARRUTHERS, D. H. ELWYN AND J. ASHMORE, *Am. J. Physiol.*, 203 (1962) 919.
152 SÖLING, H. D., R. KATTERMAN, H. SCHMIDT AND P. KNEER, *Biochim. Biophys. Acta*, 115 (1966) 1.
153 SPECTOR, A. A., K. JOHN AND J. E. FLETCHER, *J. Lipid Res.*, 10 (1969) 56.
154 SPENCER, W. A., AND G. DEMPSTER, *Can. J. Biochem. Physiol.*, 40 (1962) 1705.
155 SPITZER, J. J., in K. RODAHL AND B. ISSEKUTZ (Eds.), *Fat as a Tissue*, McGraw-Hill, New York, 1964, pp. 215–227.
156 SPITZER, J. J., AND M. GOLD, *Ann. N.Y. Acad. Sci.*, 131 (1965) 235.
157 SPITZER, J. J., AND F. J. HOHENLEITNER, *J. Lipid Res.*, 2 (1961) 396.
158 STEIN, O., R. O. SCOW AND Y. STEIN, *Am. J. Physiol.*, in the press.
159 STEIN, O., Z. SELINGER AND Y. STEIN, *J. Atheroscler. Res.*, 3 (1963) 189.
160 STEIN, O., AND Y. STEIN, *J. Cell Biol.*, 33 (1967) 319.
161 STEIN, O., AND Y. STEIN, *J. Cell Biol.*, 36 (1968) 63.
162 STEIN, Y., AND B. SHAPIRO, *Am. J. Physiol.*, 196 (1959) 1238.
163 STEIN, Y., AND O. STEIN, *J. Atheroscler. Res.*, 2 (1962) 400.
164 STEINBERG, D., P. J. NESTEL, E. R. BUSKIRK AND R. H. THOMPSON, *J. Clin. Invest.*, 43 (1964) 167.
165 STEINBERG, D., AND M. VAUGHAN, *Handbook of Physiology*, Section 5, *Adipose Tissue*, Am. Physiol. Soc., Washington, D. C., 1965, pp. 335–347.
166 STEINBERG, D., M. VAUGHAN, P. J. NESTEL, O. STRAND AND S. BERGSTROM, *J. Clin. Invest.*, 43 (1964) 1533.
167 STRAND, O., M. VAUGHAN AND D. STEINBERG, *J. Lipid Res.*, 5 (1964) 544.
168 STRUCK, E., J. ASHMORE AND O. WIELAND, *Biochem. Z.*, 343 (1965) 107.
169 TRACHT, M. E., M. S. GOLDSTEIN AND E. R. RAMEY, *Am. J. Physiol.*, 187 (1956) 2.
170 URGOITI, E. J., B. A. HOUSSAY AND C. T. RIETTI, *Diabetes*, 12 (1963) 301.
171 VAUGHAN, M., *J. Lipid Res.*, 2 (1961) 293.
172 VAUGHAN, M., *J. Biol. Chem.*, 237 (1962) 3354.
173 VAUGHAN, M., J. E. BERGER AND D. STEINBERG, *J. Biol. Chem.*, 239 (1964) 401.
174 VAUGHAN, M., AND D. STEINBERG, *J. Lipid Res.*, 4 (1963) 193.
175 VAUGHAN, M., AND D. STEINBERG, *Handbook of Physiology*, Section 5, *Adipose Tissue*, Am. Physiol. Soc., Washington, D. C., 1965, pp. 239–251.
176 VOLLE, R. L., AND G. B. KOELLE, in L. S. GOODMAN AND A. GILMAN (Eds.), *The Pharmacological Basis of Therapeutics*, 3rd ed., Macmillan, London, 1967, pp. 578–595.
177 WADSTROM, L. D., *Nature*, 179 (1957) 259.
178 WASSERMAN, K., AND H. S. MAYERSON, *Am. J. Physiol.*, 165 (1951) 15.
179 WENKE, M., D. LINCOVÁ, J. ČEPELIK, C. ČERNOHORSKÝ AND S. HYNIE, *Ann. N.Y. Acad. Sci.*, 139 (1967) 860.
180 WERTHEIMER, E., M. HAMOSH AND E. SHAFRIR, *Am. J. Clin. Nutr.*, 8 (1960) 705.
181 WERTHEIMER, E., AND B. SHAPIRO, *Physiol. Rev.*, 28 (1948) 451.
182 WIEDEMANN, M. J., *Proc. Intern. Union Physiol. Sci.*, VI (1968) 229.
183 WILLIAMSON, J. R., E. T. BROWNING, P. SCHOLZ, R. A. KREISBERG AND I. B. FRITZ, *Diabetes*, 17 (1968) 194.
184 WILLIAMSON, J. R., AND P. E. LACY, *Handbook of Physiology*, Section 5, *Adipose Tissue*, Am. Physiol. Soc., Washington, D. C., 1965, pp. 201–210.
185 WINDMUELLER, H. G., *J. Biol. Chem.*, 239 (1964) 530.
186 WINDMUELLER, H. G., AND R. I. LEVY, *J. Biol. Chem.*, 243 (1968) 4878.
187 WIRSÉN, C., *Handbook of Physiology*, Section 5, *Adipose Tissue*, Am. Physiol. Soc., Washington, D. C., 1965, pp. 197–199.

Chapter I

Assimilation, Distribution and Storage

Section C

The Function of the Plasma Triglycerides in Fatty Acid Transport

D. S. ROBINSON

External Staff, Medical Research Council, Department of Biochemistry, University of Oxford (Great Britain)

1. Introduction

This review begins with a summary of the major conclusions which may be drawn from the material presented in it. This form of presentation has been adopted because, while the functions of the plasma triglycerides in the transport of fatty acids can be described in general terms, it is not possible, at the present time, to reach definite conclusions about several specific matters which need to be considered. Without a clear, if somewhat oversimplified, picture before him at the outset, the reader could find himself in a position of confusion when such topics are being dealt with. This would reflect on the author's exposition of his subject, rather than on the reader's ability to comprehend it, and seemed best avoided at all costs.

(a) The plasma triglycerides

Triglycerides are present in the plasma in association with other lipids and with proteins in lipoprotein complexes. These can best be considered in the

References p. 105

present context as providing a mechanism for the transport of the water-insoluble triglycerides in an aqueous medium. Most of the plasma triglycerides are carried in two lipoprotein classes—the chylomicrons and the very-low-density lipoproteins (VLDL).

As part of their structure, both the chylomicrons and the VLDL contain two other lipoproteins which also exist independently in the plasma; the high-density or α-lipoproteins (HDL) and the low-density or β-lipoproteins (LDL). The VLDL can, in fact, be regarded as consisting of an association of HDL and LDL components, and perhaps additional protein, together with the load of triglyceride which is being transported. Chylomicrons resemble the VLDL in several respects and probably have a similar structure. They are, however, considerably larger and of a lower density since the proportion of triglyceride which they contain is greater. They are of sufficient size to produce a turbidity in any plasma which contains them in appreciable numbers.

Chylomicron triglycerides contain predominantly fatty acids of the diet and enter the plasma from the intestine *via* the thoracic duct. The triglycerides of the VLDL, on the other hand, are released into the plasma from the liver and the origin of the fatty acids which they contain depends on the nutritional status of the animal. Changes which occur after the entry of the chylomicrons and VLDL into the plasma—for example, exchange of triglycerides between lipoprotein classes—may significantly complicate the above classification.

The mechanism of release of triglycerides from intestinal and liver cells is poorly understood at the present time. It is particular important to know to what extent association of the triglycerides with the other lipid and protein components of the chylomicrons and VLDL must occur before the release of the triglycerides and how far such components function over and over again in the transport of triglyceride loads in the plasma.

The concentration of triglycerides in the plasma at any time reflects a balance between rates of entry and removal of triglyceride fatty acids from the circulation. Thus, during the ingestion of a meal which contains appreciable quantities of fat, the rate of entry increases and the plasma concentration rises to a level at which a higher rate of removal balances the higher rate of influx. The proportion of the total plasma triglyceride which is present in the chylomicrons is high at such times.

In addition to such short-term effects of the dietary intake of fatty acids on the plasma triglyceride concentration, alterations in the composition of the diet, or in hormonal balance, may produce more sustained changes. These may be due either to alterations in the rate of influx of triglycerides into the circulation from the liver or to changes in the rate of removal of triglyceride fatty acids from the blood by the extrahepatic tissues.

(b) *Triglyceride fatty acid transport*

(i) *In the fasting state*

In the fasting animal, fatty acids are mobilized from adipose tissue and carried in the bloodstream as free fatty acids (FFA). A considerable proportion of the circulating FFA is used directly by the body tissues as a source of metabolic energy. A part, however, is taken up by the liver. Uptake of FFA by this organ exceeds its immediate energy needs and the excess is either oxidized partially to ketone bodies or converted to triglycerides, both of which are then released again from the liver into the blood, the latter in the VLDL.

Removal of the triglyceride fatty acids of the VLDL from the blood occurs by virtue of the hydrolysis of the triglycerides in the capillary beds of the extrahepatic tissues by the enzyme, clearing factor lipase (EC 3.1.1.3) (or lipoprotein lipase). The FFA produced, like those mobilized directly from adipose tissue, pass readily out of the blood into the tissues. The eventual fate of triglyceride fatty acids taken up by the extrahepatic tissues in the fasting state is presumably, therefore, the same as that of the circulating FFA.

In the fasting state, the activity of clearing factor lipase in adipose tissue is low. Thus, the plasma triglyceride fatty acids do not return to their tissue of origin. Instead they are directed predominantly towards the muscular tissues and oxidation.

The significance of the plasma triglyceride fatty acids in meeting the metabolic requirements of the tissues in the fasting state is difficult to assess. However, it seems that, in the rat, they may satisfy some 10–20% of the total caloric needs in the resting animal.

(ii) *Under conditions of enhanced fatty acid mobilization*

Increased mobilization of fatty acids from adipose tissue occurs under a variety of conditions—for example, during stress or, experimentally, after the administration of certain hormones which increase the rate of lipolysis of the triglycerides stored in adipose tissue. Under such conditions, the concentration of triglycerides, as well as of FFA, in the plasma may be

References p. 105

significantly increased. Such increases in concentration are probably due, at least in part, to greater influx of FFA into the liver, with consequent increased hepatic triglyceride formation and output of VLDL triglycerides into the plasma. However, a contributing factor may be decreased activity of clearing factor lipase in the extrahepatic tissues, leading to a diminution in the rate of uptake of the triglyceride fatty acids from the blood.

Mobilization of fatty acids from adipose tissue is increased during exercise. However, there is evidence that the proportion of the mobilized FFA that enter the liver is reduced and increased hepatic triglyceride output into the plasma during exercise has not been demonstrated. Similarly enhanced output of triglycerides from the liver has not been shown unequivocally in uncontrolled diabetes, despite the increased fatty acid mobilization from adipose tissue. The rise in plasma triglyceride concentration in this condition could be due, at least in part, to decreased efficiency of triglyceride fatty acid removal from the blood.

(iii) *In the fed state*

When the caloric intake is in excess of the immediate metabolic needs of the tissues and the diet contains significant amounts of fat, there is an increased influx of chylomicron triglycerides into the circulation. This is usually sufficient to produce turbidity in the plasma, the so-called alimentary lipaemic response to the ingestion of a fatty meal.

Removal of the triglyceride fatty acids from the blood depends again on the hydrolysis of the triglycerides by clearing factor lipase action in the extrahepatic tissues. Unlike the situation in the fasting state, the activity of clearing factor lipase in adipose tissue is high in the fed animal. A considerable proportion of the plasma triglyceride fatty acids is, therefore, taken up by this tissue to replenish triglyceride stores depleted during fasting. However, uptake by the muscular tissues also continues, to provide fatty acids for immediate oxidation and, possibly also, for the replenishment of muscle triglyceride stores.

Some of the chylomicron triglyceride fatty acids also appear in the liver. At the present time the mechanism of this uptake is in dispute. Direct removal of chylomicron triglycerides by the liver may take place. Alternatively, uptake of FFA produced by hydrolysis of the chylomicron triglycerides at extrahepatic sites could occur. It is assumed that a proportion of the chylomicron triglyceride fatty acids taken up by the liver is released again into the plasma in the VLDL.

In animals that are fed diets rich in carbohydrate but poor in fat, a proportion of the carbohydrate is converted to fatty acids in the body. Such conversion occurs in the liver and in adipose tissue. Fatty acids formed in the liver by lipogenesis are converted to triglycerides and released into the plasma in VLDL. Because mobilization of the adipose tissue fatty acids into the blood is normally low in the fed state, no significant net influx of FFA into the liver from the plasma occurs. Thus the VLDL triglyceride fatty acids are predominantly those that have been synthesized in the liver.

The fate of the VLDL triglyceride fatty acids in the fed animal is essentially the same as that of chylomicron triglyceride fatty acids and a considerable proportion is taken up and stored in adipose tissue. The triglycerides thereof are, therefore, replenished both from fatty acids synthesized *in situ* and from fatty acids formed in the liver and delivered to adipose tissue in the plasma triglycerides.

(c) The directive function of clearing factor lipase

The extent to which the constituent fatty acids of the plasma triglycerides are taken up by a tissue will depend on the clearing factor lipase activity of that tissue relative to its activity in other tissues. The activity of the enzyme at particular tissue sites can change rapidly and independently as circumstances alter. Hence, the enzyme will have a directive role in the distribution of the plasma triglycerides. Evidence for this role is provided by the reciprocal changes in muscle and adipose tissue clearing factor lipase activity which occur with changes in nutritional state and which are associated with corresponding reciprocal changes in the extent of triglyceride fatty acid uptake by these tissues. A further example is provided by the mammary gland. Activity of clearing factor lipase in this organ is high only when uptake of triglyceride fatty acids from the blood occurs during lactation.

It is not yet clear whether such alterations in the clearing factor lipase activity of particular tissues involve changes in the amount of the enzyme present or whether they are due to the interconversion of different forms of the enzyme. Moreover, although the activity of the enzyme seems to be under hormonal control, the factors involved in regulating the activity in particular tissues remain to be elucidated. In adipose tissue, there is evidence that the concentration of cyclic AMP in the tissue may be one such regulatory factor. This raises the possibility that the reserves of both the energy stores of the body, triglyceride as well as glycogen, may be determined by the tissue cyclic AMP concentration.

References p. 105

2. The plasma lipoproteins

The plasma lipoproteins have been the subject of several recent reviews[1-7]. Attention is directed here mainly towards those matters that are particularly relevant to the function of the lipoproteins as carriers of triglyceride in the plasma.

Four main classes of the plasma lipoproteins are now generally distinguished in man: these are, the high-density or α-lipoproteins (HDL), the low-density or β-lipoproteins (LDL), the very-low-density or pre-β-lipoproteins (VLDL) and the chylomicrons. It is usually assumed that a similar classification is valid for other animal species. However, it should be recognized that the total concentration of lipoproteins, the lipid distribution within each lipoprotein class, and the relative proportions of HDL, LDL and VLDL in the plasma varies from species to species[8].

(a) The high- and low-density lipoproteins (HDL and LDL)

These two lipoprotein classes contain most of the phospholipid and cholesterol of the plasma, but only a small proportion of the triglyceride. Some of their characteristic features in man are shown in Table I. It is important to emphasize that variations in the concentration and in the composition of the plasma HDL and the LDL, as these are commonly isolated by ultracentrifugal fractionation, occur in different human populations and in different disease states[3-6].

TABLE

CHARACTERISTICS OF HIGH-DENSITY AND

Lipoprotein class	Average plasma concentration (mg/100 ml)		Molecular weight	Size (Å)	S_f class	Hydrated density
	Male	Female				
High-density	300	460	$1-4 \cdot 10^5$	50 × 300	—	1.09–1.14
Low-density	440	390	$1-3 \cdot 10^6$	150 × 350	0–20	1.01–1.05

^a The information in this table is taken from review articles which should be consulted for fuller details[1,3,7,9,17]. The plasma concentrations given are for middle-aged American men and women with no overt metabolic abnormalities. The S_f classification is one that is widely used and gives an indication of the rate of flotation of a lipoprotein in solutions of a specified density in the analytical ultracentrifuge[7]. As the S_f value increases the rate of flotation rises and the density of the lipoprotein decreases.

(b) The very-low-density lipoproteins (VLDL)

In the fasting state, or when the diet contains no significant quantity of fat, most of the plasma triglycerides are normally present in the plasma in lipoproteins of lower density (0.94–1.006) than the HDL and LDL. Such lipoproteins are referred to as very-low-density lipoproteins (VLDL). The proportions of protein, and of lipids other than triglycerides, in the VLDL are correspondingly lower than in the HDL or LDL.

The nature of the VLDL, as they exist in the plasma in man, has been intensively studied in recent years. Fredrickson and colleagues[9,10] have assembled evidence which suggests that they are, in fact, composite structures made up of HDL and LDL components, together with a triglyceride load which is being transported from the liver to the extrahepatic tissues. The concept is largely based on immunological and chemical evidence for the presence in the VLDL of HDL and LDL proteins. The lipid composition of the VLDL is generally consistent with such a composite structure. The existence of inherited conditions in man in which the plasma HDL or LDL are either absent, or present in an altered form, and in which triglyceride transport is markedly distorted, provides further evidence.

Though the concept has been adopted in the present review, it must be stressed that it does not take into account all the experimental findings[11–13a] and may need modification in the future. Moreover, while HDL and LDL components have been detected in VLDL of the rat[14,15], substantial evidence for the VLDL structure is only available for man.

I

LOW-DENSITY LIPOPROTEINS IN HUMAN PLASMA[a]

Electrophoretic mobility	Approximate percentage composition				
	Triglyceride	Phospholipid	Cholesterol ester	Cholesterol	Protein
α_1	3	22	22	3	50
β_1	6	25	39	8	22

The VLDL do not represent a single macromolecular species. In fact, several sub-groups, differing in their density, can be distinguished. However, such sub-groups appear to differ from one another primarily in the proportions of triglyceride which they contain[7,16,17]. The VLDL may, there-

References p. 105

fore, be considered on present evidence as comprising a class of lipoproteins in which the HDL and LDL constituents are associated with varying amounts of triglyceride. Those of lowest density have about 70% of their lipid in this form.

The VLDL can be separated from the HDL and LDL in the ultracentrifuge because of their lower density (S_f values—see Table I—are between 20 and 400 by definition). They are also distinguishable from these lipoproteins by electrophoresis. In human plasma on starch electrophoresis they behave as α_2-globulins[5] and on paper electrophoresis, under appropriate conditions, they move as a discrete band in front of the LDL (or β-lipoproteins)[18]. For this reason they are also referred to as pre-β-lipoproteins. They are considerably larger than either the HDL or the LDL, with estimated diameters of between 250 and 750 Å and molecular weights of the order of 5–10 million[7].

The usual concentration of VLDL triglyceride in the plasma in man after an overnight fast is about 70 mg/100 ml (range 25–150 mg/100 ml)[9,19]. In individuals on a mixed diet, the plasma triglyceride concentration rises during the course of a day, with the influx into the plasma of triglycerides of primarily dietary origin contained in chylomicrons (Section 2c, p. 59). However, VLDL can still be isolated from the plasma at such times and their triglyceride content remains high, though the contribution which they make to the total plasma triglycerides is much less than in the fasting state[9,17].

When individuals are fed diets rich in carbohydrate but low in fat, the concentration of VLDL triglyceride in the plasma normally rises over a period of several days, increases of the order of 200 mg/100 ml being commonly observed[20-23]. Usually such rises are not maintained for more than a few months but, in certain individuals, the hypertriglyceridaemia is more marked and persistent[10,24-26]. In such cases, lipoprotein complexes large enough to scatter light are present in the plasma. The molecular weight of such "particles" has been estimated to be of the order of 100 million and their diameter as greater than 750 Å. They can be readily separated from the VLDL by ultracentrifugation as a fraction of $S_f > 400$. In the present state of our knowledge, they are best considered as being VLDL with which abnormally large quantities of triglyceride have associated[27].

Subjects susceptible to such excessive rises of plasma triglyceride concentration on high-carbohydrate diets frequently also show abnormalities of carbohydrate utilization characteristic of the pre-diabetic condition[9,10,25,]

[28,29] and the disorder has been reported to be common in young and middle-aged individuals with ischaemic heart disease[30-32]. Hypertriglyceridaemia, associated with a high concentration of triglycerides in the VLDL, is also more generally linked with an increased incidence of ischaemic heart disease[19,33-37] (see also ref. 38) and may represent a stage in the development of maturity-onset diabetes[38a].

VLDL triglyceride concentration in the plasma in man is also increased in uncontrolled diabetes[39-44] and in various other disease states[3,6,39].

(c) The chylomicrons

In most animal species significant quantities of fatty acids are normally present in the diet. The bulk of these, after their absorption from the intestinal lumen and passage across the intestinal cell, enter the thoracic duct, and finally the plasma, as triglycerides. They are carried in lipid particles to which the general name, chylomicrons, was first given in 1920[45,46].

(i) In the thoracic duct

Chylomicrons in the thoracic duct form a stable suspension of lipid particles in the lymph. A variety of techniques has been employed in their isolation, the most widely used being those which depend upon flotation of the lipid particles by centrifugation and their subsequent washing free of contaminating lymph constituents[47]. Because chylomicrons are heterogeneous in size and in density[48,49], and because the lymph also contains VLDL which may be difficult to separate completely from the smaller chylomicrons by centrifugation, purification by flotation techniques generally involves a compromise between conditions leading to a "pure" preparation, achieved by the exclusion of smaller and denser chylomicrons, and those leading to preparations contaminated with other lymph lipoproteins.

Despite these procedural difficulties, it is clear that the bulk of the triglycerides are present in chylomicrons of a diameter between 1000 and 5000 Å. Studies with the electron microscope support such a size distribution[50,51]. Larger particles, which are sometimes seen in isolated preparations, most likely arise by aggregation and coalescence of chylomicrons as a result of their removal from their protein environment in the lymph during washing[49]. No variations of particle size due to differences in the composition of the dietary fatty acids have yet been unequivocally established[52].

Some 10% of the lipid of thoracic duct chylomicrons is present as cholesterol, cholesterol esters and phospholipid[47]. The latter component seems to be responsible for the stability of the particles in lymph[53] and is probably oriented as a stabilizing layer at the oil–water interface around a central core of triglyceride[54,55]. The appearance of chylomicrons in the electron microscope as discrete spherical particles with a homogeneous core surrounded by a rim of electron-dense material[50,51,56] is consistent with such a structure, as is the finding that the phospholipid:triglyceride ratio increases with diminishing chylomicron size[49].

Very small amounts of protein (about 0.5–1% of the total mass) are present in chylomicrons isolated from thoracic duct lymph. In general, the more efficiently the chylomicrons have been freed from contaminating lymph components, the smaller the amount of protein that is found[47]. It seems, in fact, that a part of the chylomicron protein is lymph-lipoprotein protein adsorbed at the surface and that this can be removed during the isolation procedure. Nevertheless, it also appears that a proportion of the chylomicron protein probably has its origin in the intestinal cell and forms an integral part of the structure of the particles (Section 2e, p. 68).

The nature of the protein in isolated chylomicron preparations has been difficult to establish[57]. More than one protein is present and, from the results of end-group analysis, immunological and peptide-mapping studies, a close relationship between certain of the chylomicron proteins and particular protein fractions of the plasma HDL and LDL, which are present in the lymph and which could be absorbed at the surface of the chylomicrons, has been observed[3,58–61,94].

(ii) In the plasma

Chylomicrons isolated from plasma have a higher protein content than lymph chylomicrons and differ from them in their electrophoretic properties[47,53,57,62]. These changes are probably a result of the absorption of more lipoproteins, or constituents thereof, on to the particles after their entry into the plasma where the lipoprotein concentration is greater than it is in the lymph.

HDL and LDL, as well as other proteins, are present in isolated plasma chylomicrons[3,9]. Some may have already been associated with the particles while they were in the lymph. However, from experiments in which lymph chylomicrons have been incubated with plasma lipoproteins *in vitro*[53,63,64,109] it appears that association after entry into the plasma involves particularly the HDL[64,109] (see also refs. 62, 65).

Lymph chylomicrons, incubated in plasma *in vitro*, gain cholesterol as well as protein, but they lose phospholipid[64,66,67]. Such findings suggest that association of plasma HDL at the chylomicron surface may involve the displacement of some of the phospholipid therefrom. Exchange of the phospholipid, cholesterol and protein components of the plasma lipoproteins with these same components at the chylomicron surface also occurs during such incubations.

Studies on the composition of plasma chylomicrons are complicated by the fact that, in man, they can be separated into at least two groups, similar in their behaviour in the ultracentrifuge but differing in their behaviour on starch-block electrophoresis or on columns of polyvinylpyrollidone[68-72]. One such group, the "primary particles", have a triglyceride fatty acid composition closely resembling that of the lymph chylomicrons and these predominate in the plasma early during fat absorption. The second group, "secondary particles", have a triglyceride fatty acid composition which does not closely resemble that of the lymph chylomicrons at the early stages of the absorption of a fatty meal. As absorption proceeds, not only do they become more plentiful in the plasma but their fatty acid composition changes towards that of the chylomicrons in lymph.

The origin of "secondary particles" is unclear. Direct conversion of the plasma VLDL into "secondary particles" by net transfer of triglycerides from the chylomicrons could be envisaged. However, in the plasma of individuals fed high-carbohydrate diets, lipid particles which are thought to arise from VLDL by the acquisition of more triglyceride (Section 2b, p. 57), can be separated from "secondary particles" and differ from them in their detailed lipid composition[16,18,27,72]. A more reasonable possibility is that association of lymph chylomicrons with additional plasma lipoprotein components after their entry into the circulation (*vide supra*) takes some time. Thus, at any moment during fat absorption, the plasma will contain lymph chylomicrons with and without additional lipoprotein and these could be the "primary" and "secondary" particles. The conditions under which "secondary particles" can be produced *in vitro* are consistent with such an origin[73,74]. The differences between "primary" and "secondary" particles with regard to triglyceride fatty acid composition could reflect the occurrence of further lipid transfer or exchange such as has been shown to occur between lipoproteins *in vitro*[66,75].

It may be noted here that an additional complication is introduced into studies on plasma chylomicrons by the process of triglyceride fatty acid removal from the plasma (Section 4, p. 81). This could well lead to the formation in the plasma of a range of low-density-lipoprotein complexes of varying triglyceride content and size derived from both the VLDL and the chylomicrons[5,76]. Thus, during the absorption of dietary fatty acids, the plasma may contain, in addition to VLDL and "primary" and "secondary" chylomicrons, lipoprotein complexes derived from the VLDL and chylomicrons by triglyceride loss. Triglyceride exchange between these various complexes could lead to further heterogeneity.

(d) Formation and release of very-low-density lipoproteins (VLDL)

Though some plasma lipoprotein may be synthesized in the intestine (Section 2e, p. 68), there is general agreement that the liver is the chief site of formation. Evidence that plasma lipoprotein triglyceride, cholesterol and phospholipid are mainly formed in the liver has been available for several years[6,8,77]. Synthesis of the protein components therein has been shown to occur more recently[3].

Consideration of the mechanism of formation of the HDL and the LDL in the liver may be regarded as being outside the scope of this article. However, if the concept of the VLDL as complexes of HDL and LDL acting together to transport triglycerides in the plasma is essentially correct (Section 2b, p. 57), the question of whether triglyceride release from the liver occurs together with, or independently of, the formation and release of HDL and LDL must be considered. Association with HDL and LDL could occur in the plasma after triglyceride release or there could be alternative possibilities. The nature of the problem has been well stated by Trams and Brown[78].

Before reviewing the evidence relating to this matter, it must be stressed that most of the studies to be described have been carried out in the rat. While lipoproteins having many of the characteristics of the HDL, LDL and VLDL as described in man are present in this species[14,15,79], the total lipoprotein concentration is lower than in most human populations, the HDL form a much higher proportion of the total lipoproteins, and the plasma triglyceride concentration in the fasting state is lower[79-81].

The view that triglyceride release from the liver depends on the release of other lipid and protein components of the plasma lipoproteins is supported by the following findings:

(1) Under appropriate conditions, net triglyceride release by the rat liver can be demonstrated *in vitro* using either tissue slices or the isolated perfused

organ[14,81-85]. In such experiments the triglycerides appear in the media in complexes in which cholesterol, phospholipid and the protein components of the HDL and LDL can be identified[14,81,86-88]. These observations are clearly consistent with the view that the release of triglyceride from the liver is linked to that of the other lipid and protein components of the lipoproteins, though they do not exclude the possibility that association may occur after, rather than before, release.

(2) Triglyceride release from the liver is inhibited by the administration of compounds such as puromycin, carbon tetrachloride and ethionine which also inhibit hepatic protein synthesis[89,91]. The actions of such compounds on hepatic protein formation are admittedly in no way specific. However, the administration of orotic acid also inhibits hepatic triglyceride release and, in this case, the effect on the hepatic proteins seems to be mainly restricted to an inhibition of LDL formation[14,92]. Thus, the action of this compound suggests that triglyceride release is linked specifically with LDL formation in the liver. Support for this view is provided by the existence in man of the inherited condition of abeta-lipoproteinaemia[9] where again a specific failure in LDL formation is associated with a block in hepatic triglyceride release*.

(3) Studies with the electron microscope show that the liver normally contains lipid-rich complexes which are located in the channels of the endoplasmic reticulum and in vesicles and vacuoles of the Golgi apparatus. Such complexes can readily be distinguished from larger droplets of lipid free within the cell cytoplasm. When livers are perfused *in vitro* with media containing free fatty acids as a triglyceride precursor, the number of the complexes within these structures increases dramatically within 2 min[96,97]. Within 5 min, large numbers of the complexes are seen in vacuoles opening into the space of Disse and within the space of Disse itself. Finally, they appear in the perfusate where they can be shown to be of similar density and size to the VLDL of the plasma. During the perfusion, the lipid droplets within the cytoplasm also increase in number and size. A similar sequence of events can be shown to occur *in vivo*[98].

---

* It has been suggested that after orotic acid administration in animals[93] and in abeta-lipoproteinaemia in man[94,95], there may be present in the plasma a protein which is immunologically similar to LDL-protein but as an apoprotein—that is, without its associated phospholipid and cholesterol. If this were so, there could be in these situations a failure of association of the protein and lipid components of the LDL in the liver, or synthesis of a defective form of LDL-protein—rather than a block in LDL-protein synthesis. However, in other studies in abeta-lipoproteinaemia in man[9] and in animals treated with orotic acid[14] no evidence for the presence of such an apoprotein has been obtained.

References p. 105

Certain features of these morphological studies are of particular interest. First, essentially the same sequence of events occurs whether or not the perfusion medium contains plasma lipoproteins[96,97]. Second, puromycin, an inhibitor of protein synthesis, does not affect the formation of triglycerides in the liver, or the increase in the number of lipid droplets within the cytoplasm, but it does prevent release of the complexes into the space of Disse and the perfusion medium[97]. Third, when triglyceride release is increased by raising the FFA concentration in the perfusion medium, the amount of protein released rises correspondingly. Moreover, under such conditions, there is a rise in amino acid incorporation into the protein of the very-low-density complexes which appear in the medium[99].

Studies such as these suggest that triglyceride release by the liver may involve the following steps:

(1) Formation of triglyceride in association with the endoplasmic reticulum.

(2) Release of the triglyceride into the channels of the endoplasmic reticulum in association with other lipid classes and protein in lipoproteins of very-low-density.

(3) Movement of these complexes through the Golgi region of the cell. The nature of the involvement of the Golgi apparatus is unclear but it is possible that minor carbohydrate components that are present in the plasma lipoproteins are acquired there[100].

(4) Passage of the complexes to the cell surface and their release from the cell.

In such a scheme, the formation of triglyceride-rich lipoproteins is a prerequisite for the release of the triglycerides from the endoplasmic reticulum—presumably so that they can be transported in an aqueous medium. However, it is important to emphasize that such lipoproteins need not be identical with the VLDL as these are isolated from the plasma. Moreover, due account must be taken of observations showing that, though release of triglyceride by the perfused liver can occur in the absence of plasma lipoproteins, it is nevertheless markedly stimulated when they are present in the perfusion medium[88]. Release of triglycerides from liver microsomes[101] and from liver slices[83] is also stimulated in the presence of plasma lipoproteins. These findings suggest that the scheme as outlined may need to be elaborated to allow the triglyceride-rich lipoprotein complexes liberated from the endoplasmic reticulum to undergo further association with pre-formed plasma lipoprotein if release from the liver cells is to occur at a normal rate. Possibly release into the space of Disse, and thence into the plasma[102,103], might be facilitated by this. In such circumstances, the plasma VLDL would be more properly identified with the end-products of such further association. Since it appears that release of triglyceride from the endoplasmic reticulum may involve LDL components (*vide supra*), then HDL, possibly together with more LDL, might be concerned in the secondary association outside the liver cells. The description of an inherited condition in man in which, in the absence of normal HDL formation in the liver, there is no blockage of triglyceride release but the triglyceride is transported in the plasma in abnormal lipoproteins, is consistent with the normal involvement of the HDL in such secondary association[9].

A feature of such a modified scheme is that it could explain differences that have been observed between the turnover times of the protein and triglyceride constituents of the plasma lipoproteins *in vivo*. Thus, the circulating half-lifes of the protein moieties of injected lipoproteins have been calculated to be of the order of a few days in man and in certain experimental animals[104-109]. That of the VLDL triglyceride fatty acids in most experimental animals is, on the other hand, measured in minutes[110-113] and, even in man, where it may be appreciably longer, it still appears to be much less than that of the VLDL protein[114-116]. These findings have usually been interpreted as indicating that triglyceride and lipoprotein protein release from the liver are *not* linked directly to each other[78,89]. However, if triglyceride release occurs in association with only small amounts of lipoprotein protein and further association occurs subsequently, the disparity in turnover times could be explained.

The problem can be illustrated in a different way. In the rat, hepatic triglyceride output appears to be about 100 mg/h/kg body weight in the fasting state (Section 3b, p. 71) and the protein content of the circulating VLDL is not less than 10% of the triglyceride content[79]. If this protein was all released with the triglyceride from the liver, hepatic VLDL protein output would be of the order of 10 mg/h/kg body weight. But the total amount of lipoprotein protein in rat plasma (VLDL+LDL+HDL) is only about 30 mg/kg body weight[79,117]. An output of the order of 10 mg/h would therefore require an average turnover time of about 3 h which is much less than that observed*. If the protein content of the triglyceride-rich complexes as

---

* The amount of VLDL protein in the rat is about 1 mg/kg body weight[79,117]. Thus, if no exchange of VLDL protein with other plasma lipoprotein protein occurred, a VLDL protein output from the liver of 10 mg h/kg body weight would mean that the total plasma VLDL protein would be replaced in 0.1 h.

References p. 105

these are released is only 1% of the triglyceride content, on the other hand, the hepatic output of new lipoprotein protein would only be of the order of 1 mg/h/kg body weight and the turnover times would be compatible with those found *in vivo*.

It may be noted that there is not necessarily any disagreement between the relatively rapid labelling of the plasma lipoprotein protein that has been observed to follow the administration of radioactive amino acids[117-119] and the long turnover time of the plasma lipoprotein protein as determined *in vivo* (*vide supra*). On the basis of the scheme that has been outlined, one would expect small amounts of newly-synthesized lipoprotein protein to be released from the liver, with associated triglyceride, into a large plasma pool of lipoprotein which was itself turning over relatively slowly.

Experiments in rats, in which triglyceride release from the liver has been blocked following carbon tetrachloride administration, have shown an apparent fall in the plasma VLDL protein concentration to 50% of normal levels within 1 h[79]. This could be taken as evidence for a very rapid turnover of VLDL protein in this species. However, triglyceride removal from the circulation (Section 4, p. 81) is presumably unimpaired in such conditions. Consequently, since hepatic release is blocked, the triglyceride content of the VLDL will fall and their density will increase. As a result, they will be separated in a different lipoprotein class in the ultracentrifuge. Whether or not the VLDL protein concentration is actually changed, an apparent fall in concentration will, therefore, be observed. Such a consequence of inhibition of hepatic triglyceride release may need to be taken into account in the interpretation of other experiments in which an apparent fall in VLDL "synthesis" occurs following the administration of agents which inhibit hepatic triglyceride release.

In addition to complete HDL and LDL, there is some evidence that human plasma contains the protein moieties of these lipoproteins without their associated lipid components[3,94,120]; though, in the case of the HDL, some phospholipid remains associated with the protein. In studies with the perfused rat liver, it has been reported that such plasma lipoprotein "apoprotein" facilitates hepatic triglyceride release[93,121]. On the basis of this, it has been suggested that triglyceride release normally involves entry of pre-formed lipoprotein "apoprotein" into the liver cells. Therein the full lipid complement of the lipoprotein, including triglyceride, is acquired and VLDL are released.

Such a concept is consistent with the differences noted above in the circulating half-lifes of the triglyceride and protein components of the lipoproteins. On the other hand, it is not readily reconciled with the evidence that the release of triglycerides from the liver requires the simultaneous formation of lipoprotein protein. It seems possible, however, that the triglyceride-rich complexes released from the endoplasmic reticulum, already associated with small quantities of LDL, could subsequently combine with further quantities of lipoprotein "apoprotein", as well as with intact lipoprotein; particularly should the "apoprotein" concerned turn out to be that of the HDL. Whether circulating "apoprotein" can accept phospholipid and cholesterol and be converted to intact lipoprotein without entering the liver cell is a separate consideration which also cannot be excluded*.

Although the precise mechanism of triglyceride release from the liver is clearly not yet fully understood, it seems that the concept of the HDL and LDL acting as transporters of triglycerides in the VLDL is fairly well established. However, the extent to which the lipid and protein constituents thereof function over and over again in this capacity remains obscure. On present evidence, it seems that the HDL may have a particularly important function in this respect, while the LDL may be primarily concerned in the release process.

There is little evidence that triglyceride release and transport from the liver is ever limited, under normal physiological circumstances, by the availability of HDL or LDL. However, the blockage of triglyceride release from the liver which occurs in abeta-lipoproteinaemia[9] and that produced by hepatotoxic agents which interfere with lipoprotein formation[89,91,122], together with evidence that triglyceride release from the liver may be inhibited in choline deficiency because of a reduction in lipoprotein phospholipid formation[123-125], emphasizes the importance of these carrier components of the plasma lipoproteins in triglyceride transport. Variations in triglyceride output by the liver in different physiological circumstances (Section 3b, p. 71) could, therefore, necessitate alterations in the rates of formation of such components. Some evidence that such alterations may occur is already available[108,119,126] but further studies are warranted. Thus, the high plasma lipoprotein concentration in man, as compared to most other animal species, and the raised plasma lipoprotein concentrations found in particular individuals, could be related to a need for greater amounts of

* The source of plasma apoprotein is unknown. However, it may be derived from VLDL or chylomicrons after the triglyceride load of these complexes has been removed from the circulation (Section 4c, p. 87).

References p. 105

HDL or LDL to facilitate the transport of increased amounts of triglycerides from the liver. Altered plasma lipoprotein concentrations characteristic of particular disease states and those brought about by particular dietary regimes or drug treatments could also be secondary, in some cases, to changes in the rate of hepatic triglyceride output.

(e) *Formation and release of chylomicrons*

At least part of the phospholipid of lymph chylomicrons has its origin in the intestinal cell and it seems probable that formation of the chylomicrons depends on the introduction of the phospholipid as a stabilizing component at the surface of the particles[55]. Variations in chylomicron size in the thoracic duct lymph (Section 2ci, p. 59) could then be related to phospholipid availability in the intestinal cell. Thus, at peak periods of fat absorption, larger particles, which would require less phospholipid at their surface, would be formed.

Part of the protein in lymph chylomicrons also appears to be formed in the intestinal cell. The chylomicron protein becomes strongly labelled when radioactive amino acids are included in the diet[127] and, when intestinal preparations are incubated with labelled amino acids *in vitro*, particles of a similar density to chylomicrons, containing highly-labelled protein, are formed[128]. Striking similarities, though not absolute identity, between such protein and the proteins of washed lymph chylomicrons have been demonstrated[129].

Such findings suggest that a similar sequence of events to that postulated to occur during triglyceride release from the liver (Section 2d, p. 62) could be involved in triglyceride release from the intestine. In support of such a view are the observations that triglyceride release from the intestine, as well as from the liver, is inhibited by ethionine and puromycin administration[130,131]. Again, in abeta-lipoproteinaemia, not only is there no release of triglyceride from the liver, in the absence of LDL formation, but chylomicron release from the intestine is also inhibited[132,133].

Orotic acid administration blocks triglyceride release from the liver but does not inhibit the release of triglycerides from the intestine[14]. Though this could indicate that the two processes occurred by different mechanisms it more probably reflects a greater effectiveness of orotic acid action in the liver because of its concentration there. The inhibitory action of orotic acid on triglyceride release by the liver is believed to derive from its inhibition of LDL formation (Section 2d, p. 62) and a consequence of this is a marked reduction in the concentration of LDL in the plasma of orotic acid-treated animals. It is particularly significant, therefore, that the concentration of LDL in the thoracic-duct lymph after orotic acid administration is higher than in the plasma[134]. This indicates that the intestine, as well as the liver, releases LDL—a conclusion consistent with studies showing that some incorporation of labelled amino acids into lymph and plasma lipoproteins continues in hepatectomized animals[135]—and suggests, moreover, that chylomicron formation continues after orotic acid feeding, simply because LDL formation in the intestine is not inhibited.

In animals on fat-free diets, triglyceride release from the intestine is markedly reduced but does not cease completely (Section 3a, p. 69). It can be shown that in such circumstances most of the triglyceride is carried in the lymph in complexes which resemble the VLDL in their density and which contain most of the lymph LDL[134]. When the rate of triglyceride release from the intestinal cells is increased by feeding a fatty meal then these VLDL-like complexes in the lymph are replaced by the chylomicrons. The situation appears to correspond, therefore, to that in individuals fed carbohydrate-rich diets when the VLDL in the plasma are replaced by lipid particles, still derived from the liver, but much richer in their triglyceride content (Section 2b, p. 57). In fact, on present evidence it seems that the processes of triglyceride release from the intestine and the liver may differ only in so far as the lipoprotein complexes released by the liver normally carry a much smaller triglyceride load than do the chylomicrons released by the intestine during the absorption of a fatty meal.

It has been suggested (Section 2c, p. 59) that additional lipoprotein components associate with the chylomicrons after their entry into the lymph and the plasma. This could clearly correspond to the secondary association that has been postulated to occur after the release of triglyceride-rich complexes from the liver cells (Section 2d, p. 62). In both cases the HDL may be particularly concerned in this secondary association (Sections 2c and 2d).

3. Entry of triglycerides into the blood

(a) *From the intestine*

The rate of entry of triglycerides into the blood from the intestine varies directly with the dietary intake of fatty acids. Ingestion of a meal containing

References p. 105

appreciable amounts of fat increases the rate of entry and leads to a rise in the concentration of chylomicron triglycerides in the plasma that persists for the several hours during which the absorption of the meal continues. Because the chylomicrons are sufficiently large to scatter light this rise in concentration is normally accompanied by an increase in the plasma turbidity—the so-called alimentary lipaemic response to fat ingestion.

In man, fat may constitute some 20% of the dietary intake and provide up to 40% of the caloric needs. The time taken for the absorption of meals containing fat in these amounts is such that chylomicrons can usually be recovered from the plasma from the time of the first meal of the day onwards. The plasma triglyceride concentration is lowest before breakfast and rises to a peak a few hours thereafter. It is raised again by the midday meal and, after some decline, increases again after the evening meal[9].

The timing and duration of the tidal flows of dietary triglycerides into the plasma that are associated with each meal may be modified by a variety of factors such as the nature of the dietary fatty acids and the presence or absence of other dietary components which affect the rate of fat absorption[136-140].

Increases in the duration and magnitude of the alimentary lipaemic response are associated with increasing age and with ischaemic heart disease[141,142]. However, such changes do not seem to be due primarily to alterations in rates of entry of chylomicron triglycerides into the plasma. Rather they derive from the fact that the plasma triglyceride concentration is already raised in the fasting state in such conditions (Section 2b, p. 58) so that a normal influx of chylomicron triglycerides causes the plasma concentration to reach an abnormally high level[143-145]. Enhancement of the alimentary lipaemic response in diabetes[146] and in carbohydrate-induced hyperlipaemic states[26,27] may have a similar explanation (Section 2b, p. 58; see also Section 4f, p. 97).

An acute dietary load of carbohydrate is reported to lower the alimentary lipaemic response to a standard fat meal in man[147]. Again, however, possible effects of the carbohydrate on the rate of absorption of the dietary fatty acids appear to be of less importance than its effect on the rate of triglyceride entry into the plasma from the liver (Section 3bi, p. 75), which could itself lead to a change in the plasma triglyceride concentration, or on the rate of triglyceride fatty acid removal from the circulation (Section 4fi, p. 98). Changes in response brought about by exercise[148,149] and those resulting from the operation of a variety of environmental and genetic factors[150-154] may also be best explained in terms of effects on the rate of triglyceride fatty acid removal (Section 4f, p. 97).

In certain individuals the ingestion of a normal dietary fat load leads to a massive alimentary lipaemic response which persists for much longer than normal. Such a state is definitely associated with impaired removal of triglyceride fatty acids from the blood (Section 4c, p. 87).

Although chylomicron triglyceride fatty acids in the thoracic duct lymph resemble the dietary fatty acids fairly closely in their composition[46,138,155], a minor proportion appear to be derived from sources other than the diet[57,140,155]. Small quantities of triglyceride are also present in lipoprotein complexes of very-low-density in the intestinal and thoracic duct lymph of animals that have been fed fat-free diets or that have been starved for a prolonged period[156,157]. These findings suggest that the intestinal cell releases into the lymph some triglycerides containing fatty acids not directly derived from the diet. Such fatty acids could be synthesized by the intestine or they could be derived from the circulating plasma FFA, from exfoliated mucosal cell lipids, or from bile lipids[140,158,159].

The extent of such release is difficult to assess[156,157]. However, it has been recently estimated that in the rat on a fat-free diet about 10% of the total triglyceride entering the plasma may arise in this way[134].

(b) *From the liver*

The fatty acids of the triglycerides that are released into the bloodstream by the liver are derived from a variety of sources. In this section an attempt is made to assess the significance of these sources in different metabolic situations. No distinction is made between different fatty acid species with respect to their incorporation into the hepatic triglycerides and their subsequent secretion into the plasma. This should be borne in mind since there is evidence that the saturated and monounsaturated fatty acids, which predominate in the plasma triglycerides, are treated somewhat differently from the polyunsaturated fatty acids[160-167]. Nor has the regulation of triglyceride formation at the level of the enzymes concerned in the process been considered. The reader is referred for such information to the article in this volume by Marinetti[168].

Knowledge of the metabolic pathways open to fatty acids in the liver has been obtained by following the fate of labelled FFA that are taken up by the organ *in vivo* and *in vitro*. In studies *in vivo* radioactive FFA is injected into the bloodstream and the distribution of the label in different lipid fractions in the liver and in the plasma is determined at different times thereafter. Studies with the perfused liver or with liver slices *in vitro* allow, in addition, measurements of the appearance of radioactivity in CO_2 to be made.

Such studies show that the rates of transfer of labelled FFA to sites of oxidation or esterification in the liver are extremely rapid: thus, within 2 min of the injection of radioactive FFA into the bloodstream, the labelled fatty acids recovered from the liver are almost all in esterified form[110,169] and maximum rates of labelled CO_2 production are reached within a few minutes[170]. Radioactive triglyceride fatty acids do not reappear in the plasma until 10–15 min after such injections, however, and peak-specific activities of this fraction are not reached for 30–40 min in the rat and rabbit[110,111,158,161,171,172], for 40–100 min in the dog[112,160] or for 1–3 h in man[114–116,166,173]. These times have usually been taken to represent the period during which triglycerides, newly formed from the injected FFA in association with the endoplasmic reticulum, are passing through the various stages of transport to the exterior of the liver cell which precede their appearance in the plasma VLDL (Section 2d, p. 62). However, note should be taken of a recent study which suggests that, in the rat at least, the FFA may be first converted to a non-triglyceride esterified form[169]. The time needed for subsequent conversion of this ester to triglycerides may be responsible for part of the time-lag in the appearance of labelled triglyceride fatty acids in the plasma.

A proportion of an injected dose of labelled FFA is converted to phospholipids in the liver cell[82,110,171]. Incorporation into phospholipids is also observed when liver slices are incubated with FFA[83,164] or when livers are perfused with FFA[164,174,175]. Moreover, after the injection of a dose of radioactive FFA, the plasma lipoprotein phospholipids, as well as the triglycerides, become labelled[114,161,171,176]. These observations are significant for two reasons. First, they redirect attention to the fact that triglyceride release by the liver involves the simultaneous release of associated phospholipids and that phospholipids produced in the liver are important constituents of all the plasma lipoprotein fractions (Section 2, p. 56). Secondly, since phospholipids form a considerable proportion of the total liver lipids and their fatty acid moieties can be rapidly renewed by the Lands mechanism[177,178], the hepatic phospholipids will presumably contribute FFA to the hepatic pool(s) which serves as a source of the plasma triglycerides*.

The liver also contains considerable quantities of lipid in the form of triglyceride. In the intact liver cell, this triglyceride is believed to be contained largely in the cytoplasmic lipid droplets described in studies with the electron microscope (Section 2d, p. 62). When the liver is homogenized, it comprises the major constituent of the "floating fat" layer which separates on centrifugation. When labelled FFA are injected into the circulation, they enter this triglyceride pool but complete equilibration between it and other smaller triglyceride pools in the microsomes and mitochondria is relatively slow[82,110]. It is usually, therefore, considered as a storage pool, into which and from which fatty acids are continually passing. However, it is noteworthy that, in certain circumstances—for example, following the administration of hepatotoxic agents (Section 2d, p. 62)—the size of this pool may increase markedly over relatively short time periods and in man considerable quantities of fatty acids may pass through it[114]. Moreover, net breakdown of hepatic triglycerides has been observed when mobilization of adipose tissue fatty acid is prevented[179,180]. Such findings suggest that the flow of fatty acids through the hepatic triglyceride pool may be greater than is usually envisaged, and that it may, under certain circumstances, provide a source of metabolic energy for the peripheral tissues. At the present time hepatic lipases which act on the liver triglycerides have not been fully characterized[181-187]. It is of interest, however, that glucagon increases hepatic lipase activity and that insulin-deficiency has a similar effect[188-190].

Fatty acids are synthesized in the liver when the diet is rich in carbohydrate and poor in fat (Section 3biii, p. 78). It is generally assumed that the pathways open to such synthesized fatty acids are essentially the same as those available to FFA entering the liver from the plasma.

The extent of triglyceride output by the liver in different metabolic conditions has generally been assessed by one of two methods. In the first, doses of a labelled triglyceride precursor (FFA or glycerol) have been given intravenously and the specific activity and total radioactivity of the liver and plasma lipids have been determined at intervals thereafter. The values obtained have then been interpreted in terms of model systems constructed on the basis of estimated fatty acid fluxes in the liver through the pathways

---

* Though the fatty acids of cholesterol esters also become labelled when radioactive FFA are injected intravenously, the rate and the extent of labelling is low. Fatty acid flow through hepatic cholesterol esters seems unlikely to be of any great quantitative significance and it has been neglected here.

References p. 105

discussed above. Such models are necessarily based on a number of assumptions and, while certain of the divergent findings that have been reported can be attributed to valid differences between animal species, others are no doubt accounted for by the different techniques and models used. For a description of the various studies the reader is referred to the original papers[110-115,172,173,191,192], and, in particular, to a detailed recent analysis of the problems of interpretation[169]. The second method depends upon the fact that the plasma triglyceride concentration is the result of a balance between rates of triglyceride fatty acid entry into and removal from the circulation (Section 4, p. 81). Removal can be prevented by the use of surface-active substances—the most commonly used is the non-ionic detergent Triton WR 1339—which associate with the circulating triglycerides in such a way as to prevent the normal removal mechanisms from operating (Section 4c, p. 87). Measurement of the rate of increase of the plasma triglyceride concentration then provides a measure of the rate of triglyceride influx. The possibility that the rate of influx might be affected, either by the detergent itself, or by the accumulation of triglyceride in the plasma which it causes, has been examined[89,193]. It seems that neither of these factors need affect the usefulness of the method significantly. However, since the removal from the circulation of all triglyceride fatty acids in the plasma is blocked by the administration of such detergents, the method can only provide a measure of hepatic triglyceride release when release from the intestinal cells is negligible by comparison: that is, in the fasting state or when the diet contains no fat.

Measurements of hepatic triglyceride output by each of these techniques are complicated by two additional factors. One derives from the possible heterogeneity of the plasma triglyceride pool (Section 4a, p. 82; footnote p. 82). Exchange of triglycerides between the different plasma lipoprotein fractions can occur[75,110,116] and the existence of such heterogeneity may complicate, in particular, the interpretation of some of the isotopic studies. A second factor is the possibility of the return of a proportion of the plasma triglyceride fatty acids to the liver. This might occur in one of two ways. Either triglyceride in the plasma might be taken up as such by the liver again and be hydrolysed therein, or the constituent fatty acids of the plasma triglycerides might be released by hydrolysis in the extrahepatic tissues, exchange with the plasma FFA pool, and reenter the liver in this way (Section 4aii, p. 84). The importance of such recycling processes is not clear but they have usually been assumed to occur in the isotope studies.

The effect which Triton injection might have on any direct uptake of triglycerides by the liver is unknown. Certainly, it would presumably inhibit the hydrolysis of circulating triglycerides in the extrahepatic tissues and hence prevent any hepatic uptake of FFA derived from such triglycerides (Section 4c, p. 87).

(i) In the fasting state

In fasted animals, the plasma FFA, deriving from the mobilized triglyceride fatty acids of adipose tissue[194], are the major source of the plasma triglyceride fatty acids[115,158,195,196]. The pattern of flow of fatty acids through the liver in the fasting state can, therefore, be taken to approximate to that shown in simplified schematic form in Fig. 1. About one-third of the FFA entering the plasma from adipose tissue are normally taken up by the liver in the resting state (Route A) and are either oxidized (Route B) or esterified in the endoplasmic reticulum (Route C) and released into the circulation again as triglyceride fatty acids in the VLDL (Route D). Hepatic lipogenesis (Route I) is minimal in the fasting animal[197,198]. The flux of fatty acids through the liver phospholipids (Routes E and F) and the stored liver triglycerides (Routes G and H) is not yet clear. Such flows have been

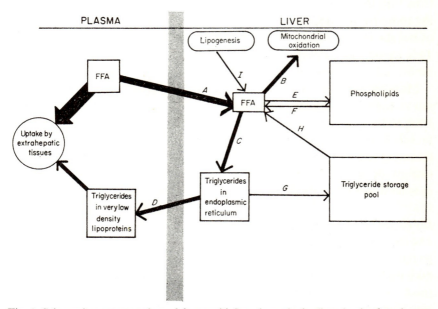

Fig. 1. Schematic representation of fatty acid flux through the liver in the fasted state.

References p. 105

shown as small in Fig. 1 and, moreover, no direct transfer of fatty acids from phospholipids or stored triglycerides to the triglycerides in the endoplasmic reticulum is indicated. Both of these views may need reassessment as a result of further work[114,169]. Possible reuptake by the liver of some of the triglyceride fatty acids released from it, either before or after their hydrolysis in the extrahepatic tissues, is not shown in Fig. 1.

Though quantitative values for hepatic triglyceride output in the fasting state must be treated with caution, agreement between the two techniques used to measure it is reasonably good where they have been applied to the same experimental animal. Thus, for the fasted male rat, hepatic outputs of triglyceride of 100 (ref. 172), 70 (ref. 89), 60–140 (ref. 80), 100 (ref. 199), and 110 (ref. 193) mg/h/kg body weight have been recorded in different studies. In the fasted female rat the output appears to be somewhat higher at about 150 mg/h/kg body weight[193]. If the fatty acids of such triglycerides were all oxidized by the extrahepatic tissues (Section 5, p. 102), outputs of this order would satisfy some 10–20% of the caloric needs of rats requiring between 5 and 6 kcal/h/kg body weight[200]. It may be noted that some 10–20% of the total fatty acids mobilized from adipose tissue in the fasted rat may recirculate in the plasma as triglyceride fatty acids[193].

In man the only figures presently available for hepatic triglyceride output are those from isotopic studies. The calculated values for individuals fasted overnight and at rest are of the order of 10–40 mg/h/kg body weight[114,115]. This would be sufficient to provide from 125 to 500 kcal/day for a 70-kg individual and, therefore, could still satisfy some 10–20% of the caloric needs.

Carbohydrate fed to fasting rats, or to human subjects that have been fasted overnight, inhibits fat mobilization and causes a fall in the plasma FFA and triglyceride concentrations[80,201,202]. In the rat the last effect may be satisfactorily explained in terms of decreased hepatic triglyceride output following reduced plasma FFA influx into the liver[80]. However, an increased rate of removal of triglyceride fatty acids from the plasma could be a secondary contributory factor (Section 4fi, p. 98).

(ii) Under conditions of enhanced fatty acid mobilization

Administration of appropriate hormones can cause increased mobilization of fatty acids from adipose tissue[194]. In susceptible species, increases in the liver and plasma triglyceride concentrations, as well as in the plasma FFA level, are observed[44,203-206]. The simplest explanation of these effects in terms of Fig. 1 is that increased influx along Route A leads to increased triglyceride formation (Route C), secretion (Route D) and hepatic storage (Route G). In support of this view are the experimental findings that increases occur in both secretion and storage of triglycerides by the perfused liver when the perfusate FFA concentration is increased[164,207,208]. However, it should be pointed out that the increases in plasma triglyceride concentration observed *in vivo* may require sustained administration of the hormones and frequently occur only several hours after the rise in plasma FFA concentration[44,203]. The possibility that the hormones may also cause a progressive impairment in triglyceride fatty acid removal from the plasma, and that this may contribute to the hypertriglyceridaemia, certainly cannot be excluded (Section 4e, p. 95, and 4fii, p. 99).

Increased mobilization of adipose-tissue fatty acids also occurs in exercise, in various stress conditions, and in certain diseases such as diabetes[194]. Hepatic triglyceride output might, therefore, be expected to be increased in such conditions.

Prolonged heavy exercise, however, usually provokes a fall in the plasma triglyceride level[209,210]. This may be attributable partly to increased triglyceride utilization by the muscular tissues (Section 5, p. 102) and partly to reduction in the blood flow to the liver and consequent decreased hepatic FFA uptake and VLDL triglyceride output[209,211-213]. It has been calculated that during heavy exercise in man the hepatic output of fatty acids can account for no more than 5% of the body's energy needs[214].

In acute and chronic stress, rises in plasma triglyceride concentration can occur[44]. However, as well as increased hepatic triglyceride release, impaired triglyceride fatty acid removal from the plasma (Section 4e, p. 95, and 4fii, p. 98) may be a contributing factor in such conditions.

Uncontrolled diabetes is frequently accompanied by raised plasma VLDL triglyceride concentrations (Section 2b, p. 57). Again, however, in addition to enhanced hepatic triglyceride output, the possibility of decreased removal of triglyceride fatty acids from the plasma has to be considered (Section 4fii, p. 99). Studies with the perfused livers of alloxan-diabetic animals have, in fact, provided somewhat paradoxical findings[86,87,208]. At equal FFA concentrations in the perfusate, FFA uptake was enhanced in the diabetic liver as compared to the control, but triglyceride, phospholipid and cholesterol release into the perfusate was inhibited. Various explanations for these findings have been considered[86-88] but, at present, increased hepatic triglyceride release cannot be considered an established feature of the diabetic condition[215,216].

References p. 105

(iii) In the fed state

Normally when the caloric intake in the form of ingested foodstuffs is greater than the caloric expenditure by the tissues, mobilization of fatty acids from adipose tissue is minimal and the plasma FFA concentration is low. In these circumstances, although FFA in the liver and plasma pools will exchange freely, net uptake of FFA from the plasma by the liver seems likely to be small[194]. If, in addition, the diet is rich in carbohydrate but free of fat, the rate of lipogenesis in the liver will be increased[197,198,217-220], while the rate of hepatic fatty acid oxidation will be reduced[174,221,222,*]. Under such circumstances, therefore, the pattern of flow of fatty acids depicted in Fig. 1 for the fasted animal may be expected to change to that shown in Fig. 2. Thus, fatty acids synthesized from dietary carbohydrate (Route I) will substitute in large measure for the influx of plasma FFA (Route A) and the oxidation of fatty acids by Route B will be reduced.

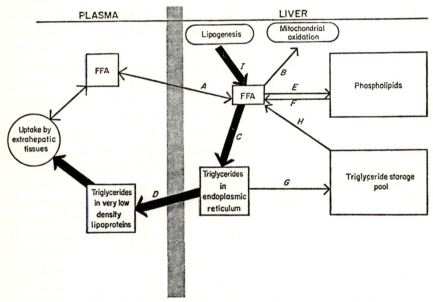

Fig. 2. Schematic representation of fatty acid flux through the liver in the fed state in animals on a high-carbohydrate, low-fat diet.

* Adaptation from a low rate to a high rate of hepatic lipogenesis when fasted animals are fed a high-carbohydrate diet may take several hours[80]. This can explain why the initial effect of feeding carbohydrate may be a lowering of the plasma triglyceride concentration (Section 3bi, p. 75).

Since the liver remains the main source of the plasma triglycerides in animals on fat-free diets[110,111,171,195], fatty acids synthesized within the liver will, therefore, provide the major triglyceride precursor *via* Routes C and D[88,223]. The flux of fatty acids through the phospholipid and stored triglyceride pools in the liver is considered in Fig. 2 to be unchanged when a fat-free diet is fed but further information on this point is needed. Re-uptake by the liver of some of the triglyceride fatty acids released into the circulation may occur in the fed, as in the fasting state, but again the extent of such re-uptake is not known.

A direct comparison of the extent of hepatic triglyceride output in the fasted state with that in the fed state in animals on fat-free and carbohydrate-rich diets has only been made in the female rat using the Triton method[193]. Output increased from 150 to 250 mg triglyceride/h/kg body weight. Values of 90 (ref. 89), 90 and 160 (ref. 224), and 170 (ref. 223) mg/h/kg body weight have been recorded for hepatic triglyceride output in male rats on such diets. In one investigation *in vitro*, using perfused livers from young male rats that had been fed a high-carbohydrate diet, values for net triglyceride release approximating those recorded *in vivo* have been obtained[88]. Only about half the newly-synthesized fatty acids appeared as triglyceride in the perfusate in this study. Perfusate FFA and hepatic and perfusate esterified lipids presumably contributed the remainder.

In man a high-carbohydrate diet leads to a rise in the plasma triglyceride concentration and, in particular subjects, abnormally high levels may be reached (Section 2b, p. 57). Although the evidence is indirect, it appears probable that increased hepatic lipogenesis, leading to increased release of triglycerides from the liver, is the primary factor contributing to the hyper-triglyceridaemia[173,191,225,226], though the relative inefficiency in man of VLDL triglyceride removal from the circulation may magnify the effect (Sections 4a, p. 82, and 4fii, p. 99). An adaptive increase in the efficiency of removal could account for the fact that the increase in plasma triglyceride concentration on high-carbohydrate diets is frequently a temporary phenomenon (Section 2b, p. 57).

The possibility that an increased incidence of ischaemic heart disease may be specifically linked to high dietary intakes of sucrose has recently been raised[227] (see also ref. 38). It is of interest, therefore, that diets containing carbohydrate in the form of fructose cause the plasma triglyceride concentration to rise to a higher level than do diets containing the equivalent amount of carbohydrate as glucose. Such an effect is probably related to a

References p. 105

more pronounced action of dietary fructose in increasing hepatic triglyceride formation and release[31,228-235].

Normally the diet contains fat as well as carbohydrate and, in these circumstances, the extent of lipogenesis in the tissues is reduced. In fact, when the proportion of fat is as high as it is in many human diets (Section 3a, p. 69), this reduction may be very marked indeed[197,236-238]. Under such conditions, it seems probable that the main supply of fatty acids to the liver in the fed state is provided from the dietary intake. Some dietary fatty acids could enter the liver directly by the portal route[45]. Of more significance, however, is the possibility that some of the chylomicron triglyceride fatty acids could be taken up by the liver after their entry into the blood from the thoracic duct. A proportion of such fatty acids could then be resecreted into the plasma in the VLDL triglycerides. Such cycling of a fraction of the dietary fatty acids through the liver would not represent net hepatic release of triglyceride fatty acids but it could allow the selection of polyunsaturated fatty acids present in the diet for incorporation into the plasma and liver phospholipids and cholesterol esters. Unfortunately, the question of the extent of uptake of chylomicron triglyceride fatty acids by the liver, either direct or indirect, is still controversial (Section 4aii, p. 84). Moreover, the techniques available to measure hepatic triglyceride release are not readily applicable in the fed state when the diet contains significant quantities of fat (*vide supra*). Consequently the extent of any cycling of dietary fatty acids through the liver is far from clear on present evidence. Studies that have been carried out in the rat suggest that net hepatic triglyceride release may be small under such conditions, being reduced to take account of the increased influx of chylomicron triglycerides into the plasma from the intestine[193]. However, further studies are clearly needed.

(iv) In various special conditions

The injection of nicotinic acid, chlorpropamide or salicylate in man causes inhibition of fatty acid mobilization and reduction of the plasma triglyceride concentration[44,239]. The latter effect is most simply explained in terms of a fall in hepatic triglyceride output secondary to reduced hepatic uptake of FFA. Possible effects of these substances on the rate of triglyceride fatty acid removal from the circulation, which could also affect the plasma triglyceride concentration, remain to be investigated.

The administration of carbon tetrachloride, ethionine, orotic acid or puromycin to experimental animals causes a fall in the plasma triglyceride concentration and a rise in the hepatic triglyceride content with the eventual development of a fatty liver. As has already been indicated (Section 2d, p. 62), such effects are probably the result of inhibition of triglyceride release from the liver due to a block in plasma lipoprotein formation. The fatty livers produced by white phosphorus administration[240], and those found in choline deficiency[123-125] and amino acid deficiency[241,242], as well as those of protein malnutrition and advanced cirrhosis in man, may be explained on a similar basis.

The fatty liver produced by alcohol administration in experimental animals seems likely, on the other hand, to be the result of increased hepatic triglyceride formation[243]. Though the factors involved are not fully understood, triglyceride release, far from being blocked, is probably occurring at a high rate in such animals[240]. An increased rate of triglyceride, and lipoprotein, release from the liver may also occur in nephrosis[3,122], though a decrease in the efficiency of triglyceride fatty acid removal from the blood could also play a role in producing the enhanced plasma triglyceride concentration which is a feature of this condition[3,141].

An increased rate of hepatic lipogenesis is reported to characterize various types of experimental obesity and it has been suggested that this could lead to increased hepatic triglyceride output, hypertriglyceridaemia and increased fatty acid deposition in adipose tissue[218,244-246]. In man, relationships between obesity and high serum triglyceride concentrations have been recorded[249,250], and correlations between excessive body weight and increased incidence of ischaemic heart disease are well established[4,247,248]. A link with the situation in the obese animal clearly cannot be excluded.

Finally, it may be noted that variations in the extent of hepatic lipogenesis occur at different stages of development[251,252]. For example, lipogenesis is negligible when the diet is extremely rich in fat during suckling. Such variations may be expected to be accompanied by corresponding changes in hepatic triglyceride output.

4. Removal of triglyceride fatty acids from the blood

The concentration of triglyceride fatty acids in the plasma normally reflects a balance between particular rates of influx and efflux: changes in concentration may therefore be initiated by alterations in either entry or removal rates. Factors concerned in determining the rate of triglyceride fatty acid entry into the blood have been considered in the previous section. The process

References p. 105

of triglyceride fatty acid removal from the blood and its regulation are discussed here.

(a) The mechanism of removal

(i) By the extrahepatic tissues

Investigations with the electron microscope have shown that the luminal surface of the blood capillaries in most extrahepatic tissues is lined by a continuous layer of endothelial cells[253], and that, when chylomicrons are injected intravenously, they are entirely confined within the vascular lumen[254-258]. When the endothelial cells are caused to separate after vascular injury[254], or when they are already separated as in newborn animals[259], chylomicrons still do not penetrate to the extravascular tissue spaces but are held up by the barrier which is then provided by the basement membrane of the endothelial cells. Such studies suggest that some change in the structure of the chylomicrons needs to occur before their component triglyceride fatty acids are removed from the blood.

Evidence to support this view is provided by studies on the rates of removal of the triglyceride fatty acids from the circulation. When chylomicrons are injected intravenously in animals, the triglyceride fatty acids thereof disappear from the blood with a half-life of only a few minutes[260]. The circulating half-life of VLDL triglyceride fatty acids is of a similar order of magnitude (Section 2d, p. 68)*. Such rapid rates of removal are in sharp contrast to the rates of removal of the soluble plasma proteins which, though they do leave the plasma in the extrahepatic tissues, do so at a very much slower rate[266]. They can also be contrasted with the slower rates of removal of minor lipid components of the chylomicrons, such as the cholesterol esters. When both the triglyceride fatty acids and the cholesterol ester fatty

---

* In man, there appears to be considerable discrepancy between the rates of removal of injected triglyceride fatty acids in chylomicrons on the one hand and in VLDL on the other. The half-lifes of the former are similar to those in animals[144,145,261] whereas those of the latter are reportedly much longer[114,191] (but see also ref. 262). A possible complicating factor with respect to the studies with VLDL triglycerides, however, is the existence of heterogeneity within the lipoprotein triglyceride fractions. In animal species a small proportion of the plasma triglyceride is in lipoproteins of higher density than the VLDL and the triglyceride fatty acids in such lipoproteins have a much slower rate of turnover[263,264]. The possibility that the VLDL triglycerides used in studies in man could be heterogeneous with respect to the rate of turnover of their constituent fatty acids needs to be borne in mind. There is already evidence for the existence of such heterogeneity in certain pathological conditions in man[265].

acids of chylomicrons are labelled with different isotopes and the chylomicrons are then injected intravenously, the rate of removal from the circulation of the former is much more rapid than that of the latter[267]. Such findings indicate that the uptake of triglyceride fatty acids by the extrahepatic tissues involves disruption of the chylomicrons and VLDL with which they are associated in the bloodstream; they suggest, moreover, that some selective process must operate in the removal process.

There is now abundant evidence to show that this selective process involves hydrolysis of the triglycerides by lipase action[76,141]. This evidence is not reviewed here in detail. However, special significance attaches to studies using chylomicrons in which the fatty acid and glycerol moieties of the triglycerides have been labelled respectively with ^{3}H and ^{14}C. At short times following the injection of such chylomicrons, the ratio ^{14}C/^{3}H in the triglycerides found in tissues such as adipose tissue and heart muscle is much lower than that in the injected triglycerides[268-271]. This indicates that lipolysis, followed by greater reutilization of the liberated fatty acid than of the glycerol, must have occurred during the transfer of fatty acids from the circulating triglycerides to those in the tissues. Similar findings have been reported in studies on the lactating rat-mammary gland[272,273], with the isolated perfused rat heart[274] and with isolated perfused rat adipose tissue. Earlier work with the latter preparation had suggested that some chylomicron triglycerides might be taken up without hydrolysis[275,276] but more recent studies have indicated that this is not the case[277].

Clearly the involvement of a lipase in the uptake of triglyceride fatty acids from the blood explains adequately the speed of the removal process since the rate of uptake and esterification by the tissues of the liberated FFA would be expected to be extremely rapid[194]. Moreover, present evidence suggests that the lipase is not likely to be saturated under normal physiological conditions. When increasing doses of chylomicron triglyceride fatty acids are injected into the plasma in the rat, the quantity removed per unit time increases up to a maximum that is only reached at plasma concentrations well above the physiological[261,278,279]. Similarly, in the dog, the removal mechanism appears adequate to deal with normal rates of entry of VLDL triglycerides into the blood[112]. In man, while the removal process does not appear to be saturated at normal chylomicron and VLDL triglyceride concentrations in the plasma[145,261], it appears that saturation may be approached with respect to the VLDL triglyceride fatty acids at the high concentrations found in some individuals on carbohydrate-rich diets

References p. 105

(Section 2b, p. 57), in so far as a small increment in triglyceride influx may then lead to a marked rise in the plasma concentration[191,226].

(ii) By the liver

The evidence which implicates lipolysis in the process of triglyceride fatty acid removal from the circulation does not exclude the possibility that some triglycerides may be taken up directly from the blood in certain tissues. On morphological grounds alone it seems most likely that such direct uptake would occur in the liver where, in contrast to other tissues, the endothelial lining of the blood capillaries or sinusoids is fenestrated[253] so that chylomicrons and VLDL can pass freely into the space of Disse and come into direct contact with the liver cells[254,280].

When animals are injected with chylomicrons or VLDL, a considerable proportion of the triglyceride fatty acids thereof do, in fact, appear in the liver[122]. Studies in which both the glycerol and fatty acid moieties of injected chylomicron triglycerides have been labelled with ^{14}C and ^{3}H respectively have shown, moreover, that the ^{14}C/^{3}H ratios in the hepatic triglycerides at early stages after such injections are the same as those in the chylomicrons[268-270,273,281]. These findings led early to the suggestion that the liver was indeed able to take up chylomicron and VLDL triglycerides without hydrolysis and further evidence to support this view was obtained in studies that were carried out in the rat with the perfused organ[84,85,174,175,207,282,283]. In fact, it was at one time considered by some investigators that virtually all chylomicron triglycerides were first taken up from the blood by the liver and that a proportion were then redistributed to the extrahepatic tissues[284].

Recently the question of hepatic uptake has been re-evaluated. It is now recognized that part of the chylomicron triglyceride found in the liver at early times after injection represents that which is "trapped" in the sinusoids and the spaces of Disse[122]. In fact, a considerable proportion of such triglyceride can be washed out of the organ again[285,286]. Moreover, experiments with the perfused rat liver have been repeated under conditions in which particular care was taken to exclude hydrolysis of the triglycerides during the perfusion[222,287]. Neither chylomicron nor VLDL triglyceride fatty acids were taken up or oxidized to a significant extent under such conditions.

These findings have, in their turn, led to still further studies[285,286,288-290]. In addition to the contribution made by trapping of injected chylomicrons in the sinusoids, it now appears that a proportion of the fatty acids found in the liver after the injection of chylomicrons could arise by hydrolysis of the triglycerides thereof in the extrahepatic tissues and recirculation of the FFA to the liver. Not all the hepatic uptake which occurs *in vivo* can yet be accounted for in this way, however. With respect to the remainder, it is still not clear whether triglyceride is taken up into the liver cells without hydrolysis or not; nor whether, if such uptake occurs, it represents a physiological process. Kupffer cells in the liver have the capacity to take up triglycerides in artificial emulsions, presumably without hydrolysis[266,283], and injected chylomicrons appear to be sequestered by these cells under certain circumstances[254,280]. When such uptake is observed it could be due to the pretreatment of the chylomicrons before their injection which may have altered their properties so as to make them more susceptible to removal by the reticulo–endothelial system[291]. Uptake of chylomicron triglycerides by liver parenchymal cells by pinocytosis has been reported[292] but other workers have not observed this[254,280]. It seems more probable, in fact, that, if uptake of triglyceride fatty acids by the liver is a significant physiological process, then it will involve lipolysis as in the extrahepatic tissues. Present evidence suggests that such lipolysis could occur at the surface of the liver parenchymal cells[187,293,294]. While rat liver is reported not to contain the lipase concerned in triglyceride fatty acid uptake by the extrahepatic tissues (Section 4c, p. 87) in an active form[187,282,295] (see also ref. 185), a role for this enzyme in triglyceride fatty acid uptake by the liver in other species cannot be excluded[296-299].

In conclusion, the role of the liver in triglyceride fatty acid removal from the blood was clearly over-estimated in earlier studies. Though its importance could well vary from species to species, it appears on present evidence that, in the rat, a maximum of about 10–15% of an injected dose of chylomicron triglyceride fatty acids may be taken up initially by the liver. In so far as part of this may represent uptake by Kupffer cells, the percentage may be lower under physiological conditions *in vivo*. Recirculation of FFA to the liver after hydrolysis of the triglycerides in extrahepatic tissues no doubt occurs *in vivo* and this process may be significant with respect to both VLDL and chylomicron triglycerides (Section 3b, p. 71).

(b) The role of particular extrahepatic tissues in triglyceride fatty acid removal

The pattern of distribution of radioactive triglyceride fatty acids in the body following their injection into the bloodstream shows that they are taken up by most of the major extrahepatic tissues, excluding the brain. Quantitative analysis of such studies carried out in whole animals is complicated by the present uncertainty regarding the importance of hepatic uptake of triglyceride fatty acids (Section 4a, p. 82). However, the results indicate that adipose tissue and the body musculature are the major sites of extrahepatic removal in the fed animal[260,263,300,301] and that uptake by the musculature is increased during exercise[271]. In the fasted animal, uptake by adipose tissue is markedly reduced *in vivo*[260,300,301] and *in vitro*[275,277,302,303].

Triglyceride fatty acid uptake and utilization by the perfused isolated heart has been studied in several investigations. In early work[274,304-307], the possibility that hydrolysis of the triglycerides had occurred in the perfusate and that uptake of FFA was being studied was not rigorously excluded. More recently this appears to have been done[308]. In the rat, rates of oxidation of perfused triglyceride fatty acids, provided as the sole exogenous substrate, are sufficient under appropriate conditions to account for over half of the concomitant oxygen utilization of the non-working heart[309]. Similar high rates of triglyceride fatty acid utilization have been reported using a heart–lung preparation[310]. When glucose, as well as chylomicron triglyceride, is available to the rat heart there is no evidence that the glucose is preferentially utilized[311].

Uptake of triglyceride fatty acids by the heart has also been studied *in vivo* but the results are not in agreement. In some investigations arteriovenous differences in plasma triglyceride concentration across the myocardium have been such as to indicate considerable uptake[312-314], but in other studies no differences have been observed[274,315,316]. It may be noted, however, that a significant proportion of the metabolic requirements of the heart in man could be met by the uptake of quantities of triglyceride fatty acids that would result in only small arteriovenous differences in total plasma triglyceride concentration.

Starvation is reported to increase triglyceride fatty acid uptake by the heart *in vivo*[317] but the results of studies with the perfused heart are not in good agreement[305,306,308]. Increased cardiac utilization *in vivo* is also reported after adrenaline infusion, after ethanol administration and in diabetes[314,318,319]. However, in studies with the perfused rat heart no increase in utilization was observed in alloxan diabetes[306].

It may be noted that when the perfused heart is used to study the effect of different physiological or pathological conditions on triglyceride fatty acid uptake, comparisons are usually made at a single perfusate triglyceride concentration. Should the plasma triglyceride concentrations differ *in vivo* under the conditions being compared, a failure to find any alteration in uptake *in vitro* would not exclude the possibility of a change in uptake *in vivo*.

An organ of particular interest in relation to triglyceride fatty acid uptake is the mammary gland. Uptake occurs only during lactation but it is extremely efficient, the amounts removed being adequate to account, at least in certain species, for most of the milk triglyceride output[320-324].

(c) The role of clearing factor lipase in triglyceride fatty acid removal

From the foregoing considerations, it is possible to state certain properties which any lipase involved in the removal of triglyceride fatty acids from the blood by the extrahepatic tissues must possess. It must be present in a wide variety of such tissues and must be active on triglycerides in chylomicrons and VLDL. Moreover, its site of action must be compatible with its suggested role in facilitating the removal from the bloodstream of the triglyceride fatty acids in the circulating chylomicrons and VLDL. Finally, a delay in their removal should occur when the activity of the enzyme is reduced and changes in the extent of their uptake by particular tissues should be accompanied by corresponding changes in the activity of the enzyme in those tissues.

A lipase which generally fulfils these requirements is that known as clearing factor lipase, lipoprotein lipase or heparin-induced lipase[76,141,325-329],*. This enzyme was first described as a factor ("clearing factor") which was present in the plasma of animals that had been injected with heparin and which was able to clear the turbidity of added chylomicron suspensions. Its identification as a lipase, liberated from the tissues by the injected heparin, led to an explanation of its clearing action in terms of hydrolysis of the chylomicron triglycerides with the liberation of FFA which were able to form water-soluble complexes with the plasma albumin. The enzyme is now known to be present in tissue extracts of adipose tissue, heart and skeletal muscle, lactating mammary gland, spleen, lung and kidney medulla and it thus fulfils the requirement of wide distribution in the tissues[141]. It is not present in the brain and triglyceride fatty acids are not taken up from the blood by this organ. In the fed animal its activity in adipose tissue and

---

* For reasons already given[76], the name "clearing factor lipase" has been preferred to "lipoprotein lipase" in this review. The term "lipoprotein triglyceride hydrolase"[330] may prove to be a suitable systematic name for the enzyme.

References p. 105

in muscle would appear to account for a high proportion of the total activity in the body, in line with the importance of these tissues in the removal of triglyceride fatty acids from the blood. During lactation the activity of the enzyme in the mammary gland may contribute a significant proportion of the body total.

Except after the injection of heparin, the level of clearing factor lipase activity in the blood is low and the plasma cannot be considered to be the normal physiological site of its action. Nevertheless, it is released into the blood within seconds of the injection intravenously of heparin, or of other much larger molecules of similar charge[331-333], and a correspondingly rapid release is observed when isolated organs are perfused with heparin *in vitro*[334,335]. These findings show that the enzyme is present at some site in the tissues immediately accessible to heparin in the vascular lumen and led, several years ago, to the suggestion that its normal site of action was at the capillary endothelial cell surface[76,141,325,336]. Hydrolysis of chylomicron and VLDL triglycerides with release of FFA at this site would evidently explain the rapid removal of triglyceride fatty acids from the circulation and would also be consistent with a number of other findings reported in the literature—for example, the presence of labelled FFA in the blood following the injection of ^{14}C-labelled chylomicron triglycerides[76] and the appearance of FFA, derived from the plasma triglycerides, in the venous outflow from a tissue rich in the enzyme such as the lactating mammary gland[320,322].

Recent studies with the electron microscope have provided important supporting evidence for such a site of action of the enzyme. Thus, it has been shown that circulating chylomicrons are sequestered at the endothelial cell surface in tissues rich in clearing factor lipase and, from histochemical evidence, it appears that their constituent triglycerides are hydrolysed at this site[254-257,259,337].

Tissue preparations of clearing factor lipase can apparently hydrolyse triglycerides completely to glycerol and FFA[338]. However, it has recently become clear that such preparations may also contain monoglyceride lipase activity[184,339-341]. This raises the possibility that hydrolysis may proceed in two stages, and by the activity of two distinct enzymes. Moreover, although the rate of monoglyceride breakdown may not normally limit the overall rate of triglyceride hydrolysis with tissue preparations of the enzyme it could do so in particular circumstances. In early studies with post-heparin plasma as an enzyme source, evidence for a partial block in hydrolysis at the monoglyceride stage was obtained[342]. More recent work has, in fact, indicated that a monoglyceride lipase is also liberated into the plasma after heparin injection[342a]. Nevertheless the possibility that hydrolysis of triglycerides at the capillary endothelial cell surface may result in the accumulation of monoglycerides cannot be excluded. Should this be so, monoglycerides as well as FFA, would need to be transported into the tissues and a clear analogy with the situation in the intestine during the absorption of a fatty meal would exist[45,46].

Post-heparin plasma is also reported to possess esterase and phospholipase activity[141,343–345]. Whether these activities reflect a relatively low specificity of clearing factor lipase or the activity of independent enzymes released by heparin cannot be decided on present evidence. However, since phospholipid is an important constituent of chylomicrons and VLDL (Sections 2d, p. 62, and 2e, p. 68), the possibility that at least part is hydrolysed during the course of the removal of the triglyceride fatty acids from the blood is clearly of considerable interest.

Although detailed consideration of the properties of clearing factor lipase is outside the scope of this article[76,141,325,326,328] its substrate specificity is of particular interest in relation to its function and deserves comment. Artificial emulsions of triglyceride are not adequate substrates for the enzyme unless they have been incubated with plasma[346–348]. During such incubations, HDL associates with the lipid particles and, as a result, the triglyceride becomes capable of being hydrolysed[62,326,327,347,349]. Lymph chylomicrons and plasma VLDL already contain HDL, or lipoprotein closely related thereto (Sections 2b, p. 57, and 2ci, p. 59), and their triglyceride is hydrolysed rapidly by clearing factor lipase in the absence of added plasma. However, lymph chylomicrons may be activated further by incubation with plasma[141] under circumstances in which they associate with additional plasma HDL (Section 2cii, p. 60).

Such specificity of clearing factor lipase strongly suggests that its physiological substrate is the triglyceride of the plasma chylomicrons or VLDL and provides, moreover, a specific function for HDL in these complexes. It seems probable, in fact, that HDL is concerned in the linkage of the enzyme to the chylomicron or VLDL[327]. Though nothing is known at present about the particular components of the HDL that may be involved in this association it may be significant that emulsions of triglyceride stabilized by phospholipid can be hydrolysed at a moderate rate by clearing factor lipase under appropriate circumstances[76].

Association of clearing factor lipase with its substrate can be shown to

References p. 105

occur *in vitro* by mixing the enzyme in solution with chylomicrons and then isolating the latter by flotation in the centrifuge[348,350]. Under such conditions, most of the enzyme is present in the chylomicron layer. Such association seems likely to be of considerable importance if the enzyme is active at the capillary endothelial cell surface since the sequestration of circulating chylomicrons and VLDL at this site could depend upon it. It could also explain the presence of low levels of clearing factor lipase in the plasma during fat absorption[325] and in the venous outflow from an organ such as the lactating mammary gland which is particularly rich in the enzyme[320]. Thus association of the enzyme with its chylomicron substrate could result not only in the sequestration of the substrate at the capillary wall surface, but also in the liberation of small quantities of the enzyme linked to chylomicrons that remained in the circulation.

The mechanism of release of clearing factor lipase from its tissue-binding site by injected heparin remains obscure[141]. There is evidence that the enzyme in the tissues contains heparin, or a related molecule, as an integral component[326,327]. Injected heparin could, therefore, release the enzyme by competition with such an endogenous moiety. It has been suggested that the endogenous heparin might also be normally involved in the linkage of the enzyme to its substrate[326,327] and that heparin may stabilize the enzyme molecule[351].

Important evidence for the function of clearing factor lipase in the removal of triglyceride fatty acids from the blood derives from studies in which a reduction in the rate of their removal is associated with a decrease in the activity of the enzyme. Of particular interest in this connection is the existence of a hypertriglyceridaemic state in man in which chylomicron triglyceride fatty acids are removed from the plasma at a much reduced rate and the level of clearing factor lipase in the tissues is abnormally low[9,352]. The injection intravenously of inhibitors of clearing factor lipase— for example, protamine sulphate, Triton WR 1339 and acidic azoproteins— also causes hypertriglyceridaemia[141,310,353]. Triton, being a surface-active agent, may act on the triglyceride substrate so as to make it inaccessible to clearing factor lipase (Section 3b, p. 71). Protamine sulphate, on the other hand, more probably acts directly on the enzyme; it may, in fact, combine with the heparin moiety and thereby destabilize the enzyme[333,348].

A direct relationship between tissue clearing factor lipase activity and triglyceride fatty acid uptake is also suggested by observations made with the isolated rat heart. When such a preparation is perfused with heparin, part of the tissue clearing factor lipase is released into the perfusate. If the preparation is then washed free of heparin and perfused next with chylomicrons, its capacity to take up and oxidize the triglyceride fatty acids thereof is much less than that of hearts not perfused with heparin[308]. In fact, the residual oxidation—approximately 20% of normal—can be entirely accounted for by the oxidation of FFA in the chylomicron preparations[309].

The clearing factor lipase activity of particular tissues in the body can vary in different physiological situations and it is of interest to consider how far such variations can be correlated with changes in the uptake of triglyceride fatty acids by these tissues. A satisfactory correlation exists with respect to the enzyme in adipose tissue and in the mammary gland. Thus, activity in adipose tissue is lowered in starvation[354-358] and in diabetes[355,359-361] and, in both these situations, there is a marked drop in the capacity of adipose tissue to take up triglyceride fatty acids *in vivo* and *in vitro*[301,361,362]. Clearing factor lipase activity can also be correlated with triglyceride fatty acid uptake in adipose tissue under other experimental conditions[302,303,362a]. In the mammary gland, the activity of clearing factor lipase is only high during lactation when triglyceride fatty acids are taken up from the blood[363,364]. During pregnancy and after weaning, the activity of the enzyme in the gland is low and uptake of triglyceride fatty acids does not occur.

Recent studies suggest that the transitory lipaemia of pregnancy, which is due essentially to a rise in the plasma triglyceride concentration, may be accounted for by changes in the clearing factor lipase activity of adipose tissue and mammary gland[365,366]. At the time when the lipaemia develops, adipose tissue clearing factor lipase activity is reduced and the rise in the plasma triglyceride concentration may, therefore, be accounted for by reduced uptake of triglyceride fatty acids by this tissue. Disappearance of the lipaemia just before parturition could then be attributed to the increased uptake of triglyceride fatty acids by the mammary gland, secondary to the appearance of high levels of the enzyme in this organ.

Correlations between triglyceride fatty acid uptake and clearing factor lipase activity are not so clear-cut in the case of heart muscle. Marked changes in its clearing factor lipase activity have been reported in different physiological and pathological situations and after a variety of experimental treatments[359,367-377]. However, the various findings are not in good agreement, nor do the changes always correlate with reported changes in the utilization of triglyceride fatty acids by the heart. For instance, though the

References p. 105

clearing factor lipase activity of the heart has been reported to be increased by fasting[369,372,377], and by an increased work load[368,369], there is conflicting evidence as to whether or not triglyceride fatty acid uptake is increased in such conditions[271,305,306,308,314,378]. Alloxan diabetes in the rat is reported to increase cardiac clearing factor lipase activity[359] and triglyceride uptake by the heart *in vivo*[319], but triglyceride fatty acid oxidation in the perfused rat heart is apparently reduced[306].

These discrepancies are not readily explained at the present time. However, apart from the difficulties of relating work *in vivo* and *in vitro* (Section 4b, p. 85), different systems have been used for the assay of clearing factor lipase in the various studies. This could explain some of the observed discrepancies. Further significance may be attached to a recent finding that very short periods of fasting produce marked changes in the activity of the enzyme in rat heart[309]. In the light of this, some of the reported variations in heart-enzyme activity, supposedly brought about by particular treatments, need to be reassessed.

The view has been advanced that clearing factor lipase may be important in determining not only the rate but also the pattern of uptake of triglyceride fatty acids from the blood[141]. Such a role for the enzyme may be of considerable significance since there appears to be essentially no method for controlling the distribution of circulating FFA to the tissues, other than by changes in tissue blood flow[194]. Changing needs of particular tissues could, however, be met by changes in the capacity to take up triglyceride fatty acids, secondary to alterations in clearing factor lipase activity. The alterations in the enzyme activity in adipose tissue and in mammary gland described above would clearly fit such a concept. An increase in heart- and skeletal-muscle clearing factor lipase activity has recently been found to coincide with the abrupt fall in the activity of the enzyme in adipose tissue which occurs on fasting for a short period in the rat[309]. This indicates that circulating triglyceride fatty acids may not only be directed away from adipose tissue in the fasting animal but that their uptake by muscular tissue may be actively increased. There is already evidence that this is the case[317], but further studies *in vivo* with more rigorous control of nutritional conditions are needed. Reciprocal changes in adipose tissue and heart-muscle clearing factor lipase activity also occur in diabetes and during exercise[359,368] and the decreased uptake of triglyceride fatty acids by adipose tissue in the diabetic animal has been shown to be accompanied by increased uptake by muscle[319,379].

(d) Localization and states of clearing factor lipase in the tissues

Evidence that the physiological function of clearing factor lipase is normally exerted at the luminal surface of the capillary endothelial cells (Section 4c, p. 87) does not exclude the possibility that the enzyme may also be found at other sites in the tissues. When heparin is injected intravenously, tissue activities of the enzyme are reduced but not to a very marked extent[329,346]. Similarly, prolonged perfusion of heparin through the rat heart[308,334], or through isolated adipose tissue[275], results in the release of only a proportion of the total tissue-enzyme activity. If heparin releases enzyme only from the endothelial cell surface, then these results indicate that the enzyme must also be situated elsewhere.

This conclusion is reinforced by the observation that clearing factor lipase is associated with fat cells isolated from adipose tissue by collagenase treatment[380]. In these studies, no enzyme activity was found in stromal vascular components isolated from the tissue (but see ref. 335). It must be emphasized that this finding should not be taken as evidence against the view that clearing factor lipase normally acts in association with the capillary endothelial cells. Thus, collagenase is an inhibitor of clearing factor lipase[381] and recent studies[382] have shown that a combination of collagenase inhibition and enzyme instability during the procedure for isolation of fat cells can lead to a loss of up to 80% of the initial enzyme activity of adipose tissue from fed rats. They have indicated further that it is the activity of enzyme outside the fat cells, which could well represent enzyme at the endothelial cell surface, which is specifically lost. Nevertheless, the finding of clearing factor lipase in isolated fat cells is clearly incompatible with the exclusive localization of the enzyme at a capillary wall site.

If clearing factor lipase is present at more than one site in adipose tissue, the possibility that it may also exist in more than one state must be considered*. Methods for the partial purification of clearing factor lipase in postheparin plasma have been described[386-388] (see also ref. 141), but purification of the enzyme in the tissues has proved difficult. In these circumstances, distinction of different enzyme states depends on the demonstration that enzyme fractions with different properties exist in the tissue. All of the enzyme in adipose tissue[382] and in heart muscle[309] has the same substrate specificity and shows the same response to inhibitors. There is, nevertheless, some

---

* The possibility that different tissue clearing factor lipases may form a group of isoenzymes is a separate consideration[383-385].

evidence that two states of the enzyme may exist. Thus, in fed rats, where the activity of the enzyme in adipose tissue is high, most of this activity is unstable to incubation at 37°. In fasted animals, on the other hand, the low activity of the enzyme in the tissue is stable under the same conditions[358]. Actinomycin treatment causes the activity of the stable enzyme in adipose tissue from fasted animals to increase markedly by an, as yet, unknown mechanism[303,358,389,390].

Present evidence suggests that the unstable form of clearing factor lipase is that which exists outside the fat cells, while stable enzyme is associated with the cells[382]. Stability of the enzyme could, in fact, simply derive from such an association. If this is so, then the transformation from the fasted to the fed state involves the appearance of clearing factor lipase outside the cells where its stability is reduced. Though it is possible that enzyme associated with the fat cells has an entirely different function, a more attractive possibility is that it is a precursor of that which appears outside the cells in the fed animal and which functions at the endothelial cell surface.

Evidence for such a precursor–product relationship is provided by studies in which intact adipose tissue from fasted animals is incubated in a serum-based medium *in vitro*[391,392]. Under appropriate conditions a progressive increase in the total clearing factor lipase activity towards that characteristic of tissue from animals in the fed state occurs and this increase in activity is largely accounted for by the appearance of unstable enzyme in the incubation medium. Similar progressive increases in activity have been observed with isolated fat cells incubated in the same medium[382].

In summary, there is evidence that clearing factor lipase in adipose tissue from fasted animals is in a stable state in association with the fat cells and that, during the transformation from the fasted to the fed condition, when the total enzyme activity of the tissue and its capacity to take up triglyceride fatty acids both increase markedly, unstable enzyme appears outside the cells. *In vivo*, such enzyme must presumably move to its proposed site of action on the luminal surface of the endothelial cells in order to fulfil its physiological function. Though any mechanism proposed for such movement must be speculative at the present time, the caveolae and vesicles, which are such prominent features of the endothelial cells, could conceivably function in the transport process[253,392a].

Studies, similar to those carried out with adipose tissue, have not been performed with other tissues. However, there is some evidence that a similar situation may exist in the heart. Thus, the rise in clearing factor lipase activity in this organ which occurs after very short periods of fasting in the rat (Section 4c, p. 87) is associated specifically with an increase in the activity of that fraction of the total enzyme which is releasable from the heart by heparin and which is presumably at the endothelial cell surface[309]. Studies on the association of clearing factor lipase in heart with particular sub-cellular fractions[369,393,394] cannot be readily interpreted at the present time.

(e) *The regulation of tissue clearing factor lipase activity*

The changes in the clearing factor lipase activity of particular tissues that have been described (Section 4c, p. 87) could result either from changes in the tissue enzyme content or from the interconversion of forms of the enzyme differing in their specific activity. Unfortunately it is not possible at present to distinguish between these possibilities. The fact that the injection of inhibitors of protein synthesis such as puromycin and cycloheximide causes a rapid fall in the activity of the enzyme in a variety of tissues[395] is consistent with the view that clearing factor lipase is normally being synthesized and broken down in the tissues with a half-life of between 1 and 2 h. In these circumstances, changes in tissue enzyme activity could evidently result from alterations in either the rate of enzyme synthesis or breakdown[392] and it is noteworthy that the increase in enzyme activity which occurs when adipose tissue from fasted animals is incubated in a serum-based medium *in vitro*, besides being inhibited by puromycin, shows other characteristics that suggest it involves the synthesis of new enzyme protein[358,392].

The findings with puromycin may be capable of another interpretation, however, since administration of this substance has been reported to increase the adenosine cyclic 3′,5′-monophosphate (cyclic AMP) content of the rat diaphragm[396]. Cyclic AMP is thought to play an important role in the regulation of glycogen synthesis and breakdown by the interconversion of enzyme forms of different specific activities[397]. It may have a similar function with respect to the interconversion of forms of the lipase concerned with the mobilization of triglyceride fatty acids from adipose tissue[398-400]. In these circumstances, a recent finding that the increase in clearing factor lipase activity on incubation of adipose tissue from fasted animals in a serum-based medium *in vitro* is inhibited by dibutyryl cyclic AMP is of considerable interest[401]. The possibility that the cyclic AMP concentration in adipose tissue may regulate the activity of the lipase concerned with the mobilization of triglyceride fatty acids stored in adipose tissue *and* the activity of clearing factor lipase concerned with the deposition of triglyceride fatty acids in the tissue is clearly an attractive one. Thus, a rise in adipose tissue cyclic AMP concentration could increase the activity of one lipase while reducing the activity of the other. If this proves to be the case then both triglyceride and glycogen storage could be determined by changes in tissue cyclic AMP concentration.

It has recently been suggested that the tissue FFA concentration might control the activity of clearing factor lipase in adipose tissue[402]. Thus, a high concentration of FFA in adipose tissue in the fasted animal, resulting from mobilization of the triglyceride stores, would inhibit clearing factor lipase and hence triglyceride fatty acid uptake. This attractive possibility was examined in the course of the studies with dibutyryl cyclic AMP[401], but it was concluded that the inhibitory action of this substance was not mediated by an induced rise in the tissue FFA concentration. Moreover, in heart muscle the increase in enzyme activity on fasting (Section 4c, p. 87) occurs when FFA concentrations in the tissue are actually raised[309]. Though these findings give little support to the notion of a regulatory role of the tissue FFA concentration, the possibility that FFA concentrations in particular tissue compartments may move in an opposite direction from those in the tissue as a whole clearly cannot be excluded.

Whatever the nature and the mechanism of action of the immediate effector of tissue clearing factor lipase, there is now abundant evidence to suggest that the activity of the enzyme is under hormonal control. The low activity of the enzyme in adipose tissue in diabetes, and the effects of hormones on the activity of the enzyme in the rat heart (Section 4c, p. 87), are consistent with such control. Further evidence is provided by the finding that the increase in clearing factor lipase activity when adipose tissue from fasted animals is incubated in a serum-based medium *in vitro* is inhibited by adrenaline, noradrenaline and adrenocorticotrophic hormone and enhanced by insulin[358,392]. These hormones also affect, though in the opposite direction, the activity of the lipase concerned with hydrolysis of the triglyceride stores of adipose tissue and may, in this way, influence the rate of triglyceride mobilization from the tissue[194]. It will evidently be of interest to see whether hormones such as growth hormone, glucagon and prostaglandin, which also affect the activity of this lipase[194], can, in addition influence clearing factor lipase activity.

Under appropriate circumstances, adrenaline, adrenocorticotrophic hormone and insulin alter the concentrations of cyclic AMP in adipose tissue in directions which are consistent with this substance being the mediator of their actions on clearing factor lipase[403]. However, it should be emphasized that the mechanism whereby clearing factor lipase activity in the tissue is reduced on fasting remains obscure. A reduction in the insulin concentration in plasma, either alone or in conjunction with a reduced glucose concentration, could conceivably play a regulatory role[401]. On the other hand, it has been suggested that the rise in the plasma concentration of growth hormone which occurs on fasting may be involved in the rise in the activity of the lipase in adipose tissue concerned with triglyceride mobilization[403a]. It could, at the same time, cause a reduction in the tissue clearing factor lipase activity.

The mechanism whereby clearing factor lipase activity is controlled in tissues such as skeletal muscle and the mammary gland has not been studied. Diabetes, exercise and fasting for a short period are all reported to cause the clearing factor lipase activity in rat-heart muscle to rise as that in adipose tissue falls (Section 4c, p. 87). It is not easy to envisage a mechanism whereby the activity of a single enzyme species in different tissues can change in different directions if the changes are controlled by alterations in hormone levels in the plasma that are common to all the tissues. However, it is of interest that prostaglandins have been recently reported to raise the cyclic AMP concentrations in certain cell types while lowering it in others[403b]. A regulatory role of some local tissue constituent, such as the tissue FFA, may also need to be considered further.

(f) Tissue clearing factor lipase activity as a determinant of the plasma triglyceride concentration

Though the mechanism for triglyceride fatty acid removal from the plasma may not be saturated at normal rates of influx (Section 4ai, p. 82), a reduction in the efficiency of removal, due to lowered tissue clearing factor lipase activity, could nevertheless lead to an increase in the plasma triglyceride concentration. Experimentally, this possibility has generally been investigated in one of two ways. In the first, the rate of removal of injected triglyceride fatty acids from the blood has been taken as an indicator of the effectiveness of the removal mechanism. In the second, the activity of clearing factor lipase in the plasma following heparin injection has been used as a measure of the activity of the enzyme in the tissues and, hence, of the efficiency of

References p. 105

triglyceride fatty acid removal. Attempts have then been made to see whether, in any particular circumstances, an inverse relationship existed between the plasma triglyceride concentration and either of these two measures of the removal process.

The interpretation of the results of such studies is difficult for a variety of reasons. The level of clearing factor lipase activity in the plasma after heparin injection, besides giving no information as to the distribution of the enzyme in different tissues, may not even be an accurate guide to the activity of the enzyme at the endothelial cell surface[141]. It depends, moreover, on the dose of heparin injected[141] and on the balance between rates of release and of inactivation of the enzyme in the plasma[404]. Alterations in the rate of inactivation in the circulation may occur independently of changes in the activity of the enzyme at its physiological site of action in the tissues. Again, the activity of the enzyme in the blood may be affected by the presence of inhibitors which may not alter the activity of the enzyme at the endothelial cell surface[76,141,405].

Rates of removal of triglyceride fatty acids from the circulation can also prove difficult to interpret. The rate of disappearance of injected labelled triglyceride fatty acids is influenced by the size of the plasma pool into which the triglyceride is introduced[144]. The possible heterogeneity of the triglycerides with respect to removal from the plasma introduces an additional complicating factor (Section 4ai; footnote p. 82) since exchange of injected labelled triglycerides with unlabelled triglycerides in complexes of a different nature may occur after injection (Sections 2cii, p. 60, and 3b, p. 71). Finally, changes in the extent of hepatic uptake and recirculation of injected triglyceride fatty acid may need to be considered (Section 4aii, p. 84).

Though, in view of these difficulties, considerable caution must clearly be exercised in attributing altered plasma triglyceride concentrations to changes in the efficiency of triglyceride fatty acid removal from the blood, it nevertheless seems worthwhile to consider the evidence which exists for such changes in particular situations.

(i) In different physiological situations

The fact that during short periods of fasting in the rat the clearing factor lipase activity of the musculature rises as that in adipose tissue falls can explain why the tissue distribution of injected triglyceride fatty acids alters in the fasted animal (Section 4c, p. 87). However, the overall effect of such changes on the rate of uptake of plasma triglyceride fatty acids by the tissues as a whole is not easy to predict and, in fact, there is no convincing evidence at present that the rates of removal of injected triglyceride fatty acids from the blood are different in fed and fasting states[110,202,317,406]. Changes in the plasma clearing factor lipase activity after heparin injection have been reported in the rat on fasting but the direction of the change depends on the heparin dosage[407,408]. Caloric restriction in man apparently results in a reduction in post-heparin plasma clearing factor lipase activity, as does the substitution of fat in the diet by carbohydrate[409-411] (*cf.* also ref. 362a).

The falls in plasma triglyceride concentration which occur with heavy exercise in man and when carbohydrate is administered after a period of starvation appear to be accounted for, at least in part, by decreased hepatic output of triglycerides (Sections 3bi, p. 75, and 3bii, p. 76). However, a contributory factor during exercise could be increased triglyceride fatty acid utilization by the muscular tissues (Section 5, p. 102), secondary to enhanced clearing factor lipase activity (Section 4c, p. 87). Moreover, feeding carbohydrate after fasting may be expected to increase adipose tissue clearing factor lipase activity[362a] while reducing that in muscle[309]. Such changes could lead to an alteration in the efficiency of triglyceride fatty acid removal from the blood and, hence, to a change in the plasma triglyceride concentration. Increased efficiency of triglyceride fatty acid removal could also contribute to the reductions in the alimentary lipaemic response to a standard fat meal observed during exercise or when carbohydrate is fed (Section 3a, p. 69)[148,149,412]. The possibility that changes in tissue clearing factor lipase activity may be responsible for the lipaemia of pregnancy has already been raised (Section 4c, p. 87).

(ii) In different pathological conditions

Familial hyperchylomicronaemia (Type 1 hyperlipoproteinaemia). The existence of this inherited hypertriglyceridaemic condition in man has already been mentioned (Section 4c, p. 87). It is characterized by an abnormally low tissue clearing factor lipase activity and by markedly reduced rates of removal of chylomicron fatty acids from the circulation. On normal diets, chylomicrons are present in the plasma in excessive amounts. Substitution of a diet poor in fat and rich in carbohydrate leads to a dramatic reduction in the plasma chylomicron count and the triglyceride concentration. Normal triglyceride concentrations are not usually achieved, presumably because the removal of VLDL triglyceride fatty acids from the plasma is also inhibited.

References p. 105

Diabetes. The existence of a high plasma triglyceride concentration in the uncontrolled diabetic patient is well established (Section 2b, p. 57). In the absence of unequivocal evidence for enhanced hepatic output of triglycerides in diabetes (Section 3bii, p. 76), attention has been directed to a possible inadequacy of triglyceride fatty acid removal from the circulation as a contributory factor. Adipose-tissue clearing factor lipase activity is low in diabetes, as is triglyceride fatty acid uptake by this tissue (Section 4c, p. 87). However, although the level of clearing factor lipase in post-heparin plasma is reduced in experimental diabetes[413] and in certain groups of diabetic patients following insulin withdrawal[42,43], no reduction has been observed in other studies in man[414-416]. It is possible that in such cases decreased activity of the enzyme in adipose tissue is compensated for by increased activity in the muscular tissues (Section 4c, p. 87).

Several studies have been carried out to see whether the removal of injected triglyceride fatty acids from the plasma is reduced in diabetes. In most[208,319,361,417], though not all[362], studies in experimental animals some reduction has been demonstrated. In man, however, the evidence is conflicting[43,418]. Problems created by differences in the plasma triglyceride pool size and by variations in the degree of insulin deprivation and, indeed, in the type of diabetic patient under study[43,361], make the interpretation of such studies far from easy.

Carbohydrate-induced hypertriglyceridaemia and ischaemic heart disease. The clearing factor lipase activity of post-heparin plasma does not seem to be reduced in individuals demonstrating carbohydrate-induced hypertriglyceridaemia[409]. Nevertheless, if the relative inefficiency of removal of VLDL triglyceride fatty acids from the blood in man is confirmed (Section 4a, p. 82), it is possible that, at the high hepatic outputs of triglycerides presumed to characterize this condition (Section 3biii, p. 78), the capacity of the extrahepatic tissues to hydrolyse these triglycerides may become a limiting factor. In this connection reports that insulin reduces the degree of carbohydrate-induced hypertriglyceridaemia in man, possibly through stimulation of adipose-tissue clearing factor lipase, are of interest[28].

Many studies have been carried out to see whether in ischaemic heart disease the elevated plasma triglyceride concentration in the post-absorptive state (Section 2b, p. 57), and the enhanced alimentary lipaemic response to a fat meal (Section 3a, p. 69), might be associated with decreased plasma clearing factor lipase activity after heparin injection. Since the subject was last reviewed[141], the results of further investigations have been reported[404,416,419,420]. The findings are difficult to assess in view of the many problems involved in such work[141] and there is no general agreement amongst the published studies. However, a partial deficiency of the tissue enzyme in at least some patients, notably those with a previous history of diabetes, certainly cannot be excluded.

Hepatic and pancreatic disease. A reduction in the clearing factor lipase activity of post-heparin plasma, which may be secondary either to low calorie intake, low insulin secretion, or to the presence of plasma inhibitors of the enzyme, is reported to be a feature of various types of pancreatic disease[141,411,421].

Early studies suggested that the clearing factor lipase activity of post-heparin plasma was increased in certain types of cirrhosis of the liver, a rise attributed to decreased destruction of the enzyme in the liver following damage to that organ[141]. Recent work suggests that this is not a general feature of hepatic cirrhosis, however. Thus, the activity of the enzyme in post-heparin plasma has been recently reported to be low rather than high[404,405]. The presence in the plasma of inhibitors of the enzyme, such as bile salts[141], could explain such reduced activity in at least certain cases.

Various metabolic diseases. Changes in plasma triglyceride concentration occur in many diseases that are essentially due to hormone imbalance[3,5,6]. In the light of the evidence that clearing factor lipase is under hormonal control (Section 4e, p. 95), the activity of the enzyme in the tissues could be altered in a number of such disease states and this could be a factor, in addition to any change in hepatic triglyceride output (Section 3b, p. 71), determining the alteration in plasma triglyceride concentration. Mention may be made of reports indirectly implicating, in addition to hormones already mentioned (Section 4e, p. 95), thyroid hormones[410], adrenocortical hormones[422-424], and oestrogens[423,425] in the regulation of plasma triglyceride concentration through effects on tissue clearing factor lipase activity.

The possibility that adrenaline and noradrenaline may play a part in regulating the activity of clearing factor lipase in the tissues has already been raised (Section 4e, p. 95). Whenever a role of the enzyme in the regulation of the plasma triglyceride concentration in a particular physiological condition or disease state is under consideration, therefore, the possibility that a change in enzyme activity could be brought about by catecholamine action should be considered. Such action could presumably lead to secondary effects both on the rate of removal of injected triglyceride fatty acids from the circulation and on the alimentary lipaemic response to a standard fat meal.

5. The fate of plasma triglyceride fatty acids in the extrahepatic tissues

Triglyceride fatty acids taken up from the blood by the extrahepatic tissues fulfil three main functions. They provide fatty acids for oxidation, for replenishment of depleted adipose-tissue triglyceride stores and, in the lactating animal, for incorporation into the milk triglycerides.

(a) Oxidation

Because hydrolysis is involved in their uptake by the extrahepatic tissues, the ultimate fate of triglyceride fatty acids removed from the blood by such tissues is essentially the same as that of FFA taken up directly from the circulation: that is, a proportion is oxidized and a proportion is incorporated into the tissue pools of esterified fatty acids[263,270,271,317,426,427].

The quantitative significance of the plasma triglyceride fatty acids in providing oxidisable substrate for the energy needs of the extrahepatic tissues is far from clear at present. Particular interest attaches to the situation in skeletal muscle because of the high proportion of the oxidative metabolism of the body which occurs there during exercise[213].

Recent studies in animals and in human subjects have confirmed the importance of the circulating plasma FFA in meeting the energy needs of muscle in the fasting state but have also shown that a substantial proportion of these needs remains to be provided from other sources[211-213,428,429]. Attention is currently being directed towards the possible significance of stores of esterified fatty acids in the muscles themselves in meeting such needs. The precise importance of these is still a matter of some controversy[430,431], but there is indirect evidence to suggest* that it may be considerable[179,432-437]: perhaps, in particular, in providing for early requirements during periods of exercise before mobilization of the adipose-tissue fatty

---

* In the perfused rat heart, endogenous triglyceride fatty acids certainly appear to be oxidized when exogenous substrates are not available in adequate amounts[438-440].

acids is increased. The hepatic output of VLDL triglycerides is probably reduced during exercise (Section 3bii, p. 76) and it seems that the contribution from this source is small under such conditions[211-213,428,429]. VLDL triglycerides could, however, contribute to the eventual replenishment of depleted muscle stores of esterified fatty acids.

During exercise in the fed state only a small proportion of the energy needs of the muscular tissues can be provided by the plasma FFA[211]. When, as is usually the case, the diet contains significant quantities of fat, an appreciable part could, however, be met by the direct uptake and oxidation of the circulating chylomicron triglyceride fatty acids. Certainly such triglyceride fatty acids appear to be actively oxidized by the body tissues, even when carbohydrate is readily available[426,427]. However, further information on the quantitative significance of their contribution under such conditions is urgently needed.

(b) Replenishment of adipose-tissue stores and formation of milk triglycerides

The function of the plasma triglycerides in providing fatty acids for replenishment of the adipose-tissue triglyceride stores and for incorporation into the milk triglycerides in the mammary gland seems clear on the basis of all the available evidence (Sections 4b, p. 85, and 4c, p. 87). Control of the extent of triglyceride fatty acid uptake at these sites through regulation of the clearing factor lipase activity does not, of course, exclude the existence of additional regulatory steps operating in the tissues during the subsequent reconversion of the fatty acids to triglycerides[339,441]. Nevertheless, it appears that primary control is exerted at the point of supply of the fatty acids and that the plasma triglycerides fulfil a function which could not be carried out by the plasma FFA*. Thus, under conditions when the body stores of triglyceride are being replenished during refeeding after a period of fasting, net movement of plasma FFA into adipose tissue seems unlikely to be of great significance. Similarly, the needs of the mammary gland during lactation for fatty acids to incorporate into the milk triglycerides can best be met by a specific mechanism which allows this organ to select fatty acids

---

* It may be noted that labelled plasma FFA are incorporated into both adipose tissue and milk triglycerides. However, this reflects only their capacity for free exchange between plasma and tissue pools and gives, by itself, no information regarding the extent of any net uptake by these tissues[329].

References p. 105

in large amounts from a plasma pool that is also available to the other body tissues.

6. Concluding remarks

The major conclusions to be drawn concerning the function of the plasma triglycerides in fatty acid transport have been outlined in the Introduction. It is nevertheless perhaps worthwhile emphasizing here the particular areas where further clarification of specific matters is required.

The question of the mechanism of triglyceride release from intestinal and liver cells has received particular attention in this article. However, the details of the process remain obscure and further information respecting the functions of the carrier lipoproteins is needed. It is particularly important to identify more rigorously the conditions in man where changes in plasma triglyceride concentration can be attributed primarily to alterations in hepatic triglyceride release and to determine how far such changes require corresponding changes in lipoprotein production by the liver.

The possibility that changes in the rate of removal of triglyceride fatty acids from the blood may be an important factor determining changes in plasma triglyceride concentration has been emphasized here. Further work is needed, however, to establish the importance of particular tissues in the uptake of triglyceride fatty acids and, in particular, to establish the quantitative significance of such acids in meeting the energy needs of the muscular tissues in the fed state.

Finally, the important question of the hormonal control of the plasma triglyceride concentration deserves further attention. Though it seems clear that this may operate through effects on either hepatic triglyceride release or triglyceride fatty acid uptake by the extrahepatic tissues, secondary to induced changes in the tissue clearing factor lipase activity, the detailed mechanism of such hormonal effects remains unknown.

REFERENCES

1 J. L. ONCLEY, in J. FOLCH-PI AND M. BAUER (Eds.), *Brain Lipids and the Leucodystrophies*, Elsevier, Amsterdam, 1963, pp. 1–17.
2 D. G. CORNWELL, in G. SCHETTLER (Ed.), *Lipids and Lipidoses*, Springer, Berlin, 1967, pp. 168–189.
3 A. M. SCANU, *Advan. Lipid Res.*, 3 (1965) 63.
4 J. W. GOFMAN, *Coronary Heart Disease*, Thomas, Springfield, Ill., 1959.
5 F. T. LINDGREN AND A. V. NICHOLS, in F. W. PUTNAM (Ed.), *The Plasma Proteins*, Vol. 2, Academic Press, New York, 1960, pp. 1–58.
6 R. E. OLSON AND J. W. VESTER, *Physiol. Rev.*, 40 (1960) 677.
7 N. K. FREEMAN, F. T. LINDGREN AND A. V. NICHOLS, *Progr. Chem. Fats Lipids*, 6 (1963) 215.
8 D. S. ROBINSON, in R. C. MACFARLANE AND A. H. T. ROBB-SMITH (Eds.), *Functions of the Blood*, Academic Press, New York, 1961, pp. 431–452.
9 D. S. FREDRICKSON, R. I. LEVY AND R. S. LEES, *New Engl. J. Med.*, 276 (1967) 34, 94, 148, 215, 273.
10 D. S. FREDRICKSON AND R. S. LEES, in J. B. STANBURY, J. B. WYNGAARDEN AND D. S. FREDRICKSON (Eds.), *The Metabolic Basis of Inherited Disease*, 2nd ed., McGraw Hill, New York, 1966, pp. 429–485.
11 J. L. GRANDA AND A. SCANU, *Biochemistry*, 5 (1966) 3301.
12 A. GUSTAFSON, P. ALAUPOVIC AND R. H. FURMAN, *Biochemistry*, 5 (1966) 632.
13 K. W. WALTON, *J. Atheroscler. Res.*, 7 (1967) 533.
13a W. V. BROWN, R. I. LEVY AND D. S. FREDRICKSON, *Circulation*, 38, *Suppl.* VI (1968) 2.
14 H. G. WINDMUELLER AND R. I. LEVY, *J. Biol. Chem.*, 242 (1967) 2246.
15 G. CAMEJO, *Biochemistry*, 6 (1967) 3228.
16 A. GUSTAFSON, P. ALAUPOVIC AND R. H. FURMAN, *Biochemistry*, 4 (1965) 596.
17 A. M. EWING, N. K. FREEMAN AND F. T. LINDGREN, *Advan. Lipid Res.*, 3 (1965) 25.
18 R. S. LEES AND D. S. FREDRICKSON, *J. Clin. Invest.*, 44 (1965) 1968.
19 L. A. CARLSON, *Acta Med. Scand.*, 167 (1960) 399.
20 J. W. GOFMAN, *Am. J. Cardiol.*, 1 (1958) 271.
21 A. ANTONIS AND I. BERSOHN, *Lancet*, i (1961) 3.
22 J. M. R. BEVERIDGE, S. N. JAGANNATHAN AND W. F. CONNELL, *Can. J. Biochem.*, 42 (1964) 999.
23 R. S. LEES AND D. S. FREDRICKSON, *Clin. Res.*, 13 (1965) 327.
24 F. T. HATCH, L. L. ABELL AND F. E. KENDALL, *Am. J. Med.*, 19 (1955) 48.
25 E. L. BIERMAN AND J. T. HAMLIN, *Diabetes*, 10 (1961) 432.
26 E. H. AHRENS JR., J. HIRSCH, K. OETTE, J. W. FARQUHAR AND Y. STEIN, *Trans. Assoc. Am. Physicians*, 74 (1961) 134.
27 E. L. BIERMAN, D. PORTE JR., D. D. O'HARA, M. SCHWARTZ AND F. C. WOOD JR., *J. Clin. Invest.*, 44 (1965) 261.
28 L. W. KINZELL AND G. SCHLIERF, *Ann. N. Y. Acad. Sci.*, 131 (1965) 603.
29 J. L. KNITTLE AND E. H. AHRENS JR., *J. Clin. Invest.*, 43 (1964) 485.
30 G. M. REAVEN, J. W. FARQUHAR, L. B. SALANS, R. C. CROSS AND R. M. WAGNER, *Clin. Res.*, 12 (1964) 277.
31 P. T. KUO, *J. Am. Med. Assoc.*, 201 (1967) 101.
32 P. J. NESTEL, *Metabolism*, 15 (1966) 787.
33 M. J. ALBRINK, J. W. MEIGS AND E. B. MAN, *Am. J. Med.*, 31 (1961) 4.
34 E. M. M. BESTERMAN, *Brit. Heart J.*, 19 (1957) 503.
35 M. J. ALBRINK, *Arch. Internal Med.*, 109 (1962) 345.

36 R. J. HAVEL AND L. A. CARLSON, *Metabolism*, 11 (1962) 195.
37 M. A. DENBOROUGH AND P. J. NESTEL, *Med. J. Australia*, 2 (1964) 91.
38 R. B. MCGANDY, D. M. HEGSTED AND F. J. STARE, *New Engl. J. Med.*, 277 (1967) 186, 242.
38a M. J. ALBRINK AND P. C. DAVIDSON, *J. Lab. Clin. Med.*, 67 (1966) 573.
39 J. CAMPBELL, *Metabolism*, 11 (1962) 762.
40 Y. MARUHAMA, Y. GOTO AND S. YAMAGATA, *Metabolism*, 16 (1967) 985.
41 M. I. NEW, T. N. ROBERTS, E. L. BIERMAN AND G. G. READER, *Diabetes*, 12 (1963) 208.
42 J. D. BAGDADE, D. PORTE AND E. L. BIERMAN, *New Engl. J. Med.*, 276 (1967) 427.
43 J. D. BAGDADE, D. PORTE AND E. L. BIERMAN, *Diabetes*, 17 (1968) 127.
44 L. A. CARLSON, J. BOBERG AND B. HÖGSTEDT, in A. E. RENOLD AND G. F. CAHILL JR. (Eds.), *Handbook of Physiology*, Section 5, *Adipose Tissue*, American Physiological Society, Washington, D.C., 1965, pp. 625–644.
45 J. M. JOHNSTON, in M. FLORKIN AND E. H. STOTZ (Eds.), *Comprehensive Biochemistry*, Vol. 18, Elsevier, Amsterdam, 1969, pp. 1–18.
46 J. R. SENIOR, *J. Lipid Res.*, 5 (1964) 495.
47 V. P. DOLE AND J. T. HAMLIN, *Physiol. Rev.*, 42 (1962) 674.
48 G. G. PINTER AND D. B. ZILVERSMIT, *Biochim. Biophys. Acta*, 59 (1962) 116.
49 A. YOKOYAMA AND D. B. ZILVERSMIT, *J. Lipid Res.*, 6 (1965) 241.
50 D. KAY AND D. S. ROBINSON, *Quart. J. Exptl. Physiol.*, 47 (1962) 258.
51 G. I. SCHOEFL, *Proc. Roy. Soc. (London)*, B, 169 (1968) 147.
52 D. B. ZILVERSMIT, P. H. SISCO AND A. YOKOYAMA, *Biochim. Biophys. Acta*, 125 (1966) 129.
53 D. S. ROBINSON, *Quart. J. Exptl. Physiol.*, 40 (1955) 112.
54 D. B. ZILVERSMIT, *J. Clin. Invest.*, 44 (1965) 1610.
55 D. B. ZILVERSMIT, *J. Lipid Res.*, 9 (1968) 180.
56 M. M. SALPETER AND D. B. ZILVERSMIT, *J. Lipid Res.*, 9 (1968) 187.
57 E. L. BIERMAN, in A. E. RENOLD AND G. F. CAHILL JR. (Eds.), *Handbook of Physiology*, Section 5, *Adipose Tissue*, American Physiological Society, Washington, D.C., 1965, pp. 509–518.
58 R. L. SWANK AND J. H. FELLMAN, *Am. J. Physiol.*, 192 (1958) 318.
59 M. RODBELL AND D. S. FREDRICKSON, *J. Biol. Chem.*, 234 (1959) 562.
60 P. ALAUPOVIC, *Proc. 1967 Deuel Conf. on Lipids: The fate of dietary lipids*, Carmel, U.S. Public Heart Service Publication No. 1742, U.S. Government Printing Office, Washington, D.C., 1968, pp. 145–152.
61 J. D. WATHEN AND R. S. LEVY, *Biochemistry*, 5 (1966) 1099.
62 A. SCANU AND I. H. PAGE, *J. Exptl. Med.*, 109 (1959) 239.
63 A. F. HOFMANN, *Am. J. Physiol.*, 199 (1960) 433.
64 W. J. LOSSOW, F. T. LINDGREN AND L. C. JENSEN, *Biochim. Biophys. Acta*, 144 (1967) 670.
65 A. SCANU AND W. L. HUGHES, *J. Biol. Chem.*, 235 (1960) 2876.
66 O. MINARI AND D. B. ZILVERSMIT, *J. Lipid Res.*, 4 (1963) 424.
67 F. CHEVALLIER AND J. PHILIPPOT, *Bull. Soc. Chim. Biol.*, 44 (1962) 809.
68 E. GORDIS, *Proc. Soc. Exptl. Biol. Med.*, 110 (1962) 657.
69 E. L. BIERMAN, E. GORDIS AND J. T. HAMLIN, *J. Clin. Invest.*, 41 (1962) 2254.
70 E. GORDIS, *J. Clin. Invest.*, 44 (1965) 1451.
71 E. L. BIERMAN, T. L. HAYES, J. N. HAWKINS, A. M. EWING AND F. T. LINDGREN, *J. Lipid Res.*, 7 (1966) 65.
72 D. PORTE JR., D. D. O'HARA AND R. H. WILLIAMS, *J. Lipid Res.*, 7 (1966) 368.
73 D. D. O'HARA, D. PORTE JR. AND R. H. WILLIAMS, *J. Lipid Res.*, 7 (1966) 264.

74 E. L. BIERMAN AND D. E. STRANDNESS JR., *Am. J. Physiol.*, 210 (1966) 13.
75 A. V. NICHOLS AND L. SMITH, *J. Lipid Res.*, 6 (1965) 206.
76 D. S. ROBINSON AND J. E. FRENCH, *Pharmacol. Rev.*, 12 (1960) 241.
77 R. G. GOULD, *Am. J. Med.*, 11 (1951) 209.
78 E. G. TRAMS AND E. A. BROWN, *J. Theoret. Biol.*, 12 (1966) 311.
79 B. LOMBARDI AND G. UGAZIO, *J. Lipid Res.*, 6 (1965) 498.
80 N. BAKER, A. S. GARFINKEL AND M. C. SCHOTZ, *J. Lipid Res.*, 9 (1968) 1.
81 M. HEIMBERG, I. WEINSTEIN, G. DISHMON AND M. FRIED, *Am. J. Physiol.*, 209 (1965) 1053.
82 Y. STEIN AND B. SHAPIRO, *Am. J. Physiol.*, 196 (1959) 1238.
83 M. HAMOSH AND B. SHAPIRO, *Am. J. Physiol.*, 201 (1961) 1030.
84 M. HEIMBERG, I. WEINSTEIN, H. KLAUSNER AND M. L. WATKINS, *Am. J. Physiol.*, 202 (1962) 353.
85 R. E. KAY AND C. ENTENMAN, *J. Biol. Chem.*, 236 (1961) 1006.
86 M. HEIMBERG, D. R. VAN HARKEN AND T. O. BROWN, *Biochim. Biophys. Acta*, 137 (1967) 435.
87 H. G. WILCOX, G. DISHMON AND M. HEIMBERG, *J. Biol. Chem.*, 243 (1968) 666.
88 H. G. WINDMUELLER AND A. E. SPAETH, *Arch. Biochim. Biophys.*, 122 (1967) 362.
89 R. O. RECKNAGEL, *Pharmacol. Rev.*, 19 (1967) 145.
90 D. S. ROBINSON AND A. SEAKINS, *Biochim. Biophys. Acta*, 62 (1962) 163.
91 E. FARBER, *Advan. Lipid Res.*, 5 (1967) 119.
92 H. G. WINDMUELLER, *J. Biol. Chem.*, 239 (1964) 530.
93 P. S. ROHEIM, S. SWITZER, A. GIRARD AND H. A. EDER, *Biochem. Biophys. Res. Commun.*, 20 (1965) 416.
94 R. S. LEES, *J. Lipid Res.*, 8 (1967) 396.
95 R. S. LEES AND E. H. AHRENS, *J. Clin. Invest.*, 47 (1958) 59a.
96 R. L. HAMILTON, D. M. REGEN, M. E. GRAY AND V. S. LEQUIRE, *Lab. Invest.*, 16 (1967) 305.
97 A. L. JONES, N. B. RUDERMAN AND M. G. HERRERA, *J. Lipid Res.*, 8 (1967) 429.
98 O. STEIN AND Y. STEIN, *J. Cell Biol.*, 33 (1967) 319.
99 N. B. RUDERMAN, K. C. RICHARDS, V. VALLES DE BOURGES AND A. L. JONES, *J. Lipid Res.*, 9 (1968) 613.
100 M. NEUTRA AND C. P. LEBLOND, *J. Cell Biol.*, 30 (1966) 137.
101 R. TZUR AND B. SHAPIRO, *J. Lipid Res.*, 5 (1964) 542.
102 K. ATERMAN, in CH. ROUILLER (Ed.), *The Liver*, Academic Press, New York, 1963, pp. 61–136.
103 CH. ROUILLER AND A.-M. JEZEQUEL, in CH. ROUILLER (Ed.), *The Liver*, Academic Press, New York, 1963, pp. 195–252.
104 W. VOLWILER, P. D. GOLDSWORTHY, M. P. MACMARTIN, P. A. WOOD, I. R. MACKAY AND K. FREMONT-SMITH, *J. Clin. Invest.*, 34 (1955) 1126.
105 J. AVIGAN, H. A. EDER AND D. STEINBERG, *Proc. Soc. Exptl. Biol. Med.*, 95 (1957) 429.
106 D. GITLIN, D. G. CORNWELL, D. NAKASATO, J. L. ONCLEY, W. L. HUGHES JR. AND C. A. JANEWAY, *J. Clin. Invest.*, 37 (1958) 172.
107 A. SCANU AND W. L. HUGHES, *J. Clin. Invest.*, 41 (1962) 1681.
108 K. W. WALTON, P. J. SCOTT, J. V. JONES, R. F. FLETCHER AND T. WHITEHEAD, *J. Atheroscler. Res.*, 3 (1963) 396.
109 R. H. FURMAN, S. S. SANBAR, P. ALAUPOVIC, R. H. BRADFORD AND R. P. HOWARD, *J. Lab. Clin. Med.*, 63 (1964) 193.
110 R. J. HAVEL, J. M. FELTS AND C. VAN DUYNE, *J. Lipid Res.*, 3 (1962) 297.
111 S. LAURELL, *Acta Physiol. Scand.*, 47 (1959) 218.

112 R. C. GROSS, E. H. EIGENBRODT AND J. W. FARQUHAR, *J. Lipid Res.*, 8 (1967) 114.
113 E. A. NIKKILÄ, M. KEKKI AND K. OJALA, *Ann. Med. Exptl. Biol. Fenniae (Helsinki)*, 44 (1966) 348.
114 J. W. FARQUHAR, R. C. GROSS, R. M. WAGNER AND G. M. REAVEN, *J. Lipid Res.*, 6 (1965) 119.
115 S. J. FREIDBERG, R. F. KLEIN, D. L. TROUT, M. D. BOGDONOFF AND E. H. ESTES JR., *J. Clin. Invest.*, 40 (1961) 1846.
116 R. J. HAVEL, *Metabolism*, 10 (1961) 1031.
117 G. R. FALCONO, B. N. STEWART AND M. FRIED, *Biochemistry*, 7 (1968) 720.
118 H. G. WILCOX, M. FRIED AND M. HEIMBERG, *Biochim. Biophys. Acta*, 106 (1965) 598.
119 W. H. FLORSHEIM, M. A. FAIRCLOTH, D. GRAFF, N. S. AUSTIN AND S. M. VELCOFF, *Metabolism*, 12 (1963) 598.
120 P. ALAUPOVIC, S. S. SANBAR, R. H. FURMAN, M. L. SULLIVAN AND S. L. WALRAVEN, *Biochemistry*, 5 (1966) 4044.
121 P. S. ROHEIM, L. MILLER AND H. A. EDER, *J. Biol. Chem.*, 240 (1965) 2994.
122 D. S. ROBINSON, *Proc. Intern. Symp. on Lipid Transport, Nashville, 1963*, Thomas, Springfield, Ill., 1964, pp. 194–201.
123 D. S. M. HAINES, *Can. J. Biochem.*, 44 (1966) 45.
124 D. S. M. HAINES AND S. MOOKERJEA, *Can. J. Biochem.*, 43 (1965) 507.
125 B. LOMBARDI, P. PANI AND F. F. SCHLUNK, *J. Lipid Res.*, 9 (1968) 437.
126 P. EATON AND D. M. KIPNIS, *J. Clin. Invest.*, 47 (1968) 27a.
127 M. RODBELL, D. S. FREDRICKSON AND K. ONO, *J. Biol. Chem.*, 234 (1959) 567.
128 K. J. ISSELBACHER AND D. M. BUDZ, *Nature*, 200 (1963) 364.
129 F. T. HATCH, Y. ASO, L. M. HAGOPIAN AND J. J. RUBENSTEIN, *J. Biol. Chem.*, 241 (1966) 1655.
130 S. M. SABESIN AND K. J. ISSELBACHER, *Science*, 147 (1965) 1149.
131 D. E. HYAMS, S. M. SABESIN, N. J. GREENBERGER AND K. J. ISSELBACHER, *Biochim. Biophys. Acta*, 125 (1966) 166.
132 H. B. SALT, O. H. WOLFF, J. K. LLOYD, A. S. FOSBROOKE, A. H. CAMERON AND D. V. HUBBLE, *Lancet*, ii (1960) 325.
133 K. J. ISSELBACHER, R. SCHEIG, G. R. PLOTKIN AND J. B. CAULFIELD, *Medicine*, 43 (1964) 347.
134 H. G. WINDMUELLER AND R. I. LEVY, *J. Biol. Chem.*, 243 (1968) 4878.
135 P. S. ROHEIM, L. I. GIDEZ AND H. A. EDER, *J. Clin. Invest.*, 45 (1966) 297.
136 H. J. DEUEL, *The Lipids*, Vol. 1, Interscience, New York, 1951.
137 H. J. DEUEL, *Progr. Chem. Fats Lipids*, 2 (1954) 99.
138 S. BERGSTRÖM AND B. BORGSTRÖM, *Progr. Chem. Fats Lipids*, 3 (1955) 351.
139 W. H. FALOR, *Proc. 1967 Deuel Conf. on Lipids: The Fate of Dietary Lipids, Carmel*, U.S. Public Health Service Publication No. 1742, U.S. Government Printing Office, Washington, D.C., 1968, pp. 125–132.
140 R. BLOMSTRAND, *Proc. 1967 Deuel Conf. on Lipids: The Fate of Dietary Lipids, Carmel*, U.S. Public Health Service Publication No. 1742, U.S. Government Printing Office, Washington, D.C., 1968, pp. 99–123.
141 D. S. ROBINSON, *Advan. Lipid Res.*, 1 (1963) 133.
142 D. F. BROWN, S. H. KINCH AND J. T. DOYLE, *J. Atheroscler. Res.*, 6 (1966) 232.
143 M. A. DENBOROUGH, *Clin. Sci.*, 25 (1963) 115.
144 P. J. NESTEL, *J. Clin. Invest.*, 43 (1964) 943.
145 P. J. NESTEL, M. A. DENBOROUGH AND J. O'DEA, *Circulation Res.*, 10 (1962) 786.
146 V. KALLIO, *Acta Med. Scand.*, 181, Suppl. 467 (1967).
147 M. J. ALBRINK, J. R. FITZGERALD AND E. B. MAN, *Metabolism*, 7 (1958) 162.

148 H. COHEN AND C. GOLDBERG, *Brit. Med. J.*, ii (1960) 509.
149 E. A. NIKKILÄ AND A. KONTTINEN, *Lancet*, i (1962) 1151.
150 M. FRIEDMAN AND S. O. BYERS, *Nature*, 217 (1968) 235.
151 M. FRIEDMAN, R. H. ROSENMAN AND S. O. BYERS, *Circulation*, 29 (1964) 874.
152 M. FRIEDMAN AND S. O. BYERS, *Am. J. Physiol.*, 213 (1967) 1359.
153 M. FRIEDMAN AND S. O. BYERS, *Am. J. Physiol.*, 213 (1967) 829.
154 M. FRIEDMAN, S. O. BYERS AND A. E. BROWN, *Am. J. Physiol.*, 212 (1967) 1174.
155 P. SAVARY AND M. J. CONSTANTIN, *Biochim. Biophys. Acta*, 144 (1967) 549.
156 R. V. COXON AND D. S. ROBINSON, *Quart. J. Exptl. Physiol.*, 47 (1962) 252.
157 J. H. BAXTER, *J. Lipid Res.*, 7 (1966) 158.
158 R. J. HAVEL AND A. GOLDFIEN, *J. Lipid Res.*, 2 (1961) 389.
159 B. K. SHRIVASTAVA, T. G. REDGRAVE AND W. J. SIMMONDS, *Quart. J. Exptl. Physiol.*, 52 (1967) 305.
160 P. J. NESTEL, A. BEZMAN AND R. J. HAVEL, *Am. J. Physiol.*, 203 (1962) 914.
161 G. GÖRANSSON AND T. OLIVECRONA, *Acta Physiol. Scand.*, 62 (1964) 224.
162 G. GÖRANSSON, *Acta Physiol. Scand.*, 64 (1965) 204.
163 G. GÖRANSSON, *Acta Physiol. Scand.*, 64 (1965) 387.
164 P. J. NESTEL AND D. STEINBERG, *J. Lipid Res.*, 4 (1963) 461.
165 Y. STEIN AND B. SHAPIRO, *Biochim. Biophys. Acta*, 34 (1959) 79.
166 P. J. NESTEL, *Metabolism*, 14 (1965) 1.
167 P. J. NESTEL AND R. O. SCOW, *J. Lipid Res.*, 5 (1964) 46.
168 G. V. MARINETTI, in M. FLORKIN AND E. H. STOTZ (Eds.), *Comprehensive Biochemistry*, Vol. 18, Elsevier, Amsterdam, 1969, pp. 117–155.
169 N. BAKER AND M. C. SCHOTZ, *J. Lipid Res.*, 8 (1967) 646.
170 P. A. MAYES AND J. M. FELTS, *Biochem. J.*, 103 (1967) 400.
171 B. BORGSTRÖM AND T. OLIVECRONA, *J. Lipid Res.*, 2 (1961) 263.
172 M. C. SCHOTZ, N. BAKER AND M. N. CHAVEZ, *J. Lipid Res.*, 5 (1964) 569.
173 C. WATERHOUSE, J. H. KEMPERMAN AND J. M. STORMONT, *J. Lab. Clin. Med.*, 63 (1964) 605.
174 B. MORRIS, *J. Physiol. (London)*, 168 (1963) 564.
175 L. A. HILLYARD, C. E. CORNELIUS AND I. L. CHAIKOFF, *J. Biol. Chem.*, 234 (1959) 2240.
176 S. LAURELL, *Acta Physiol. Scand.*, 46 (1959) 97.
177 W. E. M. LANDS, *Ann. Rev. Biochem.*, 34 (1965) 313.
178 L. L. M. VAN DEENEN AND G. H. DE HAAS, *Ann. Rev. Biochem.*, 35 (1966) 157.
179 L. A. CARLSON, S. O. FRÖBERG AND E. R. NYE, *Acta Med. Scand.*, 180 (1966) 571.
180 D. L. TROUT, H. L. BITTER AND W. W. LACKEY, *Biochem. Pharmacol.*, 16 (1967) 971.
181 H. VAVŘINKOVÁ AND B. MOSINGER, *Physiol. Bohemoslov.*, 14 (1965) 46.
182 P. BELFRAGE, *Biochim. Biophys. Acta*, 98 (1965) 660.
183 A. C. OLSON AND P. ALAUPOVIC, *Biochim. Biophys. Acta*, 125 (1966) 185.
184 Y. BIALE, E. GORIN AND E. SHAFRIR, *Biochim. Biophys. Acta*, 152 (1968) 28.
185 J. A. HIGGINS AND C. GREEN, *Biochim. Biophys. Acta*, 144 (1967) 211.
186 J. R. CARTER, *Biochim. Biophys. Acta*, 137 (1967) 147.
187 C. GREEN AND J. A. WEBB, *Biochim. Biophys. Acta*, 84 (1964) 404.
188 J. R. WILLIAMSON, B. HERCZEG, H. COLES AND R. DANISH, *Biochem. Biophys. Res. Commun.*, 24 (1966) 437.
189 P. D. BEWSHER AND J. ASHMORE, *Biochem. Biophys. Res. Commun.*, 24 (1966) 431.
190 J. R. WILLIAMSON, P. H. WRIGHT, W. J. MALAISSE AND J. ASHMORE, *Biochem. Biophys. Res. Commun.*, 24 (1966) 765.
191 G. M. REAVEN, D. B. HILL, R. C. GROSS AND J. W. FARQUHAR, *J. Clin. Invest.*, 44 (1965) 1826.

192 W. G. RYAN AND T. B. SCHWARTZ, *Metabolism*, 14 (1965) 1243.
193 S. OTWAY AND D. S. ROBINSON, *J. Physiol. (London)*, 190 (1967) 321.
194 R. O. SCOW AND S. CHERNICK, in M. FLORKIN AND E. H. STOTZ (Eds.), *Comprehensive Biochemistry*, Vol. 18, Elsevier, Amsterdam, 1969, pp. 19–49.
195 S. O. BYERS AND M. FRIEDMAN, *Am. J. Physiol.*, 198 (1960) 629.
196 P. V. HARPER JR., W. B. NEAL JR. AND G. R. HLAVACEK, *Metabolism*, 2 (1953) 69.
197 E. J. MASORO, *J. Lipid Res.*, 3 (1962) 149.
198 G. R. JANSEN, M. E. ZANETTI AND C. F. HUTCHISON, *Biochem., J.* 101 (1966) 811.
199 D. L. TROUT, *J. Pharmacol. Exptl. Therap.*, 152 (1966) 529.
200 E. C. ALBRITTON, *Standard Values in Nutrition and Metabolism*, Saunders, Philadelphia, 1964, p. 250.
201 R. J. HAVEL, *J. Clin. Invest.*, 36 (1957) 855.
202 J. H. BRAGDON, R. J. HAVEL AND R. S. GORDON JR., *Am. J. Physiol.*, 189 (1957) 63.
203 D. RUDMAN, *J. Lipid Res.*, 4 (1963) 119.
204 H. E. LEBOVITZ AND F. L. ENGEL, in A. E. RENOLD AND G. F. CAHILL JR. (Eds.), *Handbook of Physiology*, Section 5, *Adipose Tissue*, American Physiological Society, Washington, D.C., 1965, pp. 541–548.
205 R. J. HAVEL, in A. E. RENOLD AND G. F. CAHILL JR. (Eds.), *Handbook of Physiology*, Section 5, *Adipose Tissue*. American Physiological Society, Washington, D.C., 1965, pp. 575–582.
206 P. J. NESTEL, *J. Clin. Invest.*, 43 (1964) 77.
207 B. MORRIS, *J. Physiol. (London)*, 168 (1963) 584.
208 M. HEIMBERG, A. DUNKERLEY AND T. O. BROWN, *Biochim. Biophys. Acta*, 125 (1966) 252.
209 L. A. CARLSON AND F. MOSSFELDT, *Acta Physiol. Scand.*, 62 (1964) 51.
210 L. A. CARLSON AND S. O. FRÖBERG, *Metabolism*, 16 (1967) 624.
211 R. J. HAVEL, A. NAIMARK AND C. F. BORCHGREVINK, *J. Clin. Invest.*, 42 (1963) 1054.
212 R. J. HAVEL, B. PERNOW AND N. L. JONES, *J. Appl. Physiol.*, 23 (1967) 90.
213 R. J. HAVEL, L. G. EKELUND AND A. HOLMGREN, *J. Lipid Res.*, 8 (1967) 366.
214 L. B. ROWELL, E. J. MASORO AND M. J. SPENCER, *J. Appl. Physiol.*, 20 (1965) 1032.
215 R. J. HAVEL, *Proc. 1967 Deuel Conf. on Lipids: The Fate of Dietary Lipids*, Carmel, U.S.Public Health Service Publication No. 1742, U.S. Government Printing Office, Washington, D.C., 1968, p. 193.
216 T. R. CSORBA, I. MATSUDA AND N. KALANT, *Metabolism*, 15 (1966) 262.
217 H. M. TEPPERMAN AND J. TEPPERMAN, *Federation Proc.*, 23 (1964) 73.
218 J. TEPPERMAN AND H. M. TEPPERMAN, *Ann. N.Y. Acad. Sci.*, 131 (1965) 404.
219 D. W. ALLMANN, D. D. HUBBARD AND D. M. GIBSON, *J. Lipid Res.*, 6 (1965) 63.
220 G. R. JANSEN, M. E. ZANETTI AND C. F. HUTCHISON, *Biochem. J.*, 106 (1968) 345.
221 W. J. LOSSOW, G. W. BROWN JR. AND I. L. CHAIKOFF, *J. Biol. Chem.*, 222 (1956) 531.
222 P. A. MAYES AND J. M. FELTS, *Biochem. J.*, 105 (1967) 18C.
223 H. G. WINDMUELLER AND A. E. SPAETH, *J. Biol. Chem.*, 241 (1966) 2891.
224 N. BAKER AND M. C. SCHOTZ, *J. Lipid Res.*, 5 (1964) 188.
225 P. J. NESTEL AND E. Z. HIRSCH, *Australasian Ann. Med.*, 14 (1965) 265.
226 P. J. NESTEL, *Clin. Sci.*, 31 (1966) 31.
227 J. YUDKIN, *Am. J. Clin. Nutr.*, 20 (1967) 108.
228 I. MACDONALD, *Advan. Lipid Res.*, 4 (1966) 39.
229 H. BAR-ON AND Y. STEIN, *J. Nutr.*, 94 (1968) 95.
230 E. A. NIKKILÄ AND K. OJALA, *Life Sci.*, 4 (1965) 937.
231 E. A. NIKKILÄ AND K. OJALA, *Life Sci.*, 5 (1966) 89.

REFERENCES

232 F. Heinz, W. Lamprecht and J. Kirsch, *J. Clin. Invest.*, 47 (1968) 1828.
233 I. Macdonald and J. B. Roberts, *Metabolism*, 16 (1967) 572.
234 D. Zakim and R. H. Herman, *Biochim. Biophys. Acta*, 165 (1968) 374.
235 E. A. Nikkilä, *Scand. J. Clin. Lab. Invest.*, 18, Suppl. 92 (1966) 76.
236 G. R. Jansen, C. F. Hutchison and M. E. Zanetti, *Biochem. J.*, 99 (1966) 323.
237 G. R. Jansen, M. E. Zanetti and C. F. Hutchison, *Biochem. J.*, 102 (1967) 864.
238 P. Farvarger, *Advan. Lipid Res.*, 2 (1964) 447.
239 L. A. Carlson, *Ann. N.Y. Acad. Sci.*, 131 (1965) 119.
240 A. Seakins and D. S. Robinson, *Biochem. J.*, 92 (1964) 308.
241 A. E. Harper, *Am. J. Clin. Nutr.*, 6 (1958) 242.
242 D. S. Robinson and P. M. Harris, *Biochem. J.*, 80 (1961) 361.
243 B. Shapiro, *Ann. Rev. Biochem.*, 36 (1967) 247.
244 B. Hellman, *Ann. N. Y. Acad. Sci.*, 131 (1965) 541.
245 W. W. Shreeve, *Ann. N. Y. Acad. Sci.*, 131 (1965) 464.
246 G. R. Jansen, M. E. Zanetti and C. F. Hutchison, *Biochem. J.*, 102 (1967) 870.
247 A. Keys, in M. Sandler and G. H. Bourne (Eds.), *Atherosclerosis and its Origins*, Academic Press, New York, 1963, pp. 263–299.
248 T. R. Dawber, F. E. Moore and G. V. Man, *Am. J. Public Health*, 47 (1957) 4.
249 M. J. Albrink, J. W. Meigs and M. A. Granoff, *New Engl. J. Med.*, 266 (1962) 484.
250 S. Waxler and L. S. Craig, *Am. J. Clin. Nutr.*, 14 (1964) 128.
251 F. J. Ballard and R. W. Hanson, *Biochem. J.*, 102 (1967) 952.
252 C. B. Taylor, E. Bailey and W. Bartley, *Biochem. J.*, 105 (1967) 717.
253 G. Majno, in W. F. Hamilton and P. Dow (Eds.), *Handbook of Physiology*, Section 2, *Circulation*, Vol. 3, American Physiological Society, Washington, D.C., 1965, pp. 2293–2375.
254 J. E. French, in A. C. Frazer (Ed.), *Biochemical Problems of Lipids*, Elsevier, Amsterdam, 1963, pp. 296–303.
255 G. I. Schoefl and J. E. French, *Proc. Roy. Soc. (London)*, B, 169 (1968) 153.
256 F. Wassermann and T. F. McDonald, *Z. Zellforsch.*, 59 (1963) 326.
257 J. R. Williamson, *J. Cell Biol.*, 20 (1964) 57.
258 L. Napolitano, in A. E. Renold and G. F. Cahill Jr. (Eds.), *Handbook of Physiology*, Section 5, *Adipose Tissue*, American Physiological Society, Washington, D.C., 1965, pp. 109–123.
259 E. R. Suter and G. Majno, *J. Cell Biol.*, 27 (1965) 163.
260 D. S. Robinson, in R. M. C. Dawson and D. N. Rhodes (Eds.), *Metabolism and Physiological Significance of Lipids*, Wiley, New York, 1964, pp. 275–288.
261 D. Hallberg, *Acta Physiol. Scand.*, 65 (1965) Suppl. 254.
262 E. L. Bierman and J. T. Hamlin, *Proc. Soc. Exptl. Biol. Med.*, 109 (1962) 747.
263 A. Bezman-Tarcher, S. Otway and D. S. Robinson, *Proc. Roy. Soc. (London) B*, 162 (1965) 411.
264 M. W. Bates, *Am. J. Physiol.*, 212 (1967) 662.
265 L. A. Carlson and B. Olhagen, *J. Clin. Invest.*, 38 (1959) 854.
266 J. E. French, B. Morris and D. S. Robinson, *Brit. Med. Bull.*, 14 (1958) 234.
267 P. J. Nestel, R. J. Havel and A. Bezman, *J. Clin. Invest.*, 42 (1963) 1313.
268 B. Borgström and P. Jordan, *Acta Soc. Med. Upsalien.*, 64 (1959) 185.
269 T. Olivecrona, *J. Lipid Res.*, 3 (1962) 439.
270 T. Olivecrona and P. Belfrage, *Biochim. Biophys. Acta*, 98 (1965) 81.
271 N. L. Jones and R. J. Havel, *Am. J. Physiol.*, 213 (1967) 824.
272 C. E. West, E. F. Annison and J. L. Linzell, *Biochem. J.*, 104 (1967) 59P.
273 O. W. McBride and E. D. Korn, *J. Lipid Res.*, 5 (1964) 459.
274 J. Scheuer and R. E. Olson, *Am. J. Physiol.*, 212 (1967) 301.

275 M. RODBELL AND R. O. SCOW, in A. E. RENOLD AND G. F. CAHILL JR. (Eds.), *Handbook of Physiology*, Section 5, *Adipose Tissue*, American Physiological Society, Washington, D.C., 1965, pp. 491–498.
276 M. RODBELL, *J. Biol. Chem.*, 235 (1960) 1613.
277 R. O. SCOW, in H. C. MENG AND D. H. LAW (Eds.), *Proceedings of International Symposium on Parenteral Nutrition*, Thomas, Springfield, Ill., 1970.
278 D. HALLBERG, *Proc. 1967 Deuel Conf. on Lipids: The Fate of Dietary Lipids*, Carmel, U.S. Public Health Service Publication No. 1742, U.S. Government Printing Office, Washington, D.C., 1968, pp. 194–200.
279 M. SIMPSON-MORGAN, *Proc. 1967 Deuel Conf. on Lipids: The Fate of Dietary Lipids*, Carmel, U.S. Public Health Service Publication No. 1742, U.S. Government Printing Office, Washington, D.C., 1968, pp. 200–205.
280 O. STEIN AND Y. STEIN, *Lab. Invest.*, 17 (1967) 436.
281 Y. STEIN AND B. SHAPIRO, *J. Lipid Res.*, 1 (1960) 326.
282 B. MORRIS AND J. E. FRENCH, *Quart. J. Exptl. Physiol.*, 43 (1958) 180.
283 M. RODBELL, R. O. SCOW AND S. S. CHERNICK, *J. Biol. Chem.*, 239 (1964) 385.
284 T. OLIVECRONA, E. P. GEORGE AND B. BORGSTRÖM, *Federation Proc.*, 20 (1961) 928.
285 P. BELFRAGE, J. ELOVSON AND T. OLIVECRONA, *Biochim. Biophys. Acta*, 106 (1965) 45.
286 P. BELFRAGE, *Biochim. Biophys. Acta*, 125 (1966) 474.
287 J. M. FELTS AND P. A. MAYES, *Nature*, 206 (1965) 195.
288 M. C. SCHOTZ, B. ARNESJÖ AND T. OLIVECRONA, *Biochim. Biophys. Acta*, 125 (1966) 485.
289 J. A. ONTKO AND D. B. ZILVERSMIT, *J. Lipid Res.*, 8 (1967) 90.
290 P. BELFRAGE, *Biochim. Biophys. Acta*, 152 (1968) 266.
291 N. R. DI LUZIO AND S. J. RIGGI, *J. Reticuloendothelial Soc.*, 1 (1964) 248.
292 C. T. ASHWORTH, V. A. STEMBRIDGE AND E. SANDERS, *Am. J. Physiol.*, 198 (1960) 1326.
293 J. A. HIGGINS AND C. GREEN, *Biochem. J.*, 99 (1966) 631.
294 J. A. ONTKO, *Biochim. Biophys. Acta*, 137 (1967) 13.
295 P. A. MAYES AND J. M. FELTS, *Biochem. J.*, 108 (1968) 483.
296 M. BARCLAY, E. GARFINKEL, O. TEREBUS-KEKISH, E. B. SHAH, M. DEGUIA, R. K. BARCLAY AND V. P. SKIPSKI, *Arch. Biochem. Biophys.*, 98 (1962) 397.
297 V. S. LE QUIRE, R. L. HAMILTON, R. ADAMS AND J. M. MERRILL, *Proc. Soc. Exptl. Biol. Med.*, 114 (1963) 104.
298 R. E. CONDON, H. TOBIAS AND D. V. DATTA, *J. Clin. Invest.*, 44 (1965) 860.
299 J. BOBERG, L. A. CARLSON AND L. NORMELL, *Life Sci.*, 3 (1964) 1011.
300 P. J. NESTEL, R. J. HAVEL AND A. BEZMAN, *J. Clin. Invest.*, 41 (1962) 1915.
301 R. J. HAVEL, in A. E. RENOLD AND G. F. CAHILL JR. (Eds.), *Handbook of Physiology*, Section 5, *Adipose Tissue*, American Physiological Society, Washington, D.C., 1965, pp. 499–507.
302 A. BEZMAN, J. M. FELTS AND R. J. HAVEL, *J. Lipid Res.*, 3 (1962) 427.
303 A. S. GARFINKEL, N. BAKER AND M. C. SCHOTZ, *J. Lipid Res.*, 8 (1967) 274.
304 H. K. DELCHER, M. FRIED AND J. C. SHIPP, *Biochim. Biophys. Acta*, 106 (1965) 10.
305 A. GOUSIOS, J. M. FELTS AND R. J. HAVEL, *Metabolism*, 12 (1963) 75.
306 R. A. KREISBERG, *Am. J. Physiol.*, 210 (1966) 379.
307 M. F. CRASS AND H. C. MENG, *Biochim. Biophys. Acta*, 125 (1966) 106.
308 M. B. ENSER, F. KUNZ, J. BORENSZTAJN, L. H. OPIE AND D. S. ROBINSON, *Biochem. J.*, 104 (1967) 306.
309 J. BORENSZTAJN, *Thesis*, Oxford University, 1968.
310 M. W. SIMPSON-MORGAN, *J. Physiol. (London)*, 199 (1968) 37.

REFERENCES

311 J. A. ONTKO AND P. J. RANDLE, *Biochem. J.*, 104 (1967) 43C.
312 F. B. BALLARD, W. H. DANFORTH, S. NAEGLE AND J. R. BING, *J. Clin. Invest.*, 39 (1960) 717.
313 Y. GOTO, T. HASEGAWA, H. TAKITSUKA, A. KANDA, K. MORI, T. KATAYAMA, K. YOSHIDA, S. NAGANO, H. WAKAMURA, S. HORI AND M. KATO, *Japan. Circulation J.*, 28 (1964) 301.
314 T. J. REGAN, C. B. MOSCHOS, P. H. LEHAN, H. A. OLDEWURTEL AND H. K. HELLEMS, *Circulation Res.*, 19 (1966) 307.
315 S. CARLSTEN, B. HALLGREN, R. JAGENBURG, A. SVANBORG AND L. WERKÖ, *Scand. J. Clin. Lab. Invest.*, 13 (1961) 418.
316 A. F. WILLEBRANDS, *Clin. Chim. Acta*, 10 (1964) 435.
317 J. H. BRAGDON AND R. S. GORDON, *J. Clin. Invest.*, 37 (1958) 574.
318 T. J. REGAN, G. KOROXENIDIS, C. B. MOSCHOS, H. A. OLDEWURTEL, P. H. LEHAN AND H. K. HELLEMS, *J. Clin. Invest.*, 45 (1966) 270.
319 F. A. GRIES, S. POTTHOFF AND K. JAHNKE, *Diabetologia*, 3 (1967) 311.
320 J. M. BARRY, W. BARTLEY, J. L. LINZELL AND D. S. ROBINSON, *Biochim. J.*, 89 (1963) 6.
321 O. W. MCBRIDE AND E. D. KORN, *J. Lipid Res.*, 5 (1964) 453.
322 E. F. ANNISON, J. L. LINZELL, S. FAZAKERLEY AND B. W. NICHOLS, *Biochem. J.*, 102 (1967) 637.
323 R. F. GLASCOCK, V. A. WELCH, C. BISHOP, T. DAVIES, E. W. WRIGHT AND R. C. NOBLE, *Biochem. J.*, 98 (1966) 149.
324 A. K. LASCELLES, D. C. HARDWICK, J. L. LINZELL AND T. B. MEPHAM, *Biochem., J.*, 92 (1964) 36.
325 D. S. ROBINSON AND J. E. FRENCH, *Quart. J. Exptl. Physiol.*, 42 (1957) 151.
326 E. D. KORN, *Methods Biochem. Anal.*, 7 (1959) 145.
327 E. D. KORN, in *Digestion, Absorption Intestinale et Transport des Glycérides chez les Animaux Supérieurs*, Centre National de la Recherche Scientifique, Paris, 1961, pp. 139–150.
328 S. W. LEVY, *Rev. Can. Biol.*, 17 (1958) 1.
329 D. S. ROBINSON, *Proc. 1967 Deuel Conf. on Lipids: The Fate of Dietary Lipids, Carmel*, U.S. Public Health Service Publication No. 1742, U.S. Government Printing Office, Washington, D.C., 1968, pp. 166–188.
330 R. J. HAVEL, *Proc. 1967 Deuel Conf. on Lipids: The Fate of Dietary Lipids, Carmel*, U.S. Public Health Service Publication No. 1742, U.S. Government Printing Office, Washington, D.C., 1968, p. 211.
331 D. S. ROBINSON AND P. M. HARRIS, *Quart. J. Exptl. Physiol.*, 44 (1959) 80.
332 J. R. E. FRAZER, R. R. H. LOVELL AND P. J. NESTEL, *Clin. Sci.*, 20 (1961) 351.
333 D. S. ROBINSON, P. M. HARRIS AND C. R. RICKETTS, *Biochem. J.*, 71 (1959) 286.
334 D. S. ROBINSON AND M. A. JENNINGS, *J. Lipid Res.*, 6 (1965) 222.
335 S. J. HO, R. J. HO AND H. C. MENG, *Am. J. Physiol.*, 212 (1967) 284.
336 R. J. HAVEL, *Am. J. Clin. Nutr.*, 6 (1958) 662.
337 M. S. MOSKOWITZ AND A. A. MOSKOWITZ, *Science*, 149 (1965) 72.
338 E. D. KORN, *J. Biol. Chem.*, 236 (1961) 1638.
339 M. VAUGHAN AND D. STEINBERG, in A. E. RENOLD AND G. F. CAHILL JR. (Eds.), *Handbook of Physiology*, Section 5, *Adipose Tissue*, American Physiological Society, Washington, D.C., 1965, pp. 239–251.
340 M. YAMAMOTO AND G. I. DUMMOND, *Am. J. Physiol.*, 213 (1967) 1365.
341 D. P. WALLACH, *J. Lipid Res.*, 9 (1968) 200.
342 L. A. CARLSON AND L. B. WADSTROM, *Clin. Chim. Acta*, 2 (1957) 9.
342a H. GRETEN, R. I. LEVY AND D. S. FREDRICKSON, *Circulation*, 38, Suppl. VI (1968) 8.

343 W. C. VOGEL AND E. L. BIERMAN, *J. Lipid Res.*, 8 (1967) 46.
344 W. C. VOGEL AND E. L. BIERMAN, *Proc. Soc. Exptl. Biol. Med.*, 127 (1968) 77.
345 R. INFANTE, J. POLONOVSKI AND O. DONOU, *Biochim. Biophys. Acta*, 144 (1967) 490.
346 E. D. KORN, *J. Biol. Chem.*, 215 (1955) 1.
347 E. D. KORN, *J. Biol. Chem.*, 215 (1955) 15.
348 E. D. KORN AND T. W. QUIGLEY, *J. Biol. Chem.*, 226 (1957) 833.
349 A. SCANU, *J. Biol. Chem.*, 242 (1967) 711.
350 D. S. ROBINSON, G. H. JEFFRIES AND J. C. F. POOLE, *Quart. J. Exptl. Physiol.*, 40 (1955) 297.
351 D. S. ROBINSON, *Quart. J. Exptl. Physiol.*, 41 (1956) 195.
352 W. R. HARLAN, P. S. WINESETT AND A. J. WASSERMAN, *J. Clin. Invest.*, 46 (1967) 239.
353 J. D. BROOME, *J. Exptl. Med.*, 119 (1964) 83.
354 C. H. HOLLENBERG, *Am. J. Physiol.*, 197 (1959) 667.
355 J. PÁV AND J. WENKEOVÁ, *Nature*, 185 (1960) 926.
356 A. CHERKES AND R. S. GORDON, *J. Lipid Res.*, 1 (1959) 97.
357 D. S. ROBINSON, *J. Lipid Res.*, 1 (1960) 332.
358 D. R. WING AND D. S. ROBINSON, *Biochem. J.*, 106 (1968) 667.
359 J. I. KESSLER, *J. Clin. Invest.*, 42 (1963) 362.
360 J. D. SCHNATZ AND R. H. WILLIAMS, *Diabetes*, 12 (1963) 174.
361 D. F. BROWN, K. DAUDISS AND J. DURRANT, *Diabetes*, 16 (1967) 90.
362 D. F. BROWN AND T. OLIVECRONA, *Acta Physiol. Scand.*, 66 (1969) 9.
362a N. POKRAJAC AND W. J. LOSSOW, *Biochim. Biophys. Acta*, 137 (1967) 291.
363 D. S. ROBINSON, *J. Lipid Res.*, 4 (1963) 21.
364 O. W. MCBRIDE AND E. D. KORN, *J. Lipid Res.*, 4 (1963) 17.
365 S. OTWAY AND D. S. ROBINSON, *Biochem. J.*, 106 (1968) 677.
366 R. O. SCOW, S. S. CHERNICK AND M. S. BRINLEY, *Am. J. Physiol.*, 206 (1964) 796.
367 T. F. KELLEY, *Proc. Soc. Exptl. Biol. Med.*, 127 (1968) 337.
368 E. A. NIKKILÄ, P. TORSTI AND O. PENTTILÄ, *Metabolism*, 12 (1963) 863.
369 A. A. ALOUSI AND S. MALLOV, *Am. J. Physiol.*, 206 (1964) 603.
370 A. DURY, *Proc. Soc. Exptl. Biol. Med.*, 107 (1961) 299.
371 R. SZABÓ, M. TÉNYI AND L. VARGA, *Med. Pharmacol. Exptl.*, 7 (1962) 363.
372 E. A. NIKKILÄ, P. TORSTI AND O. PENTTILÄ, *Life Sci.*, 4 (1965) 27.
373 D. GRAFNETTER, J. GRAFNETTEROVÁ, E. GROSSI AND P. MORGANTI, *Med. Pharmacol. Exptl.*, 12 (1965) 266.
374 S. MALLOV AND A. A. ALOUSI, *Am. J. Physiol.*, 212 (1967) 1158.
375 S. MALLOV AND F. CERRA, *J. Pharmacol. Exptl. Therap.*, 156 (1967) 426.
376 J. I. KESSLER AND E. SENDEROFF, *J. Clin. Invest.*, 41 (1962) 1531.
377 C. H. HOLLENBERG, *J. Clin. Invest.*, 39 (1960) 1282.
378 R. A. KREISBERG, *Am. J. Physiol.*, 210 (1966) 385.
379 S. S. NAIDOO, W. J. LOSSOW AND I. L. CHAIKOFF, *Experientia*, 23 (1967) 829.
380 M. RODBELL, *J. Biol. Chem.*, 239 (1964) 753.
381 N. POKRAJAC, W. J. LOSSOW AND I. L. CHAIKOFF, *Biochim. Biophys. Acta*, 139 (1967) 123.
382 V. J. CUNNINGHAM AND D. S. ROBINSON, *Biochem. J.*, 112 (1969) 203.
383 M. BRADY AND J. A. HIGGINS, *Biochim. Biophys. Acta*, 137 (1967) 140.
384 C. J. FIELDING, *Biochim. Biophys. Acta*, 159 (1968) 94.
385 C. R. HOLLETT, *Arch. Biochem. Biophys.*, 108 (1964) 244.
386 E. M. MOSCHIDES AND H. B. EIBER, *J. Lab. Clin. Med.*, 63 (1964) 425.
387 A. N. PAYZA, H. B. EIBER AND I. DANISHEFSKY, *Biochim. Biophys. Acta*, 111 (1965) 159.
388 B. BASKYS, E. KLEIN AND W. F. LEVER, *Arch. Biochem. Biophys.*, 99 (1962) 25.

389 G. R. EAGLE AND D. S. ROBINSON, *Biochem. J.*, 93 (1964) 10C.
390 M. C. SCHOTZ AND A. S. GARFINKEL, *Biochim. Biophys. Acta*, 106 (1965) 202.
391 M. R. SALAMAN AND D. S. ROBINSON, *Biochem. J.*, 99 (1966) 640.
392 D. R. WING, M. R. SALAMAN AND D. S. ROBINSON, *Biochem. J.*, 99 (1966) 648.
392a M. A. JENNINGS AND H. W. FLOREY, *Proc. Roy. Soc. (London) B*, 167 (1967) 39.
393 J. D. SCHNATZ, J. W. ORMSBY AND R. J. WILLIAMS, *Am. J. Physiol.*, 205 (1963) 401.
394 S. L. GARTNER AND G. V. VAHOUNY, *Am. J. Physiol.*, 211 (1966) 1063.
395 D. R. WING, C. J. FIELDING AND D. S. ROBINSON, *Biochem. J.*, 104 (1967) 45C.
396 M. M. APPLEMAN AND R. G. KEMP, *Biochem. Biophys. Res. Commun.*, 24 (1966) 564.
397 E. W. SUTHERLAND, I. ØYE AND R. W. BUTCHER, *Rec. Progr. Hormone Res.*, 21 (1965) 623.
398 M. A. RIZACK, *J. Biol. Chem.*, 239 (1964) 392
399 R. W. BUTCHER, R. J. HO, H. C. MENG AND E. W. SUTHERLAND, *J. Biol. Chem.*, 240 (1965) 4515.
400 R. W. BUTCHER, J. G. T. SNEYD, C. R. PARK AND E. W. SUTHERLAND, *J. Biol. Chem.*, 241 (1966) 1651.
401 D. R. WING AND D. S. ROBINSON, *Biochem. J.*, 109 (1968) 841.
402 E. A. NIKKILÄ AND O. PYKÄLISTÖ, *Biochim. Biophys. Acta*, 152 (1968) 421.
403 R. W. BUTCHER AND E. W. SUTHERLAND, *Ann. N. Y. Acad. Sci.*, 139 (1967) 849.
403a J. N. FAIN, V. P. KOVACEV AND R. O. SCOW, *J. Biol. Chem.*, 240 (1965) 3522.
403b R. W. BUTCHER AND C. E. BAIRD, *J. Biol. Chem.*, 243 (1968) 1713.
404 Y. YOSHITOSHI, C. NAITO, H. OKANIWA, M. USUI, T. MOGAMI AND T. TOMONO, *J. Clin. Invest.*, 42 (1963) 707.
405 D. V. DATTA, *J. Lab. Clin. Med.*, 67 (1966) 461.
406 D. S. FREDRICKSON, D. L. MCCOLLESTER AND K. ONO, *J. Clin. Invest.*, 37 (1958) 1333.
407 M. R. SALAMAN AND D. S. ROBINSON, *Proc. Intern. Conf. Biochem. Problems Lipids, 6th, Marseilles, 1960*, Pergamon, Oxford, 1961, pp. 218–230.
408 F. C. MONKHOUSE, J. STRACHAN AND F. MCCLAIN, *Can. J. Biochem. Physiol.*, 39 (1961) 1027.
409 D. S. FREDRICKSON, K. ONO AND L. L. DAVIS, *J. Lipid Res.*, 4 (1963) 24.
410 D. PORTE, D. D. O'HARA AND R. H. WILLIAMS, *Metabolism*, 15 (1966) 107.
411 P. T. KUO, D. R. BASSETT, A. M. DIGEORGE AND G. G. CARPENTER, *Circulation Res.*, 16 (1965) 221.
412 L. H. KRUT AND R. F. BARSKY, *Lancet*, ii (1964) 1136.
413 H. C. MENG AND J. L. GOLDFARB, *Diabetes*, 8 (1959) 211.
414 M. A. DENBOROUGH AND B. PATERSON, *Clin. Sci.*, 23 (1962) 485.
415 D. P. JONES, G. R. PLOTKIN AND R. A. ARKY, *Diabetes*, 15 (1966) 565.
416 W. F. PERRY, *Clin. Chim. Acta*, 16 (1967) 189.
417 J. I. KESSLER, *J. Lab. Clin. Med.*, 60 (1962) 747.
418 E. L. BIERMAN AND J. T. HAMLIN, *Metabolism*, 12 (1963) 666.
419 P. J. NESTEL, *J. Atheroscler. Res.*, 4 (1964) 193.
420 J. SLACK, J. SEYMOUR, L. MCDONALD AND F. LOVE, *Lancet*, ii (1964) 1033.
421 S. JAKOVCIC AND D. Y. Y. HSIA, *J. Paediat.*, 62 (1963) 25.
422 F. C. MONKHOUSE AND D. G. BAKER, *Can. J. Biochem. Physiol.*, 41 (1963) 1901.
423 A. CAIRNS AND P. CONSTANTINIDES, *Can. J. Biochem. Physiol.*, 33 (1955) 530.
424 A. DURY, *Ann. N. Y. Acad. Sci.*, 72 (1958) 870.
425 E. FABIAN, A. ŠTORK, J. KOBILKOVÁ AND J. ŠPONAROVÁ, *Enzymol. Biol. Clin.*, 8 (1967) 451.
426 B. MORRIS AND M. W. SIMPSON-MORGAN, *J. Physiol. (London)*, 177 (1965) 74.
427 B. MORRIS, *Quart. J. Exptl. Physiol.*, 43 (1958) 65.

428 B. Issekutz H. I. Miller, P. Paul and K. Rodahl, *Am. J. Physiol.*, 207 (1964) 583.
429 P. Paul, B. Issekutz and H. I. Miller, *Am. J. Physiol.*, 211 (1966) 1313.
430 E. J. Masoro, L. B. Rowell, R. M. McDonald and B. Steiert, *J. Biol. Chem.*, 241 (1966) 2626.
431 E. J. Masoro, *J. Biol. Chem.*, 242 (1967) 1111.
432 E. M. Neptune, H. C. Sudduth and D. R. Foreman, *J. Biol. Chem.*, 234 (1959) 1659.
433 E. M. Neptune, H. C. Sudduth, D. R. Foreman and F. J. Fash, *J. Lipid Res.*, 1 (1960) 229.
434 M. E. Volk, R. H. Millington and S. Weinhouse, *J. Biol. Chem.*, 195 (1952) 493.
435 B. Issekutz and P. Paul, *Am. J. Physiol.*, 215 (1968) 197.
436 A. B. Maunsbach and C. Wirsén, *J. Ultrastruct. Res.*, 16 (1966) 35.
437 L. A. Carlson, S. O. Liljedahl and C. Wirsén, *Acta Med. Scand.*, 178 (1965) 81.
438 J. C. Shipp, J. M. Thomas and L. Crevasse, *Science*, 143 (1964) 371.
439 R. E. Olson and R. J. Hoeschen, *Biochem. J.*, 103 (1967) 796.
440 R. M. Denton and P. J. Randle, *Biochem. J.*, 104 (1967) 416.
441 N. J. Kuhn, *Biochem. J.*, 105 (1967) 225.

Chapter II

Biosynthesis of Triglycerides

G. V. MARINETTI

Biochemistry Department, University of Rochester, School of Medicine and Dentistry, Rochester, N.Y. (U.S.A.)

This chapter will discuss the biosynthesis of triglycerides by mammalian systems. The biochemical process to be considered is how the fatty acids are esterified to glycerol and what control mechanisms are known about this process. The biosynthesis and oxidation of fatty acids and the manner in which triglycerides are absorbed and transported in the animal body are covered by others in this book (see Chapters I and VIII). An exhaustive literature search of this topic is not intended in this chapter.

1. Introduction

It is noteworthy that triglycerides represent a major and prime fuel of most animals. The chemical structure and properties of triglycerides confer upon them a unique role in the energy metabolism of higher animals. The hydrocarbon nature of triglycerides enables them to yield on combustion more

Abbreviations: ATP, adenosine triphosphate; ADP, adenosine diphosphate; AMP, adenosine monophosphate; CTP, cytidine triphosphate; CMP, cytidine monophosphate; NAD(NADH), diphosphopyridine nucleotide; P-O-P, pyrophosphate, P_i, orthophosphate; FA, fatty acids; FFA, free fatty acids; NEFA, non-esterified fatty acids bound non-covalently to albumin; PA, phosphatidic acid; TG, triglycerides; DG, diglycerides; MG, monoglycerides; CoA–SH, coenzyme A; GP, glycerol phosphate; GPC, glycerylphosphoryl choline; FDP, fructose 1,6-diphosphate.

Fatty acids will be designated by the number of carbon atoms and the number of double bonds they contain. Thus 18:0 represents a fatty acid having a linear chain of 18 carbon atoms and no double bonds. In this system the location of the double bonds is not specified and only linear (non-branched) fatty acids are considered.

References p. 153

energy and water per unit weight than carbohydrates or proteins, as shown in Table I.

The triglycerides can be stored in large amount as potential fuel in adipose tissue. They cannot be stored in high concentration in the liver without serious consequences of liver damage. The triglycerides are also stored in a

TABLE I

THE HEAT AND WATER OF COMBUSTION OF FAT, CARBOHYDRATE AND PROTEIN

	Heat (kcal)	Water (g)
1 g fat	9.0	1.1
1 g carbohydrate	4.0	0.7
1 g protein	4.0	0.8

nearly anhydrous state. This storage form of fuel (*i.e.* large diffuse relatively anhydrous storage depots) is remarkably different from the storage of glucose as glycogen. Glycogen is stored as a hydrated polymer and primarily in the liver and muscles, but the amount which can be stored is very limited.

There are interesting cases in nature which demonstrate the unique storage form and energy potential of triglycerides[1]. Seals are known to eat ravenously for 10 months of the year, laying down hundreds of pounds of fat as triglycerides. Then for 2 months during the mating season they do not eat but rather burn up the stored fat and can lose up to 200 pounds during this time. Before salmon embark on their remarkable upstream swim for spawning they store fat as fuel. Many birds lay down fat before migrating. The camel stores large quantities of fat and can travel many miles without food or water. Hibernating animals lay down large quantities of fat before the hibernating season.

Brown adipose tissue is a unique tissue which has a special thermogenic function in the newborn[2,3]. Its anatomical location and its oxidative capacity make it act as an "electric blanket" for the newborn, especially for the region of the heart. Brown adipose tissue differs from white adipose tissue in having a higher content of mitochondria which show uncoupling of oxidative phosphorylation and in having appreciable glycerol kinase (EC 2.7.1.30) activity[4]. Brown adipose tissue oxidizes triglyceride fatty acids to generate heat in order to protect the newborn from exposure to cold temperature. As the newborn matures and this protection becomes less important the amount of brown fat diminishes or disappears.

The above examples show the way in which nature has equipped some animals to store fat as fuel to be used for a variety of needs.

With man and possibly with other mammals the storage of large amounts of excess fat is not biologically necessary and in fact may be harmful. Animals (including man) which eat meals daily on a somewhat routine basis have biochemical systems which are oscillatory in nature as a result of repeating sequences of intervals of feeding and non-feeding. After a meal, the body is geared to digesting food and storing the excess food as glycogen and triglycerides. It is known that excess glucose is converted to fatty acids and glycerol and stored as triglyceride. The reverse process does not occur to any appreciable extent. (Only the glycerol portion of triglycerides or the terminal 3 carbons of odd-chain fatty acids can be converted to glucose in mammals.) That is, excess dietary fat is not stored as glycogen but rather is still stored as fat although the dietary fat is modified to fit the needs of the animal. The modification comes about primarily by reshuffling the absorbed fatty acids so that the triglycerides finally synthesized have a fatty acid pattern characteristic for the species.

The enzyme reactions for the synthesis of triglycerides are known in fairly great detail. The cellular localization of many of the enzymes is also known. The major areas for future research will be concerned with the biosynthesis of individual molecular species of triglycerides having different fatty acid patterns, with the control mechanisms which regulate the synthesis and degradation of triglycerides, and with the quantitative aspects of lipogenesis. The mechanisms and physical–chemical aspects of conversion of triglycerides to lipoproteins and the mechanism of permeation of lipids through cell membranes or incorporation into membranes also represent vital areas for future research. Several avenues of approach will and should be used including studies with whole animals, isolated organs, isolated cells, tissue slices, tissue homogenates, cellular particles and isolated enzymes. Each system offers unique advantages and the summation of the data from all of these systems will aid in the elucidation of lipid metabolism. Isotopes will continue to serve a vital function in elucidating the mechanism of lipogenesis but care must be taken not to over-generalize with data obtained in one system with one type of labeled precursor.

There are a variety of different fatty acids in nature and thus different molecular species of triglycerides can exist.

References p. 153

The number of arrangements of n different fatty acids on glycerol to yield triglycerides can be calculated as follows:

If the α- and α'-positions have different fatty acids, the total number of arrangements will be equal to n^3. This is based on permutation theory which states that the number of permutations of n objects taken r at a time is n^r if the n objects can be repeated. Since glycerol has 3 hydroxyl groups, $r=3$. Hence the total number of permutations with n different fatty acids will be n^3.

When the α,α'-positions have different fatty acids, the center carbon atom becomes an asymmetric center and therefore D,L-isomers must be considered. The total number of permutations considering optical isomers will be $2n^3 - n^2$.

Assuming a simple situation where $n=3$, the total number of permutations where $\alpha = \alpha'$, will be $3^3 = 27$. However if α does not equal α', then D,L-isomers must be considered, and the total number of permutations will be $2 \times 3^3 - 3^2 = 45$. Hence there are 18 additional isomers due to stereoisomerism.

A more realistic but conservative number for the major fatty acids in mammals would be 10 (16:0, 18:0, 20:0, 22:0, 24:0, 16:1, 18:1, 18:2, 18:3, 20:4). The number of permutations including optical isomers will be $2 \times 10^3 - 10^2 = 1900$. However this number will be less depending on how much specificity there is in the enzyme systems which esterify the fatty acids to the glycerol to yield the triglycerides. In the case of intestine this specificity is low since the monoglyceride pathway prevails. In the case of liver and adipose tissue where the glycerol phosphate pathway predominates, the specificity is higher. In any case there will still be a large number of individual triglycerides in these tissues and the problem of separating them and studying their individual rates of synthesis poses a very formidable task. Some work along these lines has already begun. Recently different classes of triglycerides have been separated by TLC* on silver nitrate treated silica gel plates or silver nitrate treated silica-gel paper[5,6]. Kuksis has also separated the triglycerides by GLC[7]. The triglycerides are separated on the basis of their total carbon number of the fatty acids. Rat-liver triglycerides were resolved into approximately 12 peaks, and milk-fat triglycerides of man into 14 peaks. Each peak, however, contains more than one component. These newer methods will undoubtedly be used for the study of the metabolism of individual triglycerides.

* Thin-layer chromatography.

2. Digestion, absorption and transport of triglycerides

These topics will be covered very briefly since they are treated in greater detail elsewhere (Chapter I). The triglycerides are hydrolyzed in the intestinal lumen by the enzyme pancreatic lipase (EC 3.1.1.3). The bile salts, together with lysolecithin and monoglycerides aid in the emulsification of the triglycerides so that the enzyme catalysis will be more efficient. The degree of hydrolysis varies yielding free fatty acids, soaps, 1,2-diglycerides, 2-monoglycerides and free glycerol. Some triglycerides may escape hydrolysis and are probably absorbed intact in the form of small micelles. The enzyme pancreatic lipase acts on the 1,3-linked fatty acids of the glycerides to yield 2-monoglycerides and 1,2- or 2,3-diglycerides. The diglycerides are further hydrolyzed to yield 2-monoglycerides.

The mono- and diglycerides and fatty acid soaps are emulsified with bile salts to form micelles which are absorbed into the cell. Within the cell triglycerides are resynthesized and then combine with protein or lipoproteins (containing some phospholipid and cholesterol) to yield chylomicrons. β-Lipoproteins are also formed. The B-protein of the β-lipoprotein is important in this process since if it is missing, as in the rare disease, abetalipoproteinemia[8], essentially no chylomicrons are produced. The need for protein is also shown by studies in which rats are treated with puromycin. In this case triglycerides are absorbed but do not appear as chylomicrons[9]. The chylomicrons and β-lipoproteins enter the lymphatic system and then go into the vascular system where they find their way to all the tissues. The major tissues involved in the uptake of these plasma lipoproteins are adipose tissue and liver. The "clearing" of chylomicrons probably involves two processes. The major one appears to be mediated by lipoprotein lipase which hydrolyzes the triglycerides on the lipoproteins to yield glycerol and free fatty acids and a "delipidized" protein. The free fatty acids readily combine non-covalently with albumin (and other plasma proteins) to yield fatty acid–albumin complexes. These complexes have been given several names, but NEFA (non-esterified fatty acids) seems to be the one most widely used. In some manner not yet well understood, the fatty acids enter the adipose tissue cell or liver cell (or other tissue) and are either oxidized to yield energy or converted back to triglycerides and stored as fuel. It is possible that some lipoproteins are taken up directly by the cells of the body. The quantitative significance of this mechanism is unknown.

References p. 153

3. Activation of fatty acids

Fatty acids are transported in the plasma as NEFA. On reaching the tissues the fatty acids are converted to acylthiolesters of CoA by the action of three different thiokinase enzymes[10-12]. Acetic acid kinase has high activity for acetic acid, has little activity for propionic acid and essentially no activity for butyric acid or other longer chain acids. Another kinase (the general fatty acid kinase) acts on C_4–C_{12} fatty acids but has highest activity with the C-8 acid. A third kinase (the long-chain fatty acid kinase) acts mainly on long-chain C_8–C_{18} fatty acids. These kinases have been isolated from liver, heart and yeast. Acetic acid kinase and the general fatty acid kinase have been purified.

The reaction scheme proposed by Berg[13] is as follows:

$$RCOO^- + ATP \rightleftharpoons R-\overset{\overset{O}{\|}}{C}-AMP + P-O-P \quad (a)$$

$$R\overset{\overset{O}{\|}}{C}-AMP + CoASH \rightleftharpoons R\overset{\overset{O}{\|}}{C}-S-CoA + AMP \quad (b)$$

Ingraham and Green[14] have postulated an enzyme-bound complex of CoASH, AMP and Mg^{2+} as an intermediate and suggest a Mg^{2+} chelate of acyl adenylate which is bound to the enzyme such that acyl-CoA is the only free product which is observed. The thiokinase reaction is reversible. With 0.001 M heptanoate at pH 8.0, 38°, $2 \cdot 10^{-4}$ M CoASH and $2.5 \cdot 10^{-3}$ M Mg^{2+}, the equilibrium constant = 1.1.

Short-chain fatty acids, C_4–C_6, may be activated by reaction with succinyl CoA[3]. The reaction, catalyzed by a thiophorase enzyme, is shown below:

$$\text{Succinyl-S-CoA} + R\overset{\overset{O}{\|}}{C}-O- \rightleftharpoons \text{succinate} + R\overset{\overset{O}{\|}}{C}-S-CoA$$

The thiophorase system may be more important in microorganisms, while the thiokinase system appears to be the main one in animals.

4. Biosynthesis of triglycerides in liver

Tietz and Shapiro[15] were among the first investigators to study the mechanism of triglyceride biogenesis. They found that ATP was required for the incorporation of fatty acids into glycerides of rat-liver homogenates. Weiss, Kennedy and Kiyasu[16] studied the synthesis of triglycerides in chicken liver particles (mitochondria plus microsomes) and in separated mitochondria and microsomes. They found microsomes to be 2–3 times as active as mitochondria. A study of several different diglycerides (Table II) showed

TABLE II

EFFECT OF VARIOUS GLYCERIDES ON THE CONVERSION OF [1-^{14}C]PALMITOYL-CoA TO LIPID[a]

Acceptor	Stimulation of lipid formation[b]	Stimulation of lecithin synthesis
D-1,2-Diolein	[100]	[100]
L-1,2-Diolein	53	20
D,L-1,2-Diolein	62	—
D,L-1-Monoolein	7	0
D,L-Palmitoyl-β-oleyl diglyceride	44	39
D,L-1,2-Dioctanoin	39	0
D,L-1,2-Dilaurin	63	0
D-1,2-Dipalmitin	0	0

[a] Taken from Weiss, Kennedy and Kiyasu[16].
[b] Presumed to be triglyceride.

that D-1,2-diolein gave the highest yield of labeled triglyceride from palmityl-CoA. L-1,2-Diolein had half the activity of D-1,2-diolein. It was of interest that D,L-1,2-dioctanoin and D,L-1,2-dilaurin were fairly good substrates for synthesis of triglycerides with palmityl-CoA but these diglycerides were not converted to lecithin when [^{14}C]CDP-choline was used. They also observed that L-1,2-diolein was more effectively converted to triglycerides than to lecithin. Another unusual finding was the inability of D-1,2-dipalmitin to be converted to either triglycerides or lecithin. These workers recognized the hazard in interpretation of their results because of the difference in solubility of the various diglycerides. The state of dispersion of the various diglycerides would be expected to affect their rate of reaction with enzymes. The saturated diglycerides are more difficult to disperse than the unsaturated diglycerides.

References p. 153

This may in part explain why D-1,2-diolein was active whereas D-1,2-dipalmitin was not.

The reactions leading to triglyceride synthesis in liver were postulated as follows:

(a) glycerol + ATP → L-α-GP

(b) L-α-GP + acyl-CoA ⇌ mono-acyl-GP $\xrightleftharpoons{\text{acyl-CoA}}$ L-α-PA

(c) L-α-PA → D-1,2-DG + P_i

(d) D-1,2-DG + acyl-CoA ⇌ TG

This process for triglyceride synthesis has come to be known as the glycerol phosphate pathway. Reaction (a) is catalyzed by the enzyme glycerokinase and requires ATP and Mg^{2+} ions. It yields the stereospecific isomer L-α-glycerol phosphate. The reaction of L-α-GP with acyl-CoA (reaction b) has been studied by Lands et al.[17,18]. These workers used guinea-pig liver microsomes to form diacyl-GP from GP with the CoA esters of stearate and linoleate. Mono-acyl-GP (lysophosphatidic acid) did not accumulate as an intermediate since it apparently was very rapidly acylated to diacyl-GP (phosphatidic acid). The two acylations are controlled by separate enzymes since the first acylation but not the second was found to be sensitive to SH-inhibitors. 40% of the stearate and 52% of the linoleate was located at the 2-position of PA, indicating nearly random acylation. Hence these reactions do not determine the asymmetric distribution of fatty acids known to exist in most naturally occurring mammalian glycerides.

Very little is known about the fatty acid composition of natural phosphatidic acid since this phospholipid occurs in very small amount in mammalian tissues and is difficult to isolate. Hübscher and Clark[19] have reported that liver phosphatidic acid has 64-77% linoleic acid and 10-17% oleic acid.

The hydrolysis of phosphatidic acid (reaction c) is catalyzed by the enzyme phosphatidic acid phosphatase[20], to yield diglycerides and P_i. The diglycerides are then esterified with acyl-CoA to yield triglycerides (reaction d). Hill and Lands[21] have examined the rates of acylation of L-α-GP and 1-acyl-GPC in rat-liver microsomes in an attempt to explain the difference in fatty acid distribution in triglycerides and phospholipids of the same tissue. It is known that mammalian-tissue phospholipids have a higher content of long-chain polyunsaturated fatty acids at the 2-position than do the triglycerides[22-24]. Mattson and Lutton[24] and Savary et al.[25] analyzed the fatty acid distribution in vegetable oils and found certain fatty acids to exhibit a non-random distribution. Saturated fatty acids were preferentially located at the 1- and 3-positions whereas polyunsaturated fatty acids favored the 2-position. In another study Mattson and Volpenhein[26] found fatty acids with chain length greater than 18 carbon atoms (regardless of whether they are saturated or unsaturated) to be esterified almost exclusively at the 1,3-position. As a result of this distribution the 2-position contains a higher proportion of 18:1, 18:2 and 18:3 acids. Hill and Lands[21] conclude from a comparison of the relative rates of esterification of 20:2, 20:3, 20:4 and 20:5 fatty acids to 1-acyl-GP and to 1-acyl-GPC, that the acyl-GPC–acyl transferase enzyme is more important in determining the final distribution of fatty acids on phospholipids such as lecithin than are the enzymes which acylate GP. The formation of acyl-GPC however would require the action of tissue phospholipases. In essence, these workers propose a reshuffling mechanism for determining the asymmetric distribution of fatty acids in lecithins in which lecithins are made with a random distribution of fatty acids *via* the glycerol phosphate pathway and then the combined action of tissue phospholipases and acyl-GPC–acyl transferases reforms lecithins having a non-random fatty acid pattern.

It has also been postulated that certain diglycerides are used for triglyceride synthesis and other diglycerides are used for phospholipid synthesis. This may occur by enzyme specificity and/or by cell compartments.

Another way to alter the fatty acid pattern of newly synthesized triglycerides and phospholipids is by reshuffling the fatty acids by transesterification as indicated below[27].

    (a) lysolecithin + cholesterol ester $\rightleftharpoons$ lecithin + cholesterol
    (b) lysolecithin + triglyceride $\rightleftharpoons$ lecithin + diglyceride
    (c) lysolecithin + diglyceride $\rightleftharpoons$ lecithin + monoglyceride

These reactions require at least two types of enzymes to bring about the reshuffling of the fatty acids, a phospholipase or lipase to yield lysolecithin or diglycerides and monoglycerides, and a transacylase to interchange acyl groups *via* transesterification. The quantitative significance of these reactions also remains to be established.

The simultaneous synthesis of triglycerides, diglycerides and phospholipids by rat-liver cell free systems has been studied by Marinetti et al.[28,29] using [^{14}C]glycerol. These studies have shown that approximately 10 mM ATP was required for optimal lipogenesis. The effect of 10 mM ATP in stimulating the synthesis of diglycerides and triglycerides is shown by the autoradiograph in Fig. 1. The effect of 10 mM and 40 mM ATP on the labeling of neutral glycerides is given in Fig. 2. ATP at 10 mM concentration increased the labeling of both diglycerides and triglycerides but the labeling reached a

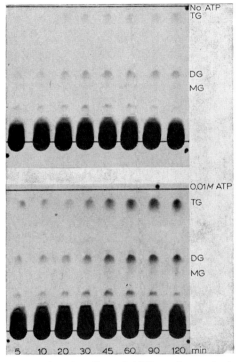

Fig. 1. Autoradiograph showing the effect of ATP on the synthesis of diglycerides and triglycerides. [^{14}C]Glycerol was incorporated into the lipids of rat-liver homogenates without added ATP (upper autoradiograph) and with added ATP at 10 mM concentration (lower autoradiograph). At 5, 10, 20, 30, 45, 60, 90 and 120 min of incubation 20-μl aliquots of the homogenate were removed for chromatographic analysis. Taken from Marinetti et al.[28].

plateau after 1 h. ATP at 40 mM concentration produced an initial inhibition in the labeling of these neutral glycerides but after 1 h the rate of labeling rose sharply. Although not shown in Fig. 2, ATP at 100 mM concentration strongly inhibits neutral glyceride labeling over a 2-h period.

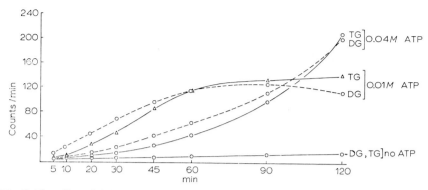

Fig. 2. The effect of ATP concentration on the synthesis of diglycerides and triglycerides. [^{14}C]Glycerol was incorporated into the lipids of rat-liver homogenates. The lipids were separated by chromatography and analyzed for radioactivity. Taken from Marinetti et al.[28].

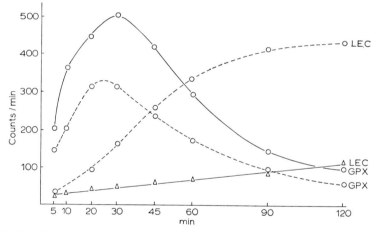

Fig. 3. The effect of magnesium ions on the synthesis of lecithin and lipid GPX. [^{14}C]-Glycerol was incorporated into the lipids of rat-liver homogenates. The lipids were separated by chromatography and analyzed for radioactivity. —, 0.01 M ATP; ----, 0.01 M ATP + 0.01 M MgCl$_2$. Taken from Marinetti et al.[28].

The effect of magnesium ions on lipid labeling from [^{14}C]glycerol in liver homogenates is seen in Figs. 3 and 4. The rate curves in Fig. 3 show the marked stimulation in labeling of lecithin and a concomitant decrease in labeling of lipid GPX by 10 mM MgCl$_2$. The inhibition of labeling of diglycerides and triglycerides is seen in Fig. 4. These effects of magnesium are dependent on ATP.

References p. 153

Marinetti et al. previously reported that a highly labeled lipid intermediate was observed in liver homogenates[28,29]. Although this lipid had chromatographic properties similar to phosphatidic acid, more recent work has indicated that this lipid may be phosphatidyl glycerol phosphate (Marinetti et al. unpublished data). This lipid has therefore been designated as GPX

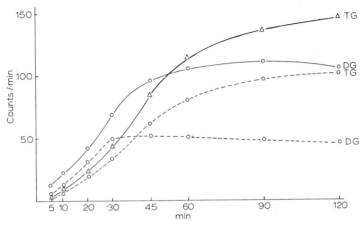

Fig. 4. The effect of magnesium ions on the synthesis of diglycerides and triglycerides. [^{14}C]Glycerol was incorporated into the lipids of rat-liver homogenates. The lipids were separated by chromatography and analyzed for radioactivity. —, 0.01 M ATP; ----, 0.01 M ATP + 0.01 M MgCl$_2$. Taken from Marinetti et al.[28].

to denote a glycerol phosphatide whose structure is still provisional. Enzyme and chemical degradation of lipid GPX together with two-dimensional chromatography on silica-gel filter paper give strong evidence that this highly labeled lipid is different from phosphatidic acid. In previous work unidirectional chromatography was used for separation of phospholipids[28,29]. Under these conditions one cannot effectively resolve phosphatidic acid from closely related glycerol phosphatides. Two-dimensional chromatography is more effective and has now been employed. Lipid GPX by two-dimensional chromatography is separated from a phosphatidic acid prepared by the action of phospholipase D (EC 3.1.4.4) on egg lecithin. Commercial phosphatidic acid was purchased from two different companies in order to have more samples for comparison. Although the commercial samples of phosphatidic acid were designated to be chromatographically pure, they both contained at least 3 components, one of which migrated like the egg phosphatidic acid prepared in our laboratory. The other two components were not identified although they may be lyso-phosphatidic acid and the phosphate ethyl ester of phosphatidic acid.

Lipid GPX does behave like a precursor to both lecithin and triglycerides in these rat-liver systems. Current studies are underway to establish its structure.

TABLE III

EFFECT OF CTP ON THE LABELING OF TRIGLYCERIDES, DIGLYCERIDES, AND LECITHIN FROM [^{14}C]GLYCEROL

	counts/min		
	TG	DG	LEC
Expt. 1			
control	3 000	4 500	20 950
plus CTP	10 100	8 700	5 050
Expt. 2			
control	4 600	4 700	24 950
plus CTP	17 800	11 250	5 900
Expt. 3			
control	2 750	2 900	19 500
plus CTP	10 950	11 000	4 700

Incubations carried out at 37° for 120 min. All systems had liver homogenate (200 mg wet weight), 10 mM ATP, 0.5 mM glycerol, and 2 μC of [1,3-^{14}C]glycerol. When CTP was added, the final concentration was 10 mM. From Marinetti (unpublished data).

In rat-liver homogenates CTP was found to stimulate the synthesis of triglycerides and diglycerides and decreased the synthesis of phospholipids[28,29]. The data in Table III also show this effect of CTP. The CTP effect was magnesium-dependent and it dominated the labeling pattern since it obliterated the Mg^{2+} stimulation of lecithin and phosphatidylethanolamine labeling. In these systems CDP-choline was incorporated into lecithin and CTP stimulated the incorporation of [^{32}P]phosphorylcholine into lecithin. These experiments serve to point out the danger in over-generalizing on results from one labeled substrate. The interplay of ATP, CTP and Mg^{2+} are believed to control in part the pathways for synthesis of triglycerides and phospholipids. These studies show that cytidine nucleotides not only influence phospholipid synthesis, but also play a role in the synthesis of triglycerides.

References p. 153

It might be anticipated that the rates of synthesis of various triglycerides having a different fatty acid pattern might vary. Evidence to support this idea has come from the determination of the specific activity of different triglycerides labeled from [^{14}C]glycerol. Thus Marinetti et al.[30] found a metabolic heterogeneity of labeled rat-liver triglycerides. The triglycerides

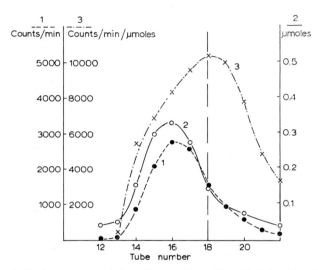

Fig. 5. Metabolic heterogeneity of rat-liver triglycerides labeled with [^{14}C]glycerol. [^{14}C]Glycerol was incorporated into the lipid of rat-liver homogenates. The triglycerides were separated on a column of silicic acid and analyzed. Taken from Marinetti et al.[30].

were isolated by column chromatography on silicic acid. The triglycerides having the highest specific activity were those which were eluted on the descending portion of the column-distribution curve (Fig. 5). These represent triglycerides which have a higher content of unsaturated fatty acids. This finding was confirmed in later studies in which the individual labeled triglycerides were separated on silver nitrate treated silica-gel paper. The more unsaturated triglycerides were more highly labeled than the more saturated triglycerides (Fig. 6).

The cellular localization of the enzymes which synthesize triglycerides has been investigated. Weiss et al.[16] found activity in chicken liver to be located in both the microsomes and cytoplasm although the former were 2–3 times as active as the latter. However these workers used frozen homogenates to prepare the cell fractions and one cannot be sure of cross-contamination of the cell particles under these conditions. Tzur and Shapiro[31] have shown a dependence of microsomal lipid synthesis (phospholipid + neutral glycerides) on added protein. A 20–40 fold increase in ester yield was obtained by supplementing microsomes with protein (albumin or serum lipoproteins).

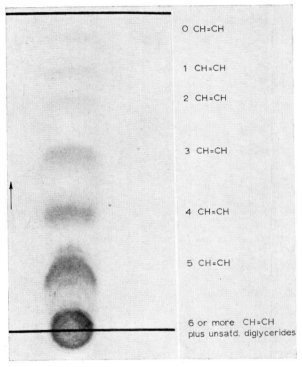

Fig. 6. Labeling of different classes of triglycerides in rat-liver homogenates. [^{14}C]Glycerol was incorporated into the lipid of rat-liver homogenates. The triglycerides were obtained by column chromatography on silicic acid and then chromatographed on silica-gel-loaded filter paper which was treated with silver nitrate. The solvent was chloroform–isopropanol 98.5/1.5(v/v). The triglycerides are resolved according to the total number of double bonds (CH = CH) in their fatty acid chains. Marinetti *et al.*, unpublished data.

Marinetti *et al.* (unpublished data) obtained the cell fractions of rat liver by differential centrifugation of a liver homogenate. The ability of these liver systems to incorporate [^{14}C]glycerol into triglycerides and diglycerides in an ATP-fortified system are shown in Table IV. The total homogenate

References p. 153

had the greatest activity. Removal of the mitochondria from a mitochondria–microsome–cytoplasm system had no appreciable effect on the synthesis of these glycerides, showing that mitochondria play a minor role in triglyceride synthesis. The cell cytoplasm was inactive. At present it is not clear why centrifugation of the homogenate at low speed to remove mainly cell nuclei and cell debris markedly decreases the incorporation of glycerol into lipid. The microsomal localization of the acylating enzymes which synthesize triglycerides in adipose tissue and intestine will be discussed in another section of this chapter.

TABLE IV

CELLULAR LOCALIZATION OF NEUTRAL GLYCERIDE SYNTHESIS IN RAT LIVER

System	counts/min[a]
Total homogenate	7500
Mitochondria + microsomes + cytoplasm	4350
Microsomes + cytoplasm	4350
Cytoplasm	0

[a] Represents total counts/min in isolated triglycerides and diglycerides from 2 μC of [1,3-^{14}C]glycerol. The systems contained 10 mM Mg^{2+}, 10 mM ATP, and 1 mM cold glycerol. The incubation time was 1 h at 37°. (Marinetti et al., unpublished data)

It is noteworthy that glycerol kinase is very active in liver[28,32,33] and is located primarily in the cell cytoplasm. Phosphatidic acid phosphatase is located in the microsomal fraction. Glycerol kinase activity is very low or absent in adipose tissue and heart[28] but some activity is found in mammary gland[34] and in intestine[35-38]. Kidney also has high activity of this enzyme[28,39].

Glycerol metabolism in rat liver, heart and kidney was studied by Marinetti et al.[28]. A time study of [1,3-^{14}C]glycerol in liver homogenate (Fig. 7) showed a very rapid conversion of glycerol to glycerol phosphate and then to dihydroxyacetone phosphate. This latter compound is subsequently converted to dihydroxyacetone. The rate curve shows that within 5 min over 90% of the glycerol is converted to glycerol phosphate. The formation of dihydroxyacetone phosphate reaches a peak between 60 and 70 min. The incorporation of glycerol into total lipid shows a continual increase over a 2-h period.

The quantitative aspects of triglyceride synthesis in liver and plasma have been investigated by Farquhar et al.[40]. These workers propose an incompletely coupled two-compartment non-recycling catenary model for the explanation of their data of turnover rates of liver and plasma triglycerides in man. The hepatic pool is claimed to be much larger than the plasma pool and has a turnover rate three times as great. Kinetic data from labeled palmitic acid was more complex to analyze than that from labeled glycerol, possibly because glycerol is removed principally by the liver and does not recycle as does palmitic acid.

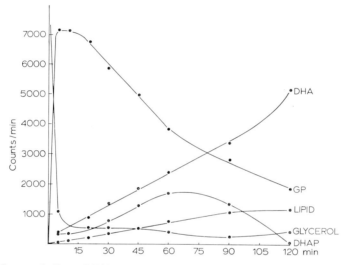

Fig. 7. The metabolism of [^{14}C]glycerol in rat-liver homogenates. [^{14}C]Glycerol (2 μC) was incubated in a liver homogenate containing 10 mM ATP. At 5, 10, 20, 30, 45, 60, 90 and 120 min, 20-μl aliquots were removed for chromatographic analysis on Whatman #1 filter paper. The solvent was propanol–conc. ammonium hydroxide–water 60/30/10 (v/v). The labeled products were first detected by autoradiography on X-ray film and then the components were cut out and counted. DHA = dihydroxyacetone, GP = glycerol-phosphate, DHAP = dihydroxyacetone phosphate. Marinetti et al., unpublished data.

A study of the endogenous triglyceride turnover in liver and plasma of the dog[41] indicated that the liver triglycerides occur in 2 compartments, part of which is metabolically inert and part of which is active. The fractional turnover rate of plasma S_f 20 triglycerides was 2–3 times that of the liver triglycerides. The average turnover rate of liver triglycerides was estimated to be about 19–21 μmoles/h/kg body weight.

References p. 153

5. Biosynthesis of triglycerides in adipose tissue

Steinberg has investigated triglyceride synthesis in homogenates of rat epididymal fat pads using [^{14}C]palmitic acid as a marker[42]. Unfortified homogenates gave very little incorporation of the labeled acid into triglycerides and had high lipolytic activity. The optimal biosynthesis of triglycerides required ATP, Mg^{2+} CoASH, α-GP, cysteine and NaF (Table V). The

TABLE V

INCORPORATION OF [1-^{14}C]PALMITIC ACID INTO NEUTRAL LIPIDS OF ADIPOSE-TISSUE HOMOGENATES

System	Relative incorporation
Complete	100
Minus ATP	0
Minus CoA	4
Minus α-GP	27
Minus Mg^{2+}	35
Minus cysteine	51
Minus NaF	32
Minus buffer	0
Heat 100°, 5 min	0

Taken from Steinberg[42].

TABLE VI

EFFECT OF SUBSTITUTIONS FOR α-GP ON THE INCORPORATION OF [1-^{14}C]PALMITIC ACID INTO TRIGLYCERIDES OF ADIPOSE-TISSUE HOMOGENATES

System	Relative incorporation
Complete	100
Minus α-GP	9
Minus α-GP and glycerol	8
Minus α-GP and monoolein	2
Minus α-GP and FDP	10
Minus α-GP and FDP and NADH	100
Minus α-GP and glyceraldehyde-3-P and NADH	70
Minus α-GP and glucose-6-P and NADH	68

Taken from Steinberg[42].

requirement for α-GP could be replaced by FDP+NADH, by glyceraldehyde-3-P+NADH, or by glucose-6-P+NADH but not by glycerol or monoolein (Table VI). These studies show that biosynthesis of triglycerides in adipose tissue goes through the glycerol-phosphate pathway and that the monoglyceride pathway is not important. It also demonstrates the lack of glycerol kinase in adipose tissue.

Steinberg et al.[42], and Tzur and Shapiro[31] have demonstrated that triglyceride synthesis in adipose tissue is associated with the cell particles, in particular, microsomes. Goldman and Vagelos[43] found that with chicken-adipose tissue, 1,3-diolein was inactive but 1,2-diolein was an excellent substrate for triglyceride formation. They found 1-oleyl-2-stearyl-diglyceride reacted more rapidly with palmityl-CoA than with oleyl-CoA but that the reverse was true with 1-stearyl-2-oleyl-diglyceride.

Steinberg[44] has discussed the problems involved in calculating the rate of synthesis of triglyceride in fat pads. The rates of esterification of triglycerides from [1-^{14}C]palmitic acid were as follows (μequiv./g/h) (a) 0.16 ± 0.014 based on the specific activity of the medium palmitic acid at zero time as that of the immediate precursor, (b) 0.25 ± 0.028 based on the arithmetic mean of the specific activity of the medium palmitic acid at zero time and at the end of 1 h, (c) 23.8 ± 2.9 based on the specific activity of the tissue fatty acids at 1 h and (d) 3.8 ± 0.85 based on the net-balance method. Steinberg favors the net-balance method as giving the most meaningful values.

The cellular localization of the acylating enzymes in rat-adipose tissue was studied by Roncan and Hollenberg[45]. Using [^{14}C]palmitate and L-α-glycerol phosphate they observed that acylation was confined to the mitochondria and microsomes but the mitochondria were more active. ATP and CoASH were obligatory cofactors. The main product was reported to be phosphatidic acid. Although the cytoplasm alone was inactive, when added to the cell particles, it led to the synthesis of triglycerides as major products. The observations of Roncan and Hollenberg are different from those of Steinberg[42], Tzur and Shapiro[31] and Goldman and Vagelos[43] who reported that the microsomes were the most active particles for synthesis of triglycerides. The conditions of the experiments were not identical and in some cases a different animal was used. Whether this difference represents a species difference or whether it represents a more fundamental difference remains to be solved.

The synthesis of different molecular species of triglycerides of adipose tissue was studied by Hollenberg[6]. Rats were fed [^{14}C]glucose and the

References p. 153

labeled triglycerides of adipose tissue were separated into 10 classes by chromatography on silver nitrate treated silica gel. These triglyceride classes were separated into the following types: SSS, SSU_1, SU_1U_1, SSU_2, $U_1U_1U_1$, SU_1U_2, $U_1U_1U_2$, SU_2U_2, $U_1U_2U_2$ and $U_2U_2U_2$ (S = saturated, U_1 = monounsaturated, U_2 = diunsaturated). Only the SSS, SSU_1 and SU_1U_1 species were obtained highly purified. The newly synthesized triglycerides resembled in kind and in proportion the triglycerides which exist in the tissue. The various triglycerides appeared to be stable (little ester interchange) and had similar turnover rates. After administration of [^{14}C]glucose the specific activity of saturated fatty acids was higher than the more unsaturated triglyceride species. However an analysis of the percent distribution of radioactivity in the various triglycerides showed more total activity in the SU_1U_2, SU_1U_1, SSU_1, SSU_2 and $U_1U_1U_2$ classes than in the SSS, SU_2U_2, $U_1U_1U_1$, $U_1U_2U_2$ and $U_2U_2U_2$ classes. The SU_1U_2 class had the highest radioactivity whereas the SSS class had the lowest percentage of the total radioactivity. Hollenberg also reported that intermolecular reshuffling of the triglyceride fatty acids in adipose tissue is slow and the small amount that may occur is believed to be due to hydrolysis and reesterification rather than transesterification[46].

Galton[47] has studied lipogenesis in human-adipose tissue. A study of 82 patients showed that fatty acyl-CoA synthetase activity was high but varied considerably. Triglycerides formed from labeled glucose had all the radioactivity in glycerol. Palmitic acid at 200 μM stimulated lipogenesis but inhibited at 700 μM. Arachidate over the same concentration range gave no inhibition of lipogenesis.

It appears that fatty acids extracted from adipose tissue do not constitute a single pool of precursor fatty acids into which medium fatty acids enter before being esterified. Steinberg[44] raised the question whether labeled medium fatty acids penetrate the adipose-tissue cell and appear there as free fatty acids prior to their incorporation into glycerides.

When adipose tissue of fed rats is incubated with labeled palmitic acid with added glucose in the medium, the amount of radioactive free fatty acids in the tissue at the end of the incubation is very small relative to the amount in triglycerides. This has led some to consider the possibility that esterification occurs at the cell membrane. However, if triglyceride synthesis is rapid, the rate of fatty acid penetration into the cell may be the rate-limiting step and free fatty acids may not accumulate.

The medium fatty acids are probably not mixing with the entire intracellular fatty acid pool prior to incorporation into glycerides. Nevertheless, a small fraction of the tissue fatty acid in the cytoplasm may be rapidly turning over and account for the relatively small amount of radioactivity in this fraction as compared to that in triglycerides.

When tissues from fasted rats are incubated without glucose in the medium, a sizable fraction (21.6%) of the labeled medium fatty acid is found in the tissue fatty acid pool[44]. The absolute as well as the relative amount of radioactivity in this fraction is increased under these conditions while the incorporation into glycerides is reduced. Incubation under anaerobic conditions in presence of fluoride decreases further the incorporation into glycerides and increases the level of cellular free fatty acid. In the presence of nitrogen and 0.1 M NaF, triglyceride synthesis is inhibited nearly 90% whereas radioactivity in the tissue free fatty acid is increased more than 3-fold. Moreover, 80% of the label taken up is present as free fatty acid. If labeled free fatty acid appearing in the tissue pool arrived there only by first being incorporated into glycerides rather than by direct penetration, less rather than more radioactivity would have been expected in the intracellular free fatty acid pool since labeling of triglycerides was less. The results suggest that labeled free fatty acids from the medium can enter the tissue free fatty acid pool and that this process is not dependent on oxidative metabolism or on concomitant triglyceride synthesis.

6. Biosynthesis of triglycerides in intestine

It is well known that glycerides are hydrolyzed in the intestinal lumen and that during passage from the lumen to the lymphatic system a resynthesis of triglycerides occurs. It was first assumed that the synthesis of triglycerides occurred by the glycerol-phosphate pathway which prevailed in liver. However, Clark and Hübscher[48,49] suggested another pathway in which monoglycerides were acylated directly to triglycerides without going through glycerol phosphate and phosphatidic acid as intermediates. This pathway, now coined the "monoglyceride path", was substantiated by Senior and Isselbacher[50] and by Johnston and Brown[51]. Further studies have shown that in intestine the monoglyceride pathway is more important than the glycerol-phosphate pathway for resynthesis of triglycerides.

Kern and Borgstrom[52] used labeled oleic acid and 1-monolein to examine the quantitative significance of the monoglyceride pathway in hamster-intestinal mucosa. Their results showed that the monoglyceride path could

References p. 153

account for 80–100% of the synthesized triglycerides. The ether analogues were acylated in these systems to yield monoalkanyl diacyl derivatives of glycerol.

Paris and Clement[53] have shown that 85% of the triglyceride synthesis in isolated intestinal loop of the rat occurs by the monoglyceride path and 15% occurs by the glycerol-phosphate path. They also found that 70% of the 2-monopalmitin added was absorbed intact and that 18% of the absorbed monoglyceride was split into fatty acid and glycerol. The liberated fatty acids were incorporated into glycerides and an appreciable amount of the free glycerol was converted to phospholipid. This indicates that glycerol kinase activity is present in intestinal mucosa, a finding which is supported by the earlier work of Clark and Hubscher[35] and Haessler and Isselbacher[36].

The specificity of the acylating enzymes which convert monoglycerides to triglycerides has also been investigated. Mattson and Volpenhein[54] found that thoracic-dust lymph triglycerides collected after feeding have the same fatty acid in the 2-position as the ingested fatty acid. Thus when β-palmityl-α,α'-dioleyl triglyceride was fed to rats, 85–90% of the palmitic acid in the resynthesized triglycerides was in the original 2-position. Skipski *et al.* have made similar observations[55].

The biosynthesis of individual molecular species of triglycerides was studied by Johnston and Rao[5] in intestinal mucosa. Individual triglycerides of the type OOO, OPO, PPO, OOP and PPP were isolated by chromatography on silver nitrate treated silica-gel plates. The fatty acid distribution was examined by using various acyl-CoA esters and free fatty acids with radioactive 1- and 2-monoglycerides (see Tables VII and VIII). The data in Table VII shows a nearly theoretical random distribution of fatty acids in the synthesized triglycerides. Using [2-^{3}H]monopalmitin and palmityl-CoA, the major triglyceride synthesized was PPP. With oleyl-CoA and [2-^{3}H]monopalmitin the major product was OPO although appreciable amounts of OOO and PPO were produced. With a mixture of palmityl- and oleyl-CoA, the major product was PPO, but some OPO and PPP were formed.

When [1-^{14}C]monopalmitin was used (Table VIII) with different acyl-CoA esters or the corresponding free fatty acids the results showed a similar random acylation. The authors concluded that the intestinal transacylase randomly acylates either the 1- or 2-position. This work demonstrates that intestinal triglycerides are unlike other natural triglycerides which have unsaturated fatty acids predominantly in the 2-position. This is due to the non-specific acylation of the intestinal transacylase[56]. In other words, tri-

TABLE VII

TRIGLYCERIDE SYNTHESIS IN INTESTINE USING [2-³H]MONOPALMITIN

		mμmoles				
		Total	OOO	OPO	PPO	PPP
Expt. 1	palmityl-CoA 1 μmole	189	—	—	1	188
	oleyl-CoA 1 μmole	194	27	126	35	6
	palmityl-CoA + oleyl-CoA 0.5 μmole each	217	13	48	104	52
Expt. 2	palmitic acid 1 μmole	282	5	5	31	241
	oleic acid 1 μmole	248	10	206	27	5
	palmitic + oleic acid 0.5 μmole each	270	10	32	94	134

Taken from Johnston and Rao[5].

TABLE VIII

TRIGLYCERIDE SYNTHESIS IN INTESTINE USING [1-¹⁴C]MONOPALMITIN

	mμmoles				
	Total	OOO	POO	PPO	PPP
Palmityl-CoA 1 μmole	218	2	—	1	215
Oleyl-CoA 1 μmole	195	15	138	31	11
Palmityl-CoA + oleyl-CoA 0.5 μmole each	213	17	39	106	51
Palmitic acid 1 μmole	353	3	2	23	325
Oleic acid 1 μmole	323	7	252	53	11
Palmitic + oleic acid 0.5 μmole each	335	10	27	107	191

Taken from Johnston and Rao[5].

glycerides synthesized by the glycerol-phosphate pathway may yield the more asymmetric pattern of fatty acid distribution.

Since in intestine there are two pathways for synthesis of triglycerides and both involve a diglyceride intermediate, it is apparent from what has been said above that the diglycerides formed from each pathway may represent different molecular species. The question also arises, are these diglycerides in different compartments in the cell or are they bound to different proteins. Johnston *et al.*[57] have suggested that in intestine these pathways are separate and that the diglycerides are in different pools within the cell. However, more work is required to confirm this idea.

References p. 153

Studies with milk fat have shown that caproate and butyrate are esterified at the 1,3-positions of glycerol whereas palmitate and myristate are preferentially located at the 2-position[58]. These data show that a general rule for fatty acid distribution in which saturated fatty acids are exclusively found in the 1,3-positions does not always hold.

7. Biosynthesis of mixed ester alkenyl or ester alkanyl glycerides

The synthesis of glycerides having saturated ether-linked hydrocarbon chains or vinyl ether-linked chains in addition to ester-linked fatty acids proceeds by the same mechanisms which exist for the synthesis of the typical triester glycerides (triglycerides). It has been shown that both α- and β-glycerol ethers are absorbed in the intestine and converted to the alkoxydiglyceride analogues of the triglycerides[52,59]. The β-ether appears to be more rapidly esterified than the α-ether.

The acylation of monoglycerides is subject to species differences[59]. In rat microsomes the 1,3-diglyceride is a major intermediate whereas in hamster microsomes very little 1,3-diglyceride accumulates.

8. Control of triglyceride synthesis

The multitude of enzyme-catalyzed reactions in living cells are under various control mechanisms. Substrate concentration can affect the rate of an enzyme reaction. Permeability through membranes undoubtedly plays a vital role in regulating substrate concentration. *De novo* synthesis of enzymes and allosteric effects on enzymes represent other important control processes. Hormones are believed to act either at the membrane level or gene level. The complex interplay of these control mechanisms remains to be determined.

Triglycerides in the fat depots are in a dynamic state. Their synthesis and deposition as fuel is favored by feeding which induces a hyperinsulin state. Excess dietary fatty acids are stored in the fat depots as triglycerides. During this process some reshuffling of fatty acids occurs and some of the dietary fatty acids may be altered by chain lengthening or chain shortening, and by chain saturation or desaturation but these alterations are limited.

It has been observed that the stored depot fat may resemble the dietary fat if the intake of dietary fat is sustained at a high level and for a long enough time period[60]. Excess dietary carbohydrate is converted to fat in the animal.

During this process glucose is converted to acetyl-CoA and glycerol phosphate which together with NADPH and ATP produce fatty acids. The fatty acids are esterified to triglycerides and phospholipids. The fatty acids formed from glucose *via* acetyl-CoA are however highly saturated or contain only 1 double bond since the animal does not have the enzymes necessary to convert acetate to linoleic, linolenic and arachidonic acids which are the polyunsaturated fatty acids containing two, three and four double bonds respectively. Since these fatty acids are obligatory dietary constituents they have been called the essential fatty acids.

Hormones affect lipogenesis[44,61,62]. Insulin is known to increase fatty acid synthesis and ultimately triglyceride synthesis. The action of insulin may be at least 2-fold. One major action is to increase the permeability of adipose tissue and muscle to glucose. It is believed that the increased glucose permeability stimulates glycolysis and the pentose-shunt pathway, the former yielding acetyl-CoA and glycerol phosphate and the latter yielding NADPH. Increased Krebs cycle oxidation will also produce ATP at a higher rate. All these processes stimulate fatty acid synthesis and triglyceride synthesis. The lowering of blood NEFA by insulin is attributed to increased synthesis of triglycerides by adipose tissue.

The second action of insulin may be at the gene level. Insulin administration has been shown to increase the activity of certain enzymes such as acetyl-CoA carboxylase (EC 6.4.2.1) and citrate-cleavage enzyme (EC 4.1.3.8) both of which are important for the synthesis of fatty acids[63,59,61]. Insulin also leads to increased activities of glucose-6-phosphate dehydrogenase (EC 1.1.1.49) and 6-phosphogluconate dehydrogenase (EC 1.1.1.43) which give more NADPH required for fatty acid synthesis.

The rate of influx or efflux of NEFA in adipose tissue is believed to be under hormonal control, in particular by insulin and epinephrine[44,61]. The net effect of these hormones is such that insulin increases influx of NEFA and epinephrine increases efflux of NEFA. These observations are secondary in explaining how insulin lowers whereas epinephrine increases plasma NEFA, since they are a consequence of the enzyme activities within the adipose-tissue cell which regulate the rate at which triglycerides are either being hydrolyzed or synthesized. Epinephrine has been shown to increase the hydrolysis of triglycerides in adipose tissue by a mechanism which appears to involve increased production of cyclic AMP which increases the activity of adipose-tissue lipase. Present day knowledge is meager indeed on the quantitative effects of the hormones on various tissues, the rapidity of these

References p. 153

effects, how much hormone is required to elicit the response and how one hormone affects the action of another hormone. For example, the lipid-mobilizing effect of epinephrine is abolished or greatly inhibited by adrenalectomy or thyroidectomy and restored by administration of cortisone or thyroxine. Thyroxine and cortisone are said to have a "permissive action" on the epinephrine effect on adipose tissue[44,64].

Epinephrine, ACTH, and glucagon suppress incorporation of labeled palmitate into glycerides of epididymal fat pads. Since these hormones increase the rate of release of fatty acids one must consider whether a dilution of the labeled fatty acid by unlabeled fatty acids can give an "apparent" decrease in glyceride synthesis. The use of labeled fatty acids for measuring triglyceride synthesis is therefore not without risk. This point has been discussed by Steinberg[44]. He describes a net balance non-isotopic method for measuring glyceride synthesis which is based on net glycerol utilization. The basic assumptions of this method are (a) glycerol released by lipolysis is not re-esterified to glycerides or oxidized to any important extent, (b) all the glycerol produced during incubation is derived from triglycerides rather than from existing mono- or diglycerides and (c) the amount of fatty acid released and re-esterified during incubation is large relative to the amount oxidized. Based on these premises one can calculate the total fatty acids released from triglycerides by determining net glycerol production (change in medium plus change in tissue). Results obtained by this method for control tissue and hormone-treated tissue showed that "lipolytic" hormones (epinephrine, ACTH) increased the rate of release of fatty acids markedly, but the rate of esterification also increased although to a much lesser degree.

Studies on the perfused parametrial fat body of the rat by Scow et al.[65] have shown that ACTH stimulated the release of glycerol and free fatty acids from this tissue, but this effect was dependent on the rate of blood flow and on the blood-albumin concentration. The release of free fatty acids was reduced when the molar ratio of fatty acid to albumin exceeded two. ACTH also increased the diglyceride content of the perfused tissue. The authors conclude that hydrolysis of diglycerides to glycerol is rate-limiting in the lipolysis of triglycerides stimulated by ACTH.

Rodbell[66] has investigated the action of hormones on lipid metabolism in isolated fat cells. He has reported that the ACTH and epinephrine stimulated lipolysis of triglycerides in isolated fat cells is 20–30 times greater than that seen in intact adipose tissue. Lipolysis stopped when the medium-albumin became saturated with fatty acids. The molar ratio of fatty acid to albumin in this case was 6–7. Under these conditions the concentration of free fatty acid in the fat cell increased. Rodbell suggests that the fatty acid concentration in the cell determines the rate of lipolysis by inhibiting the hormone-sensitive lipase.

Steinberg[44] has discussed the metabolic, hormonal and neural control of fat metabolism. When glycogen stores are very low the pool of L-α-GP is small (less than 0.2 μmoles/g). Under most circumstances the L-α-GP requirement is met by the fat tissue taking up and glycolyzing blood glucose. If blood glucose or adipose-tissue glycogen becomes limiting, then the steady-state situation favors lipolysis of the cell triglycerides. This situation would prevail in the fasted state. In the fed state triglyceride synthesis is favored since the increase in blood glucose stimulates insulin production which leads to increased influx of glucose in the fat cell. The increased glycolysis provides L-α-GP for triglyceride synthesis.

Steinberg points out that insulin, like glucose, appears to suppress release of fatty acids by accelerating esterification without altering the rate of lipolysis. The effects of insulin on glucose metabolism and fatty acid metabolism can be closely duplicated by raising the glucose concentration to high levels. Insulin alone, without glucose, does not suppress fatty acid release. Hence the effect of insulin on lipogenesis in adipose tissue is mediated principally by its enhancing glucose utilization, most likely by increasing glucose permeation through the cell membrane.

The release of free fatty acids by adipose tissue, has been shown to be stimulated by at least 11 different hormones (epinephrine, norepinephrine, ACTH, glucagon, growth hormone, adrenal glucocorticoids, etc.). It must be emphasized that some of these hormones also increase the rate of esterification of fatty acids to triglycerides but this is offset by a much greater increase in the rate of lipolysis. It is of interest that most hormones are geared to stimulate lipolysis and hence fatty acid mobilization, and that insulin alone favors the reverse process. Apparently the animal is well suited for storing triglycerides and for mobilizing this fuel when it is required. The rate of NADPH production may act as the driving force and control the synthesis of fatty acids from acetate. This facet of the problem has been discussed by Flatt and Ball[67]. They conclude that NADPH production in the pentose cycle can supply only one half the reduced NADPH required for fatty acid synthesis. The remaining half is believed to arise during oxidation of triose phosphates. Moreover, the citrate-cleavage enzyme(CCE) and malic enzyme (ME) (EC 1.1.1.38/39/40) work in concert to produce NADPH as follows:

References p. 153

(a) citrate + ATP + CoA-SH $\underset{}{\overset{CCE}{\rightleftharpoons}}$ acetyl-CoA + ADP + oxaloacetate

(b) oxaloacetate + NADH $\rightleftharpoons$ malate + NAD$^+$

(c) malate + NADP$^+$ $\underset{}{\overset{ME}{\rightleftharpoons}}$ pyruvate + CO$_2$ + NADPH

In this coupled enzyme system NADH is in effect converted to NADPH. The levels of the citrate-cleavage enzyme and the malic enzyme rise and fall in concert as a consequence of feeding and starvation[59,68].

Tepperman[69] has reviewed the role of the enzymes involved in NADPH production as it relates to lipogenesis and how this system is under control. He considers the "short" control to be related to the concentration of substrates and enzymes and the "long" control to be related to the enzyme-forming system. Although the malic enzyme levels fall during starvation and rise with feeding and hence correlate with lipogenic activity, the role of the citrate-cleavage enzyme and hexose monophosphate dehydrogenase is less clear since although their activity is increased by insulin administration and with carbohydrate feeding, these enzymes are not increased if dietary protein is withheld, yet lipogenesis is increased. These studies point to the complexity of the control systems in animals and the danger of overgeneralizing on the basis of meager evidence.

Feedback control of triglyceride synthesis has not been extensively studied. However, the work of Lynen *et al.*[70] has indicated that fatty acids or fatty acyl-CoA esters have a negative feedback effect on acetyl-CoA carboxylase and act as a throttle on synthesis of fatty acids. The rate of formation of citrate and the effect of citrate on acetyl-CoA carboxylase may represent another control process affecting the rate of formation of acetyl-CoA by the citrate-cleavage enzyme and the activity of acetyl-CoA carboxylase, which is believed to be the rate-limiting enzyme in fatty acid synthesis. Vagelos[71] has shown that acetyl-CoA carboxylase is converted from a monomer to a more active trimer by citrate but the level required to produce this effect is much higher than the "physiological" level of citrate in the cell. One can assume that conditions which affect fatty acid synthesis will also affect triglyceride synthesis.

Appreciable net synthesis of triglycerides occurs after ingesting large amounts of either carbohydrate and/or lipid. Under these conditions the levels of substrates are favorable for biosynthetic processes. Furthermore, the hormonal state of the animal, partly influenced by feeding, together with favorable substrate concentration brings about increased enzyme activities (citrate-cleavage enzyme, acetyl-CoA carboxylase, malic enzyme, phosphofructokinase (EC 2.7.1.11), pyruvate kinase (EC 2.7.1.40), glucose-6-phosphate dehydrogenase) which enhance the synthesis of fatty acids and hence triglycerides. The increased enzyme activities probably represent *de novo* synthesis of enzymes possibly at the gene level. How the substrates and hormones bring this about is now unknown, although deactivation of gene repressors has been invoked. The increased enzyme activities may also in part represent allosteric effects on enzymes by certain substrates but one can only speculate on this point with respect to triglyceride synthesis. If the substrate levels during starvation are below the amount required to obtain the V_{max} for the enzyme reaction, then in these cases, it is obvious that increasing the substrate concentration will increase the rate of the enzyme reaction. Undoubtedly all these effects act to varying degrees to control the overall rate of metabolic pathways.

The synthesis of triglycerides in liver or adipose tissue from dietary fatty acids occurs primarily by the L-α-glycerol phosphate pathway. Since normal white adipose tissue lacks glycerol kinase, the L-α-GP must be obtained through glycolysis within the fat cell since it is unlikely that hepatic L-α-GP can readily enter the blood stream or penetrate cell membranes. There is also a requirement for the production of NADPH, acetyl-CoA and ATP when the triglyceride synthesis occurs from fatty acids made endogenously from glucose. The production of NADPH for lipogenesis has been discussed by Flatt and Ball[67] whereas the production of acetyl-CoA has been discussed by Lowenstein[72]. The acetyl-CoA produced by oxidative decarboxylation of pyruvate within the mitochondria is according to Lowenstein released as citrate which is then acted on by the citrate-cleavage enzyme to regenerate acetyl-CoA in the cytoplasm. Some acetyl-CoA can be hydrolyzed in the mitochondria to form acetate which permeates the mitochondrial membrane. The acetate is then reactivated to acetyl-CoA by acetate thiokinase (EC 6.2.1.1) in the cytoplasm.

In this author's laboratory the control of lipid synthesis in liver systems has been investigated over the past several years using [^{14}C]glycerol as a marker and examining the effect of varying the concentration of ATP, CTP and Mg^{2+} ions. It was found[28] that optimum concentrations of ATP and Mg^{2+} for lipogenesis in liver homogenates were 10 mM. Increasing the

References p. 153

ATP concentration to 100 mM eliminated labelling of all lipids except an unidentified acidic phospholipid which was first believed to phosphatidic acid; further work has shown that this lipid has the properties of phosphatidyl glycerol phosphate.

Although ATP was found to be required for labeling of all lipids, the pattern of lipid labeling was influenced by Mg^{2+} and CTP. Increasing the Mg^{2+} concentration to 10 mM caused the labeling to appear mainly in the phospholipids, especially lecithin. This was done at the expense of labeling of triglycerides. However, increasing the CTP concentration to 10 mM, in the presence of both ATP and Mg^{2+}, shifted the labeling in favor of triglycerides and diglycerides.

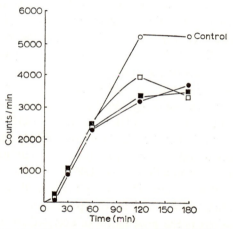

Fig. 8. The effect of puromycin on the synthesis of triglycerides in rat-liver homogenates. [^{14}C]Glycerol was incorporated into the lipids of rat-liver homogenates fortified with 10 mM ATP. The triglycerides were resolved by chromatography and counted. The control (○) homogenate did not contain puromycin. The other homogenates contained puromycin at 10^{-3} (●), 10^{-4} (□) and 10^{-5} M (■) respectively. Taken from Santora and Marinetti[73].

In order to gain information on the stabilities of the enzymes required for triglyceride synthesis, Santora and Marinetti[73] examined the effect of puromycin on the incorporation of [^{14}C]glycerol into triglycerides, diglycerides and phospholipids of rat-liver homogenate and slices. The effect of puromycin on triglyceride synthesis is shown in Figs. 8 and 9.

Puromycin inhibited triglyceride synthesis over a concentration range of

10^{-3}–10^{-5} M. The fact that 10^{-5} M puromycin gave the same inhibition as did 10^{-3} M puromycin makes it unlikely that puromycin is inhibiting the catalytic activity of one or more enzymes involved in the synthesis of triglycerides. Rather it favors the concept that the inhibition is due to a block in protein synthesis. Puromycin was also effective in inhibiting triglyceride synthesis in liver slices (Fig. 9). Although puromycin had no

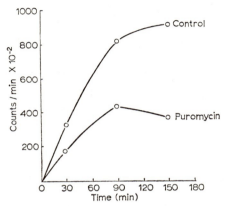

Fig. 9. The effect of puromycin on the synthesis of triglycerides in rat-liver slices. [^{14}C]-Glycerol was incorporated into the lipids of rat-liver slices fortified with 10 mM ATP. The triglycerides were resolved by chromatography and counted. The control slices did not contain puromycin. The other slices were incubated in the presence of 10^{-3} M puromycin. Taken from Santora and Marinetti[73].

detectable effect on the labeling of lecithin in liver homogenates, it gave a strong inhibition in liver slices. Moreover, puromycin had a unique effect on the labeling of lipid GPX. In both homogenates and slices puromycin caused an increase in the labeling of lipid GPX. This observation indicates that lipid GPX is a precursor to triglycerides and lecithin and that its conversion to these latter lipids is inhibited by puromycin. By inhibiting protein synthesis with puromycin and by incubating homogenates for varying periods of time before adding the [^{14}C]glycerol and then studying the rate of incorporation of label into the various lipids one could determine which enzymes are the least stable. It was found that the enzymes involved in the synthesis DG and TG were more unstable than the enzymes involved in the synthesis of phospholipids, and that glycerol kinase and the glycerol phosphate acylating enzymes were stable under these experimental conditions.

References p. 153

9. Fatty livers

In general, an impairment of carbohydrate metabolism leading to a marked diminution in glucose oxidation, or a depletion of glycogen which is sustained for a relatively long period of time, leads to an accumulation of triglycerides in the liver. This is believed to be brought about by extensive mobilization of fatty acids from the adipose tissue and their conversion to triglycerides in the liver.

Another general situation which leads to a fatty liver is an impairment of protein synthesis such that triglycerides cannot be converted to lipoproteins and transported out of the liver. Drugs and liver poisons are believed to act at this level.

The mechanism of hepatic lipid accumulation following carbon tetrachloride administration has been studied[74,75]. It has been observed that carbon tetrachloride ingestion leads to a decrease in plasma triglycerides, abolishes the marked elevation of plasma triglycerides after triton ingestion, blocks the net release of triglycerides into the perfusate of isolated perfused livers, decreases the incorporation of ingested amino acids into plasma lipoproteins, and decreases the incorporation of ingested palmitic acid into plasma triglycerides.

Excess alcohol intake leads to fatty livers in man[76]. The mechanism of how triglycerides accumulate in this situation is not known although peroxidation of microsomal lipids has been postulated to play a role[77,78].

10. Phytanic acid-containing triglycerides

In 1963 Klenk and Kahlke[79] found abnormally high levels of phytanic acid (3,7,11,15-tetramethylhexadecanoic acid) in individuals suffering from Refsum's disease. The metabolic defect is believed due to a deficiency of the enzyme required for the α-decarboxylation of phytanic acid which is obtained principally from the diet. TLC analysis of the serum lipids by Karlsson et al.[80] has revealed that the phytanic acid-containing triglycerides can be separated from the typical triglycerides. This technique can be used as a preliminary diagnosis of Refsum's disease. Laurell[81] has succeeded in separating the various molecular species of the phytanic acid-containing triglycerides obtained from a patient with Refsum's disease. In this patient the 1,2- and 1,3-diphytanic acid triglycerides were resolved and were separated from the 1,2,3-triphytanic acid triglyceride.

The phytanic acid can also be detected and analyzed by GLC.

11. Hyperlipogenesis—Obesity

The state of obesity is characterized by a large accumulation of triglyceride in the adipose tissue and may also represent marked synthesis of triglycerides. The obese syndrome is complex, being influenced by dietary, genetic, hormonal, and psychological factors[82-86]. In man, obesity is characterized as a hyperinsulin, hypothyroid state.

The control of the satiety center in the hypothalamus by glucose levels in the blood is proposed by Edelman et al.[86] as follows (a) the degree of appetite inhibition of the ventromedial area is proportional to the number of nerve cells activated (b) the generation of the action potential of the satiety neuron depends on the metabolism of a specific amount of glucose in the glial cell, and (c) there are different rates of glucose supply to and different intrinsic rates of glucose metabolism in the different neuron–glial complexes of the satiety area.

TABLE IX

COMPARISON OF METABOLIC AND REGULATORY OBESITY IN RATS[a]

	Metabolic obesity *Obese-hyperglycemic syndrome*	*Regulatory obesity* *Gold-thioglucose obesity*
Etiology and/or cause	Mendelian recessive	1 mg/g gold thioglucose
	Pancreatic dysfunction—hyperplasia of Islets of Langerhans—increase in insulin production	Hypothalamic lesion—destruction of cells in ventromedial area
Diet effect on weight gain	Maximum weight gain on high carbohydrate diet	Maximum weight gain on high fat diet
Blood-glucose level	Elevated	Normal
Blood-lipid levels	Elevated	Elevated
Lipogenesis	Increased with hyperphagia and during fasting	Increased with hyperphagia, normal otherwise
Adipose-tissue metabolism	Lipogenesis increased enormously from acetate, glucose oxidation decreased	Lipogenesis moderately increased from acetate, glucose oxidation normal
Enzymes	Increase in liver phosphorylase, glycerol kinase activity seen in adipose tissue	Normal phosphorylase, no glycerol kinase activity seen in adipose tissue

[a] Taken from Mayer[87].

References p. 153

Studies with obese-hyperglycemic mice[85,87] indicate that this disturbance is an example of a metabolic defect rather than a disturbance of the food regulatory center such as exemplified by the gold-thioglucose hypothalamic obesity. The hyperlipogenesis has been reported to be correlated with (*a*) a 5–10 fold increase of the *in vivo* incorporation of glucose into total liver fatty acids, (*b*) an increased *in vivo* incorporation of acetate into carcass and liver fatty acids even after fasting, (*c*) an increased incorporation of acetate into fatty acids and glycerides when epididymal adipose tissue is incubated in the absence of glucose, (*d*) greater capacity of fatty acid synthesis through increased activities of the citrate-cleavage enzyme, (*e*) increased esterification of palmitate when adipose tissue is incubated in absence of glucose, and (*f*) increased glycerol kinase activity in adipose tissue. In man, evidence is accumulating that obese individuals have higher insulin levels in response to glucose loads than normal individuals. When this is considered with the observations that adipose tissue in obese individuals has demonstrable glycerol kinase activity, and also appears to have an inhibitor of adipose-tissue lipase, it is easy to see that in these individuals the body is geared to storing fat.

Mayer[87,88] has discussed the differences between the metabolic obesity seen in the obese-hyperglycemic syndrome in rats and the regulatory gold-thioglucose obesity induced in rats. These differences are shown in Table IX. It is clear that obesity represents a complex disorder which is just beginning to be understood at the molecular level. Obesity which is due to a high fat diet in which preformed fatty acids are re-esterified to triglycerides represents a simpler situation than obesity which is due to a high carbohydrate diet, since in the latter case the enzymes which convert glucose to fatty acids and which produce glycerol phosphate come into play.

12. Problems in the study of lipid metabolism

Inasmuch as the various lipids within the cell are bound to protein and that this combination in all probability has an appreciable degree of specificity, and that these lipoproteins are located in specific loci in the cell, all sum up to create difficulties in the study of the metabolism of exogenous lipids which are added to cellular systems. The lipids to be added must be dispersed in a way that hopefully represents their natural state of dispersion within the cell. Since it is difficult to know precisely what this lipid–protein association is in the natural state (except with albumin-bound fatty acids) one has to guess how to disperse the lipid to be studied. Various techniques have been used to disperse the lipids but in most cases the actual physical state of the lipid is not known. The lipids may exist as micelles of varying size or as lamellae of different length and thickness. The ease of dispersion of the lipid will be markedly influenced by the chain length, degree of unsaturation, and polarity of the lipid. It is reasonable to expect that these different physical dispersions may markedly influence the permeation of these lipids through membrane barriers and their reactivity with enzymes. The techniques used to disperse lipids include sonication, slow dialysis of an alcoholic solution of lipid against water, addition of surface-active agents (such as Tweens, Triton X-100, cutscum), heating, etc. The effect of surface-active agents can create difficulties since in some cases they enhance enzyme activity and in other cases they inhibit enzyme activity. These problems continue to plague those working with lipid substrates.

The use of radioactive precursors also poses unique problems. Steinberg has discussed the difficulties in quantitating lipid synthesis using labeled fatty acids *versus* labeled glycerol[44]. The effect of multiple compartments, transesterification reactions, changing pool size and permeation through membranes all contribute to making analysis of net synthesis very difficult. Glycerol appears to be a better substrate for assessing net-triglyceride synthesis since it undoubtedly permeates the cell membrane more readily than do fatty acids or fatty acyl-CoA esters, and once in the cell can be acylated *in situ* with the endogenous fatty acids in their natural state. Moreover glycerol is more stable than fatty acids which have double bonds and which are subject to peroxidation and isomerization. For these reasons it is advisable to use more than one substrate in assessing the quantitative significance of metabolic pathways.

ADDENDUM

A recent paper by Possmayer and coworkers[89] on the positional specificity of labeled saturated and unsaturated fatty acids in phosphatidic acid from rat liver demonstrates that the majority of the saturated fatty acids are on the 1-position while the polyunsaturated fatty acids are mainly on the 2-position. The same fatty acid positioning was observed for lecithin and phosphatidylethanolamine. These results on phosphatidic acid differ from those of Lands and Hart[17] and Stoffel *et al.*[90] but are in agreement with the conclusions of Husbands and Reiser[91]. More recent work of Hill *et al.*[92]

References p. 153

has shown however, that acylation of glycerol-3-phosphate in rat-liver slices produces specific, non-random species of phosphatidic acid. Moreover, Elovson et al.[93] have now studied the initial (5–300 sec) kinetics of incorporation of labeled fatty acid and labeled glycerol into lipids of intact rats. This data strongly suggests that phosphatidic acid synthesis *in vivo* is highly specific and is believed to account for the distribution of saturated and unsaturated fatty acids on the 1- and 2-positions of 1,2-diglycerides, triglycerides and lecithins. When phosphatidic acid was isolated from livers 5 or 30 sec after injection of labeled palmitic acid, over 90% of the labeled palmitic acid was found in the 1-position. The 2-position contained predominantly unsaturated fatty acids.

REFERENCES

1. J. R. KING AND D. S. FARNER, *Ann. N.Y. Acad. Sci.*, 131 (1965) 422.
2. M. J. R. DAWKINS AND D. HULL, *J. Physiol. (London)*, 172 (1964) 216.
3. G. STEINER AND G. F. CAHILL JR., *Am. J. Physiol.*, 207 (1964) 840.
4. D. H. TREBLE AND E. G. BALL, *Federation Proc.*, 22 (1963) 357.
5. J. M. JOHNSTON AND G. A. RAO, *Biochim. Biophys. Acta*, 106 (1965) 1.
6. C. H. HOLLENBERG, *J. Lipid Res.*, 8 (1967) 328.
7. A. KUKSIS, in G. V. MARINETTI (Ed.), *Lipid Chromatographic Analysis*, Vol. I, Marcel Dekker, New York, 1967, Chapter 7.
8. D. S. FREDRICKSON, M. D. ROBERT, I. LEVY AND R. S. LEES, *New Engl. J. Med.*, 276 (1967) 32–281.
9. S. E. HICKS, D. W. ALLMAN AND D. M. GIBSON, *Biochim. Biophys. Acta*, 106 (1965) 441.
10. D. E. GREEN AND S. J. WAKIL, in K. BLOCH (Ed.), *Lipid Metabolism*, Wiley, New York, 1960, Chapter 1.
11. H. R. MAHLER, *J. Biol. Chem.*, 206 (1954) 13.
12. P. HELE, *J. Biol. Chem.*, 206 (1954) 671.
13. P. BERG, *Science*, 129 (1959) 875.
14. L. L. INGRAHAM AND D. E. GREEN, *Science*, 128 (1958) 310.
15. A. TIETZ AND B. SHAPIRO, *Biochim. Biophys. Acta*, 19 (1956) 374.
16. S. B. WEISS, E. P. KENNEDY AND S. Y. KIYASU, *J. Biol. Chem.*, 235 (1960) 40.
17. W. E. M. LANDS AND P. HART, *J. Lipid Res.*, 5 (1964) 81.
18. W. E. M. LANDS, *Ann. Rev. Biochem.*, 34 (1965) 313.
19. G. HÜBSCHER AND B. CLARK, *Biochim. Biophys. Acta*, 41 (1960) 45.
20. S. W. SMITH, S. B. WEISS AND E. P. KENNEDY, *J. Biol. Chem.*, 228 (1957) 915.
21. E. E. HILL AND W. E. M. LANDS, *Biochim. Biophys. Acta*, 152 (1968) 645.
22. H. BROCKERHOFF, *J. Lipid Res.*, 6 (1965) 10.
23. S. P. M. SLAKEY AND W. E. M. LANDS, *Lipids*, 3 (1968) 30.
24. F. H. MATTSON AND E. S. LUTTON, *J. Biol. Chem.*, 223 (1958) 868.
25. P. SAVARY, J. F. FLANZY AND P. DESNUELLE, *Biochim. Biophys. Acta*, 24 (1957) 414.
26. F. H. MATTSON AND R. A. VOLPENHEIN, *J. Biol. Chem.*, 236 (1961) 1891.
27. G. V. MARINETTI, *Biochim. Biophys. Acta*, 46 (1961) 468.
28. G. V. MARINETTI, J. F. ERBLAND AND M. BROSSARD, in R. M. C. DAWSON AND D. N. RHODES (Eds.), *Metabolism and Physiological Significance of Lipids*, Wiley, New York, 1964, pp. 71–93.
29. J. F. ERBLAND, M. BROSSARD AND G. V. MARINETTI, *Biochim. Biophys. Acta*, 137 (1967) 23.
30. G. V. MARINETTI, M. GRIFFITH AND T. SMITH, *Biochim. Biophys. Acta*, 57 (1962) 543.
31. R. TZUR AND B. SHAPIRO, *J. Lipid Res.*, 5 (1964) 542.
32. C. BUBLITZ AND E. P. KENNEDY, *J. Biol. Chem.*, 211 (1954) 951.
33. O. WIELAND AND M. SUYTER, *Biochem. Z.*, 329 (1957) 320.
34. O. W. MCBRIDE AND E. D. KORN, *J. Lipid Res.*, 5 (1964) 442.
35. B. CLARK AND G. HUBSCHER, *Nature*, 195 (1962) 599.
36. H. A. HAESSLER AND K. J. ISSELBACHER, *Biochim. Biophys. Acta*, 73 (1963) 427.
37. H. C. TIDWELL AND J. M. JOHNSTON, *Arch. Biochem. Biophys.*, 94 (1961) 546.
38. P. R. HOLT, H. A. HAESSLER AND K. J. ISSELBOCHER, *J. Clin. Invest.*, 42 (1963) 777.
39. H. KALCKAR, *Biochem. J.*, 33 (1939) 631.
40. J. W. FARQUHAR, R. C. GROSS, R. M. WAGNER AND G. M. REAVEN, *J. Lipid Res.*, 6 (1965) 119.
41. R. C. GROSS, E. H. EIGENBRODT AND J. F. FARQUHAR, *J. Lipid Res.*, 8 (1967) 114.

42 D. Steinberg, in K. Rodahl and B. Issekutz (Eds.), *Fat as a Tissue*, McGraw-Hill, New York, 1962, pp. 127–148.
43 D. Goldman and P. R. Vagelos, *J. Biol. Chem.*, 236 (1961) 2620.
44 D. Steinberg, in J. K. Grant (Ed.), *The Control of Lipid Metabolism*, Academic Press, New York, 1963, pp. 111–138.
45 D. A. K. Roncan and C. H. Hollenberg, *Biochim. Biophys. Acta*, 137 (1967) 446.
46 C. H. Hollenberg, *J. Lipid Res.*, 6 (1965) 84.
47 D. J. Galton, *J. Lipid Res.*, 9 (1968) 19.
48 B. Clark and G. Hübscher, *Nature*, 185 (1960) 35.
49 B. Clark and G. Hübscher, *Biochim. Biophys. Acta*, 46 (1961) 479.
50 J. R. Senior and K. J. Isselbacher, *J. Biol. Chem.*, 277 (1962) 1454.
51 J. M. Johnston and J. C. Brown, *Biochim. Biophys. Acta*, 59 (1962) 500.
52 F. Kern and B. Borgstrom, *Biochim. Biophys. Acta*, 98 (1965) 520.
53 R. Paris and G. Clement, *Biochim. Biophys. Acta*, 153 (1968) 63.
54 F. H. Mattson and R. A. Volpenhein, *J. Biol. Chem.*, 237 (1962) 53.
55 V. P. Skipski, M. G. Morehouse and H. J. Deuel Jr., *Arch. Biochem. Biophys.*, 81 (1959) 93.
56 M. H. Coleman, *Advan. Lipid Res.*, 1 (1963) 1–64.
57 J. M. Johnston, G. A. Rao and P. A. Love, *Biochim. Biophys. Acta*, 137 (1967) 578.
58 M. L. Blank and O. S. Privett, *J. Dairy Sci.*, 47 (1964) 481.
59 J. A. Olson, *Ann. Rev. Biochem.*, 35 (1966) 559.
60 H. A. Haessler and J. D. Crawford, *Ann. N.Y. Acad. Sci.*, 131 (1965) 476.
61 M. Vaughn, in K. Rodahl and B. Issekutz (Eds.), *Fat as a Tissue*, McGraw-Hill, New York, 1962, pp. 203–215.
62 R. G. Langdon, in K. Bloch (Ed.), *Lipid Metabolism*, Wiley, New York, 1960, Chapter 6.
63 J. Tepperman and H. M. Tepperman, *Ann. N.Y. Acad. Sci.*, 131 (1965) 404.
64 E. Shafrin, K. E. Sussman and D. Steinberg, *J. Lipid Res.*, 1 (1960) 459.
65 R. O. Scow, F. A. Stricker, T. Y. Pick and T. R. Clary, *Ann. N.Y. Acad. Sci.*, 131 (1965) 288.
66 M. Rodbell, *Ann. N.Y. Acad. Sci.*, 131 (1965) 302.
67 J. P. Flatt and E. G. Ball, in J. K. Grant (Ed.), *Control of Lipid Metabolism*, Academic Press, New York, 1963, pp. 75–77.
68 E. M. Wise Jr. and E. G. Ball, *Proc. Natl. Acad. Sci. (U.S.)*, 52 (1964) 1255.
69 J. Tepperman, *Pharmacol. Rev.*, 12 (1960) 301.
70 F. Lynen, M. Matsuhashi, S. Nauma and E. Schweizer, in J. K. Grant (Ed.), *Control of Lipid Metabolism*, Academic Press, New York, 1963.
71 P. R. Vogelos, *Ann. Rev. Biochem.*, 336 (1964) 139.
72 J. M. Lowenstein, in J. K. Grant (Ed.), *Control of Lipid Metabolism*, Academic Press, New York, 1963, pp. 57–59.
73 R. Santora and G. V. Marinetti, unpublished.
74 R. O. Recknagel and A. K. Ghoshal, *Lab. Invest.*, 15 (1966) 132.
75 B. Lombardi, *Lab. Invest.*, 15 (1966) 1.
76 Symposium, Alcohol Metabolism and Liver Disease, *Federation Proc.*, 26 (1967) 1432–1481.
77 N. R. DiLuzio and A. D. Hartman, *Federation Proc.*, 26 (1967) 1436.
78 M. Poggi and N. R. DiLuzio, *J. Lipid Res.*, 5 (1964) 437.
79 E. Klenk and W. Kahlke, *Z. Physiol. Chem.*, 333 (1963) 133.
80 K. Karlsson, A. Norrby and B. Samuelsson, *Biochim. Biophys. Acta*, 144 (1967) 162.
81 S. Laurell, *Biochim. Biophys. Acta*, 152 (1968) 75.

82 R. A. LIEBELT, S. ICHINOE AND N. NICHOLSON, *Ann. N.Y. Acad. Sci.*, 131 (1965) 559.
83 C. COHN, D. JOSEPH, L. BELL AND M. D. ALLWEISS, *Ann. N.Y. Acad. Sci.*, 121 (1965) 507.
84 W. STAUFFACHER, O. B. CRAWFORD, B. JEANRENAUD AND A. L. RENOLD, *Ann. N.Y. Acad. Sci.*, 121 (1965) 528.
85 B. HELLMAN, *Ann. N.Y. Acad. Sci.*, 131 (1965) 541.
86 M. EDELMAN, I. L. SCHWARTZ, E. P. CRONKITE AND L. LIVINGSTON, *Ann. N.Y. Acad. Sci.*, 131 (1965) 485.
87 J. MAYER, *Ann. N.Y. Acad. Sci.*, 131 (1965) 412.
88 J. MAYER, in K. RODAHL AND B. ISSEKUTZ (Eds.), *Fat as a Tissue*, McGraw-Hill, New York, 1962, pp. 329–343.
89 P. POSSMAYER, G. L. SCHERPHOF, T. M. A. DUBBELMAN, L. M. G. VAN GOLDE AND L. L. M. VAN DEENEN, *Biochim. Biophys. Acta*, 176 (1969) 95.
90 W. STOFFEL, M. E. DETOMAS AND H. G. SCHIEFER, *Z. Physiol. Chem.*, 348 (1967) 882.
91 D. R. HUSBANDS AND R. REISER, *Federation Proc.*, 25 (1966) 405.
92 E. E. HILL, D. R. HUSBANDS AND W. E. M. LANDS, *J. Biol. Chem.*, 243 (1968) 4440.
93 J. ELOVSON, B. AKESSON AND G. ARVIDSON, *Biochim. Biophys. Acta*, 176 (1969) 214.

Chapter III

Phospholipid Metabolism

GUY A. THOMPSON JR.

Department of Botany, University of Texas, Austin, Texas (U.S.A.)

1. Introduction

The universal occurrence of phospholipids in living cells is well established. A variety of evidence points to the specific localization of cellular phospholipids in lipoprotein membranes. Thus these lipids clearly play an important structural role in the cell's economy. Furthermore, considerable data suggest a dynamic function of phospholipids in controlling the movements of molecules across membranes.

Of all the compounds found in nature, none would seem more suitable for these tasks. The rather bulky hydrocarbon portions of the phospholipid molecules readily associate together by virtue of the attraction of Van der Waals forces. The polar ends of the phospholipids enable the molecules to orient themselves into sheets or micelles which can associate readily with proteins.

It is therefore easy to imagine how an amphipathic phospholipid molecule might function in membrane structure. What is less apparent is the need for a complex and very specific mixture of phospholipids in each cell type. Enough tissues have been examined to ascertain that each kind contains its own complement of phosphatides distinctive both in its distribution of polar constituents and in its fatty acid composition[1]. Effecting significant changes in these patterns even by deletion of or massive supplementation of selected dietary phospholipid precursors is extremely difficult. All our experience would point towards the requirement of each cell for some very specific sort of mosaic in which the juxtaposition of certain molecular types governs the overall physiological suitability of the structure.

References p. 197

If this is a reasonable approximation of the true situation, the turnover of membrane components and synthesis of new membranes demand precise controls on the metabolism and assembly of 5–10 major types of phospholipid, each containing a variety of specific components. This chapter will consider various aspects of this process in animal cells. Our current understanding of the principal biosynthetic pathways will be reviewed. The mechanisms available to animals for phospholipid catabolism will be summarized. Finally, I will attempt to place in perspective our existing knowledge concerning the biological exchange of whole molecules or portions thereof as a mechanism for renewing structural elements or "adjusting" membrane composition.

It is not my intent to review completely the literature of this field. Only those references necessary to provide an understanding of the various pathways will be discussed. These have been drawn almost exclusively from studies involving animal tissues, since the metabolism of bacterial and higher plant lipids is covered elsewhere in the volume. Also considered in other chapters are some details of phospholipid metabolism in brain tissue and in extracellular fluids.

For a discussion of phospholipid chemistry, the reader is referred to an article by Hanahan and Brockerhoff, Chapter III, Volume 6 (p. 83) of this series. Certain aspects of the natural occurrence and metabolism of animal phospholipids not included in the present discourse are thoughtfully considered in the essays of Van Deenen[1] and Dawson[2].

2. Biosynthesis

(a) Formation of phosphatidic acid

Phosphatidic acid is not quantitatively a major lipid component of most tissues. However, it is generally recognized as occupying a position of great metabolic importance by virtue of its role as a precursor of other lipids, including triglycerides as well as phospholipids.

Three separate pathways have been described for the formation of phosphatidic acid. Detailed evidence for the first of these was reported initially by Kornberg and Pricer[3]. A system from liver was shown to catalyze the rapid esterification of L-α-glycerophosphate utilizing the coenzyme A thioesters of long-chain fatty acids to produce phosphatidic acid.

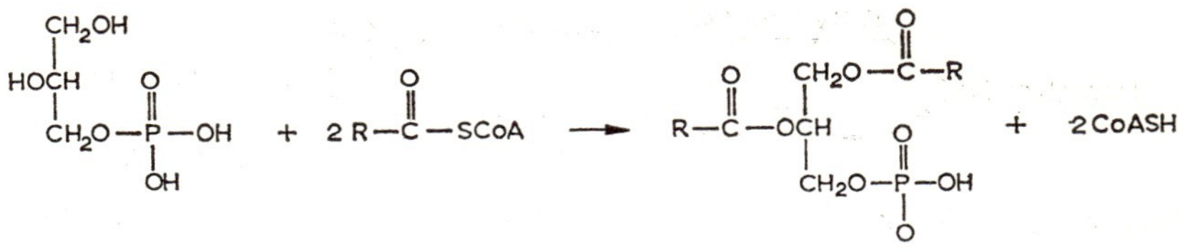

$$\begin{array}{c} CH_2OH \\ | \\ HOCH \\ | \\ CH_2O-P-OH \\ | \\ OH \end{array} \;+\; 2\,R-\overset{O}{\underset{\|}{C}}-SCoA \;\longrightarrow\; \begin{array}{c} O \\ \| \\ R-C-OCH \\ | \\ CH_2O-P-OH \\ | \\ O \end{array} \begin{array}{c} O \\ \| \\ CH_2O-C-R \\ \\ \\ \end{array} \;+\; 2\,CoASH$$

The α-glycerophosphate substrate for this reaction arises principally through the reduction of dihydroxyacetone phosphate, an intermediate of glycolysis, although some tissues, such as liver, can catalyze the direct ATP-mediated phosphorylation of free glycerol. Coenzyme A derivatives of fatty acids are formed by a soluble liver enzyme from fatty acids, CoA, and ATP.

Recent findings by Ailhaud and Vagelos[4] and by Goldfine[5] suggest that the immediate donor of acyl groups to L-α-glycerophosphate may actually be the fatty acyl–acyl carrier protein complex rather than the coenzyme A derivative indicated above. The data, recently reviewed by Majerus and Vagelos[6], include experiments showing a stimulation by acyl-carrier protein (ACP) of lysophosphatidic acid formation from L-α-glycerophosphate and palmityl-CoA and, in another system, evidence that added palmityl-ACP combines directly with glycerophosphate. So far, indications of ACP involvement in phosphatidic acid synthesis have come from the study of bacterial systems only, and its significance to animal metabolism must be established by further experimentation.

The enzyme acyl-CoA:L-glycerol-3-phosphate *O*-acyltransferase (EC 2.3.1.15), which is responsible for phosphatidic acid formation, has not been isolated in pure form, but a partially purified preparation from rat brain retains the ability to acylate both hydroxyl groups of glycerophosphate[7]. However, SH-binding reagents inhibit acylation of glycerophosphate while acylation of lysophosphatidic acid is unaffected[8], implying that separate enzymes might be involved in the two steps. The experiments of Lands and Hart[9] demonstrated that linoleate or stearate are incorporated at both the 1 and 2 positions of glycerophosphate. These results would suggest that the remarkable pattern of fatty acid specificity according to which the 1 and 2 positions of most natural phospholipids contain saturated and unsaturated fatty acids, respectively, is not established at this enzymatic step.

In addition to the above pathway for phosphatidic acid formation, two other related sequences have been reported. The reaction shown on the lower line of the following scheme has not been studied extensively, and the

References p. 197

```
                    Fatty acyl-CoA      Diglyceride      ATP
                                                        ⇌         Phosphatidic acid
     Monoglyceride                                      ⇌
                         ATP         Lysophosphatidic
                                           acid          Fatty acyl-CoA
```

observed acylation of lysophosphatidic acid may actually represent an intermediate step in the acylation of α-glycerophosphate. On the other hand, many tissues have been shown to contain the enzyme diglyceride kinase, capable of phosphorylating diglycerides in the presence of ATP. The phosphatidic acid thus produced appears to form a pool separate from that derived through acylation of glycerophosphate. The increased phosphatidic acid turnover in cells stimulated to carry out certain membrane-transport processes, such as salt or protein secretion, is effected through increased activity of diglyceride kinase and an associated enzyme, phosphatidic acid phosphatase[10]. The exact role of this phosphatidic acid "cycle" in transport phenomena of membranes remains to be clarified.

During the preparation of this chapter, a further potential source of phosphatidic acid has been described in the literature[10a]. In guinea-pig-liver mitochondria, dihydroxyacetone phosphate can react with fatty acyl coenzyme A to produce acyl dihydroxyacetone phosphate. Furthermore, when [1-^{14}C]palmitoyl-[^{32}P]dihydroxyacetone phosphate is incubated with the NADPH-supplemented mitochondrial preparation, a portion of the radioactivity can be recovered in lysophosphatidic acid, which is presumably subject to further acylation in the cell.

```
     CH₂OH                                            O
     |                      O                         ||
     C=O    O               ||                   CH₂O—C—R
     |      ||        +  R—C—SCoA      ⟶        |
     CH₂O—P—OH                                   C=O    O
            |                                    |      ||
            OH                                   CH₂O—P—OH
                                                        |
                                                        OH

                                                      ⎧ NADPH
                                                      ⎨
                                                      ⎩ NADP

                                                        O
                                                        ||
                                                   CH₂O—C—R
                                                   |
                                              HO—CH     O
                                                   |    ||
                                                   CH₂O—P—OH
                                                        |
                                                        OH
```

The quantitative significance of this pathway is not yet known.

(b) *Biosynthetic pathways utilizing phosphatidic acid as a precursor*

(i) *Formation of phosphatidylcholine*

Phosphatidylcholine or lecithin, as it is often called, is the major phospholipid component of most animal cells. Possibly for this reason, the details of its structure and biogenesis were elucidated earlier and in more detail than has been the case for most other common structural lipids. As described by Hanahan and Brockerhoff in an earlier volume of this series[11], natural lecithins generally show a very marked tendency to contain saturated fatty acids at the C-1 or α' position and unsaturated fatty acids at the C-2 or β position. Thus the structure of a typical native phosphatidylcholine would be represented as shown below.

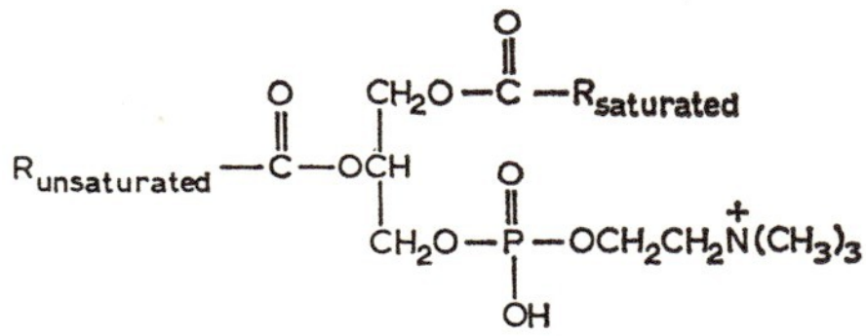

$$R_{unsaturated}-\overset{O}{\overset{\|}{C}}-O\overset{CH_2O-\overset{O}{\overset{\|}{C}}-R_{saturated}}{\underset{CH_2O-\overset{O}{\overset{\|}{P}}-OCH_2CH_2\overset{+}{N}(CH_3)_3}{\overset{|}{CH}}}$$

The major biosynthetic route to lecithin utilizes as an intermediate diglyceride, which is produced from phosphatidic acid through a reaction involving the enzyme phosphatidic acid phosphatase (EC 3.1.3.4). A second important pathway recently found to exist in certain tissues will be discussed in the section dealing with lipid interconversions.

Beginning with the common base choline, Fig. 1 outlines the steps of phosphatidylcholine formation.

The phosphorylation reaction catalyzed by ATP:choline phosphotransferase requires Mg^{2+} and is not specific for choline, although that substrate is utilized *in vitro* more rapidly than ethanolamine or dimethyl-ethanolamine[12]. Using phosphorylcholine labeled with both ^{14}C and ^{32}P, Kornberg and Pricer[13] demonstrated in 1952 the capacity of a liver enzyme to form lipid retaining the same $^{14}C:^{32}P$ ratio as measured in the starting material. The lipid was later characterized by Kennedy and Weiss[14] as phosphatidylcholine. In studying the reaction mechanism, the latter authors stumbled upon a discovery of widespread significance. Cytidine triphosphate, identified as a contaminant in their preparation of ATP, was shown to act as a cofactor in the formation of lecithin. This was the first specific requirement found for cytidine nucleotides in a major biochemical pathway. Other roles for CTP have subsequently been detected, most of them also being in some

References p. 197

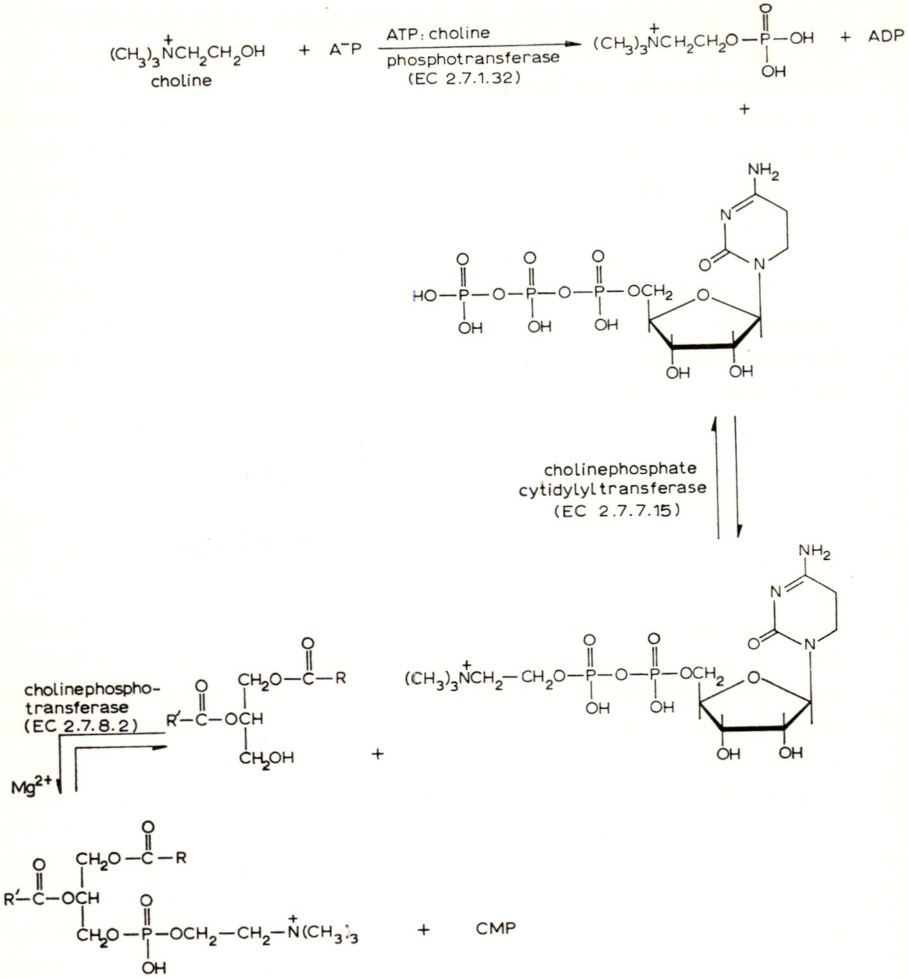

Fig. 1. Steps of phosphatidylcholine formation, beginning with choline.

phase of lipid metabolism. Characterization of CDP–choline as the active intermediate was followed by the use of synthetic CDP–choline in proving its role as the choline donor.

Recent *in vivo* experiments by Bjørnstad and Bremer[15] have established the likelihood that the terminal reaction of phosphatidylcholine biogenesis is freely reversible. Evidence for this exchange reaction

$$\text{CDP–choline} + \text{diglyceride} \rightleftharpoons \text{CMP} + \text{phosphatidylcholine}$$

includes the observation that after a period during which rats were allowed to metabolize [^{14}C]choline, the specific radioactivity of CDP–choline in liver was nearly identical to that of lecithin, while phosphorylcholine exhibited a much lower specific radioactivity. It will be valuable to ascertain whether this apparent equilibration of the diglyceride moiety of phosphatidylcholine with the free diglyceride pool involves any specificity with regard to the nature of the bound fatty acyl groups. Conceivably the reaction could be influential in establishing the characteristic fatty acid distribution found in lecithin. Diglycerides for which the enzyme had a low affinity might be utilized for other purposes, such as the synthesis of triglycerides or phosphatidylethanolamine. The reactivity of the enzyme toward several diglycerides has been tested experimentally, but significant differences in solubility of the substrates have rendered the results somewhat equivocal.

(ii) Formation of phosphatidylethanolamine

This lipid, like lecithin, accounts for a sizeable proportion of most tissue phospholipids. It contains fatty acids distributed as in phosphatidylcholine and, in fact, most natural phosphoglycerides, with a predominance of saturated acids in the C-1 position of glycerol and unsaturated acids in the C-2 position. The ethanolamine phosphatides of many animal tissues also include a high proportion of an analog in which the fatty acid at the C-1 position is replaced by an ether-linked hydrocarbon side-chain. The metabolism of these so called plasmalogens and glyceryl ether phosphatides will be described in a later section.

The pathway for *de novo* phosphatidylethanolamine biogenesis resembles

$$H_2N-CH_2-CH_2OH \xrightarrow[\text{ATP} \quad \text{ADP}]{\text{ATP:choline phosphotransferase (EC 2.7.1.32)}} H_2N-CH_2-CH_2O-\overset{\underset{\|}{O}}{\underset{OH}{P}}-OH$$

$$H_2N-CH_2-CH_2O-\overset{\underset{\|}{O}}{\underset{OH}{P}}-OH + CTP \xrightarrow{\text{Ethanolaminephosphate cytidylyltransferase (EC 2.7.7.14)}} H_2N-CH_2-CH_2O-CDP + PP_i$$

$$H_2N-CH_2-CH_2O-CDP + \underset{\underset{CH_2OH}{\overset{|}{CH_2O-\overset{O}{\overset{\|}{C}}-R}}}{R'-\overset{O}{\overset{\|}{C}}-OCH} \xrightarrow[?]{\text{Ethanolamine-phosphotransferase (EC 2.7.8.1)}} \underset{\underset{OH}{\overset{|}{CH_2O-\overset{O}{\overset{\|}{P}}-OCH_2-CH_2-NH_2}}}{R'-\overset{O}{\overset{\|}{C}}-OCH} + CMP$$

very closely that already described for phosphatidylcholine. The phosphorylation of ethanolamine is probably carried out by the same enzyme responsible for phosphorylcholine synthesis. CTP is required in the activation of phosphorylethanolamine for transfer to diglyceride by ethanolamine phosphotransferase[14]. It would appear from the work of Bjørnstad and Bremer[15] that, in liver at least, this latter reaction is much less freely reversible than is the case for the analogous step in phosphatidylcholine biosynthesis.

Phosphatidylethanolamine can also arise through the decarboxylation of phosphatidylserine. The importance of this process will be considered in the discussion of lipid interconversions.

(iii) Formation of phosphoinositides

Cytidine nucleotides also play an important role in the biosynthesis of phosphatidylinositol but, surprisingly, the mechanism of their involvement is quite unlike that found to operate in the formation of ethanolamine and choline phosphatides. In the case of phosphatidylinositol, a divalent cation-requiring, microsomal enzyme promotes the reaction of CTP with phosphatidic acid, yielding CDP-diglyceride and inorganic pyrophosphate[16,17].

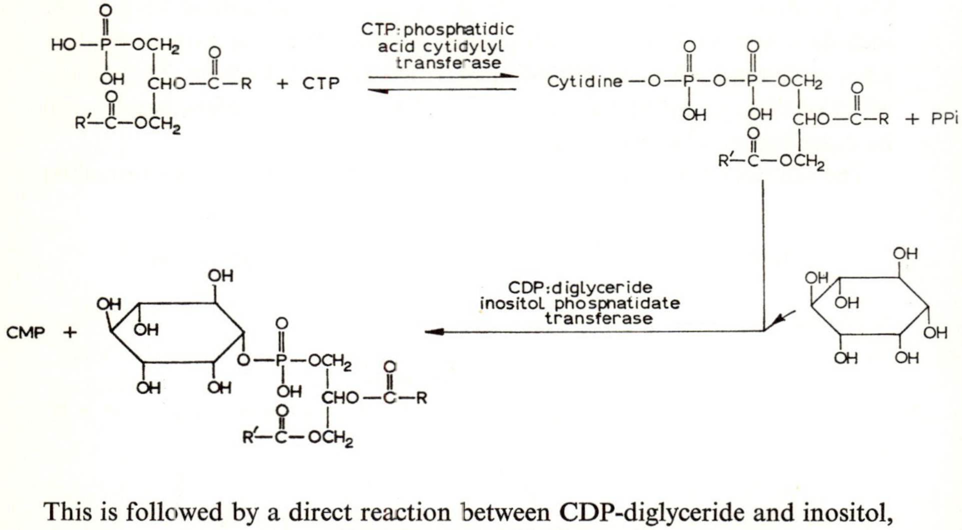

This is followed by a direct reaction between CDP-diglyceride and inositol, yielding phosphatidylinositol and CMP[16,18]. The last reaction requires manganese ions, with magnesium being a less effective substitute.

In addition to the above scheme, there are scattered indications, summarized by Hawthorne and Kemp[19], that a second biosynthetic pathway may exist, at least in brain. Such data, usually consisting of specific radioactivity measurements inconsistent with the sequence of reactions outlined in the preceding paragraph, are not yet extensive enough to provide a clear picture of the mechanisms involved.

Besides phosphatidylinositol, two additional phosphoinositides are found in certain animal tissues, particularly the nervous system. They differ from phosphatidylinositol in that the first, diphosphoinositide, contains an additional phosphate esterified to the inositol residue at position 4 while the second, triphosphoinositide (I), has phosphate atoms at both positions 4 and 5 of inositol.

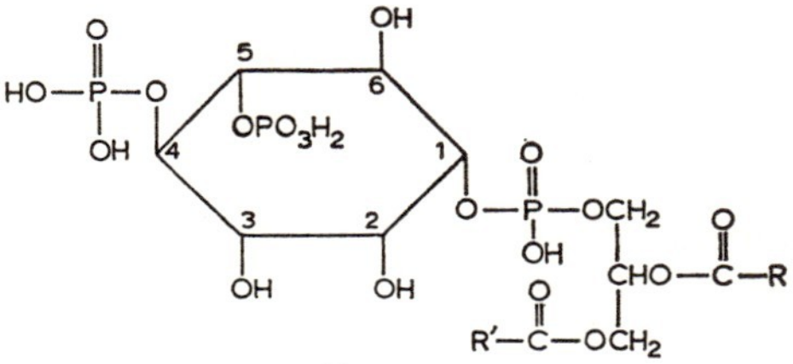

(I)

The biosynthesis of these derivatives has been studied in preparations from liver, brain, and other sources. Although such a reaction would seem logical, there is no indication that inositol mono- or di-phosphate reacts directly with CDP-diglyceride[19a]. Instead, enzymes exist which can phosphorylate phosphatidylinositol first in the 4- and then in the 5-position of inositol.

Phosphatidylinositol + ATP → diphosphoinositide + ADP

Diphosphoinositide + ATP → triphosphoinositide + ADP

The enzyme responsible for the first reaction, phosphatidylinositol kinase (ATP:phosphatidylinositol 4-phosphotransferase), requires Mg^{2+} or Mn^{2+} for activity. Unlike most enzymes of lipid metabolism, it seems to be localized somewhat specifically in the plasma membrane[20]. This finding has led to speculation that the capacity of the highly charged diphosphoinositide for binding divalent cations might function in the regulation of cellular ion transport.

The formation of triphosphoinositide is catalyzed by the enzyme diphosphoinositide kinase (ATP:diphosphoinositide 5-phosphotransferase). In rat brain the enzyme requires Mg^{2+} and is concentrated in the soluble fraction[21].

References p. 197

The distribution of triphosphoinositide among peripheral nerves and in different areas of the brain as well as its first appearance during the early stages of myelination would indicate that this lipid is primarily associated with the myelin sheath[22]. Labeling experiments with $^{32}P_i$ have shown that the turnover of the phosphate groups bound in the 4 and 5 positions of lipid inositol is very much faster than that of any other brain lipid components.

(iv) Formation of phosphatidylserine

Although phosphatidylserine accounts for 5–10% of most animal tissue phospholipids, relatively little is known concerning its metabolism. Kanfer and Kennedy[23] reported the formation of phosphatidylserine from L-serine and CDP-diglyceride in a cell-free *E. coli* system. This sequence has not yet been demonstrated in animals. Several investigators have observed a Ca^{2+}-stimulated exchange between L-serine and phosphatidylethanolamine to produce phosphatidylserine in higher animal tissues, but the quantitative importance of the interconversion is unclear. Further details of the exchange reaction are considered in Section 2c (p. 168).

(v) Formation of phosphatidylglycerol

Rat-liver and chicken-liver mitochondria contain enzymes for the biosynthesis of phosphatidylglycerol *via* the following steps[24]:

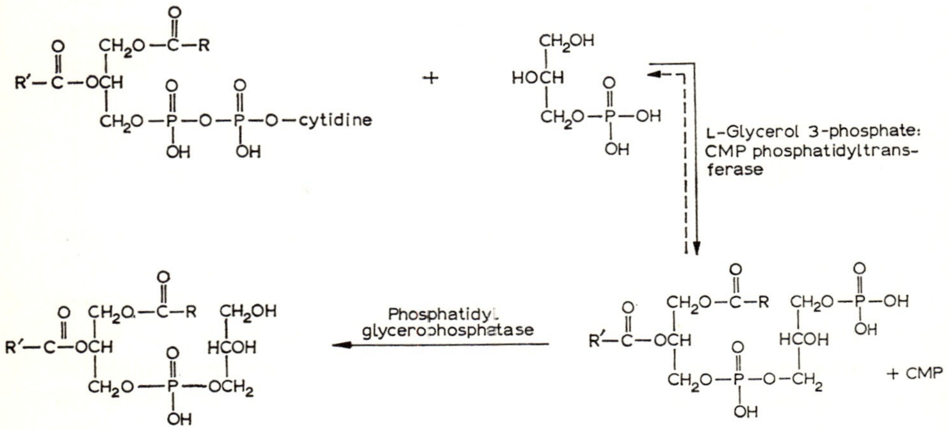

The reversibility of the first reaction in animal tissues is not yet determined, although the reverse reaction can be shown in *E. coli* preparations[25]. In test systems, the addition of sulfhydryl poisons preferentially inhibits the phosphatase reaction, and the resulting accumulation of phosphatidylglycerophosphate provides further evidence for its role as an intermediate.

(vi) Formation of cardiolipin

Cardiolipin (diphosphatidylglycerol) (I) is one of the few phospholipids which occurs almost exclusively in a single organelle of animal cells. Partly

because of its specific location in mitochondrial membranes, its metabolism carries an added dimension of interest. Unfortunately, our only detailed information concerning its biosynthesis is derived from experiments with *E. coli*, which does not contain animal-like mitochondria. Using a particulate fraction from *E. coli*, Stanacev et al.[26] demonstrated that phosphatidylglycerol is a direct precursor of cardiolipin.

An alternative possibility, that phosphatidylglycerophosphate, a precursor of phosphatidylglycerol, could react directly with 1,2-diacylglycerol to produce cardiolipin, was ruled out by the very low $^{32}P/^{3}H$ ratio of cardiolipin found after incubating the crude *E. coli* system with L-glycerol 3-$^{32}PO_4$, 2-^{3}H. Thus it seems obligatory that newly formed phosphatidylglycerophosphate be dephosphorylated prior to its conversion to cardiolipin.

References p. 197

(c) *Interconversions of phospholipids*

The reactions discussed in the previous section represent in most instances the synthesis of new phospholipid from fatty acids, L-α-glycerophosphate, and various other water-soluble precursors. In addition to these pathways, all animal cells appear to possess the capacity to convert certain intact phospholipids into others by way of reactions involving the nitrogen-base moieties. The common feature of all these processes is that although one phosphatide may thereby be replaced by another type, there is no net synthesis of phospholipid.

The most widely studied of these biological transformations involves phosphatidylcholine, phosphatidylethanolamine, and phosphatidylserine. In 1960 and 1961 work from the laboratories of Greenberg[27] and Udenfriend[28] developed the concept that in liver microsomes the conversion by decarboxylation of serine to ethanolamine and the subsequent stepwise methylation of ethanolamine to form choline take place while the nitrogenous bases are covalently bound in phospholipids. During the next few years, kinetic data provided further details which allowed the following sequence of reactions to be proposed:

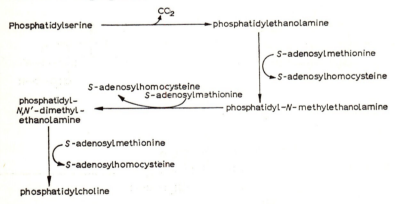

The stepwise methylation of phosphatidylethanolamine appears to be mediated by two enzymes, one catalyzing the first methyl addition and a second carrying out the two final methylations[29]. Although the inability to isolate an *in vitro* enzyme system active in performing the initial methylation step has so far permitted only circumstantial evidence for this reaction in animals, convincing support for its validity has resulted from work with mutant strains of *Neurospora*[30].

The quantitative importance of the methylation pathway of lecithin biosynthesis has been assessed by various workers. Surprisingly, among tissues of higher animals only liver derives a major fraction of its phosphatidylcholine *via* this system[15]. Considering the role of liver as a major producer of phosphatidylcholine for export to plasma and bile, this finding engendered widespread speculation that the two biosynthetic pathways might serve different cellular needs. However, our most complete information to date, derived from an *in vivo* study of [1,2-^{14}C]choline, [1,2-^{14}C]ethanolamine, [CH_3-^{14}C]- and [CH_3-^{3}H]-methionine incorporation patterns, indicates that lecithin arising from both pathways enters a common pool, which is in rapid equilibrium with plasma[15]. This explanation may be oversimplified, for indications of separate liver-lecithin pools, perhaps characterized by different fatty acid compositions, have been reported[31].

In addition to the pathway just described, interconversion of phospholipids can occur by direct base exchange. Thus free serine, choline, ethanolamine, and inositol can exchange with the existing base of a phosphatide to yield the appropriate product[32]. With the possible exception of ethanolamine, there is no ATP requirement for the exchange. As summarized by Hübscher[32], the reactions are strongly cation-dependent, all but inositol requiring Ca^{2+}. Because of the observed differences in optimum cation concentrations and in degrees of competitive inhibition among the exchangeable cases, it was concluded that a different enzyme is involved for the exchange of each base. Recently, several procedural refinements, including the replacement of tris(hydroxymethyl)amino methane, which can itself participate in the exchange reaction, by another buffering compound, have increased the likelihood that all reactions are catalyzed by a single enzyme[33]. No experimental evidence is available regarding a possible specificity with regard to the type(s) of phospholipids with which the bases may exchange.

(d) *Formation of plasmalogens and glyceryl ether phospholipids*

As analytical methods have improved during the past few years, reexamination of phospholipid patterns has provided much fresh information. Among the chief reassessments necessary has been the need to consider again the quantitative significance of phospholipids containing ether bonds. We now know that the ether phospholipids are considerably more widespread than was previously assumed. It is not unusual for them to constitute as much as 20% of the total phospholipid of a given tissue.

References p. 197

These compounds differ from the diacyl glycerophosphatides only in that the acyl group at position 1 of the glycerol backbone is replaced by an α,β-unsaturated hydrocarbon chain bound with an ether linkage in the plasmalogen class (I) and by an ether-linked hydrocarbon chain lacking α,β-unsaturation in the glyceryl ether phospholipid class (II). In both cases the most commonly found bases are ethanolamine and choline. The biochemistry of

$$\begin{array}{cc} \text{R}'-\overset{\overset{\text{O}}{\|}}{\text{C}}-\text{O}\overset{|}{\underset{|}{\text{C}}}\text{H}_2\text{OCH}=\text{CHR} & \text{R}'-\overset{\overset{\text{O}}{\|}}{\text{C}}-\text{O}\overset{|}{\underset{|}{\text{C}}}\text{H}_2\text{OCH}_2\text{CH}_2\text{R} \\ \text{CH}_2\text{O}-\overset{\overset{\text{O}}{\|}}{\underset{\text{OH}}{\text{P}}}-\text{O}-\text{N-base} & \text{CH}_2\text{O}-\overset{\overset{\text{O}}{\|}}{\underset{\text{OH}}{\text{P}}}-\text{O}-\text{N-base} \\ (\text{I}) & (\text{II}) \end{array}$$

these lipids has been effectively reviewed by Snyder[34].

It has been tacitly assumed that, aside from the unique ether bond, the metabolism of the ether lipids resembles that of their diacyl analogs. Certainly, the finding by Kiyasu and Kennedy[35] that 1-alkenyl-2-acyl glycerol reacts with CDP–choline to form choline plasmalogen would support that assumption. Furthermore, glyceryl ethers isolated from phospholipids of bone marrow given [6-^{14}C]glucose contained radioactivity almost exclusively in position 3 of the glycerol moiety[36]. This labeling pattern would suggest that in bone marrow the glycerol component arises through the glycolytic pathway, as in the case of diacyl phosphatides. The principal experimental efforts have therefore dealt with formation of the ether bond itself. Progress in understanding this step has been painfully slow.

Perhaps the most curious observation resulting from metabolic studies is the relatively sluggish rate at which potential precursors are incorporated into the vinylic ether side-chain. Thus radioactive acetate, palmitate, and palmitaldehyde label fatty acid esters of plasmalogen-rich animal tissues at a significantly faster rate than they do the ether-bound side-chain[34]. As a result, confusion exists as to whether the proper intermediate has perhaps not yet been tested or whether the turnover of plasmalogens is indeed less active than that of diacyl phosphatides. The likelihood of the former alternative is damaged by the difficulty in imagining a hydrocarbon chain precursor which would not be labeled rapidly by [^{14}C]acetate.

The saturated ether chain of glyceryl ether phospholipids, on the other hand, quickly absorbs radioactivity from [^{14}C]palmitate [37] and acetate[37,38]. Snyder et al.[38a] were successful in isolating a cell-free system from mouse preputial gland tumors which is active in synthesizing the saturated ether bond. An enzyme located in the microsomal fraction catalyzes the synthesis of glyceryl ethers from glyceraldehyde 3-phosphate and long-chain fatty alcohols. ATP, Mg^{2+}, and coenzyme A are required. Apparent intermediates in this reaction have been isolated but not yet identified.

Many cells contain both plasmalogens and glyceryl ether phospholipids. By following the fate of radioactive glyceryl ethers fed to a species of terrestrial slug, Thompson[39] detected an apparent desaturation of glyceryl ether phospholipids to yield plasmalogens. Using as substrate glyceryl ethers labeled in the hydrocarbon side-chain with ^{14}C and in the glyceryl moiety with ^{3}H, it was established that both these portions of the molecule are transferred to plasmalogens without any significant change in the $^{14}C/^{3}H$ ratio[40]. The slow rate of this transformation would explain the many literature reports of delayed plasmalogen labeling.

However, a different pattern would seem to hold in the digestive gland of the starfish. Here radioactive fatty aldehydes were more efficient precursors of plasmalogens than they were of glyceryl ether phospholipids[38]. The retention of ^{3}H located at carbon-1 of fatty aldehyde during its incorporation into plasmalogens would argue against an oxidation to fatty acid prior to utilization. It seems probable that the conflicting results obtained by various workers stem either from the presence of two or more metabolic pools of the lipids under study or from the existence of alternate biosynthetic pathways.

(e) *Formation of phosphorus-containing sphingolipids*

(i) *Formation of sphingosine*

In sphingomyelin (I), the various components of the molecule are bound, not to glycerol as in the glycerophosphatides, but rather to the nitrogenous

$$CH_3(CH_2)_{12}CH=CH-\underset{\underset{\underset{R-C=O}{|}}{NH}}{CH}-\underset{OH}{CH}-CH_2O-\underset{\underset{OH}{|}}{\overset{\overset{O}{\|}}{P}}-OCH_2CH_2\overset{+}{N}(CH_3)_3$$

(I)

base sphingosine. Carbon atoms 1 and 2 of sphingosine are derived from carbon atoms 2 and 3 of serine, and carbon atoms 3–18 arise from palmitic acid. The first details of the condensation were established in the laboratories of Brady[41,42] from experiments performed using a cell-free preparation from rat brain. Palmitaldehyde is produced by the NADPH-requiring reduction

References p. 197

of palmityl-CoA. The aldehyde intermediate then appeared to condense with

$$CH_3(CH_2)_{14}\overset{O}{\underset{\|}{C}}-SCoA \xrightarrow[]{NADPH \quad NADP} CH_3(CH_2)_{14}-\overset{O}{\underset{\|}{C}}H$$

pyridoxal phosphate-bound serine in the presence of Mn^{2+} to form dihydrosphingosine.

Recent findings from the laboratory of Stoffel[43] have revealed that an intermediate step in dihydrosphingosine formation involves the formation of 3-oxodihydrosphingosine. An enzyme from the microsomal fraction of rat liver and the yeast *Hansenula ciferrii* will utilize NADPH to reduce this compound to *erythro*-dihydrosphingosine. Experiments from the laboratory

of Snell[43a] have generally confirmed these observations and found that 3-oxodihydrosphingosine and 3-oxosphingosine accumulate in *H. ciferrii* preparations when NADPH is omitted. Thus dihydrosphingosine and sphingosine would seem to arise from their respective 3-oxo derivative. The involvement of the 3-oxo intermediate and other evidence[43b] suggest that palmityl-CoA and not palmitaldehyde is the form involved in the condensation step.

Recently, branched long-chain base analogs of sphingosine have been isolated from protista[44] and from certain bovine tissues where, in the case of kidney, they account for over 13% of total long-chain base[45]. The metabolism of these bases has not been studied, but their reported absence in rat tissues has prompted the speculation that the bovine constituent might actually be a product of rumen protozoa[45].

(ii) Formation of sphingomyelin

Sphingomyelin biosynthesis proceeds from sphingosine by a two-step process. The first reaction involves the coupling of a fatty acid residue to the amino group of sphingosine to yield ceramide (*N*-acylsphingosine). Two apparently different enzymes can catalyze the reaction. Sribney[46] has described an enzyme from liver and brain which carries out the reaction using as substrates sphingosine and fatty acyl-CoA. Free acid could not be

$$CH_3(CH_2)_{12}CH=CH-\underset{OH}{CH}-\underset{NH_2}{CH}-CH_2OH + R-\overset{O}{\underset{\|}{C}}-SCoA \longrightarrow CH_3(CH_2)_{12}CH=CH-\underset{OH}{CH}-\underset{\underset{R-C=O}{NH}}{CH}-CH_2OH$$

utilized. Both *erythro*-sphingosine, the natural isomer, and *threo*-sphingosine can be acylated.

Ceramide is also formed by a different enzyme purified from rat brain by Gatt[47]. In contrast to the system reported by Sribney, this enzyme will utilize free fatty acid, and no requirement for CoA or ATP could be demonstrated. The reaction is freely reversible.

The resulting ceramide can then react with CDP–choline under the influence of the enzyme CDP–choline:ceramide cholinephosphotransferase (EC 2.7.8.3). Sribney and Kennedy[48] found the enzyme to be widely distributed

$$CDP-choline + CH_3(CH_2)_{12}CH=CH-\underset{OH}{CH}-\underset{\underset{R-C=O}{NH}}{CH}-CH_2OH$$

$$\downarrow$$

$$CH_3(CH_2)_{12}CH=CH-\underset{OH}{CH}-\underset{\underset{R-C=O}{NH}}{CH}-CHCH_2-O-\overset{O}{\underset{\underset{OH}{\|}}{P}}-OCH_2CH_2\overset{+}{N}(CH_3)_3 + CMP$$

in animal tissues. Ceramides of dihydrosphingosine will not serve as substrates, and surprisingly, ceramides containing *erythro*-sphingosine, the natural isomer, were much less reactive than those containing *threo*-sphingosine. It has recently been shown that this unexpected finding is an artifact produced by some non-physiological property of the *in vitro* enzyme system, and that under suitable conditions both isomers are actively utilized[48a].

A possible alternate pathway for sphingomyelin synthesis has been observed in homogenates of young rat brain[49]. Sphingosylphosphorylcholine

References p. 197

could be coupled with fatty acyl-CoA to yield sphingomyelin in these preparations. Either the *erythro* or *threo* isomer is active[50]. There is no indication as yet that sphingosylphosphorylcholine is formed in sufficient quantities to make the pathway significant.

(iii) Analogs of sphingomyelin

An analog in which choline is replaced by ethanolamine is distributed widely among molluscs[51]. It is presumably formed by a pathway analogous to that operative for sphingomyelin.

Often occurring with the ethanolamine derivative in nature is the structurally similar compound ceramide aminoethylphosphonate (I), first

$$R_1-CH-CH-CH_2-O-\underset{OH}{\overset{O}{\underset{\|}{P}}}-CH_2CH_2NH_2$$
$$\underset{OH}{|}\ \underset{NH}{|}$$
$$\underset{R_2-C=O}{|}$$
(I)

characterized by Rouser et al.[52]. R_1 represents the chain of sphingosine or related bases. Certain aspects of phosphonate lipid biosynthesis will be covered in the following section.

(f) Formation of lipids containing the carbon–phosphorus bond

Since the discovery in 1959 by Horiguchi and Kandatsu[53] of 2-aminoethylphosphonic acid, the first known phosphonate in nature, these compounds have been implicated as components of several complex lipids. The biochemistry and distribution of phosphonate-containing lipids has recently been reviewed by Quin[54].

There is some evidence that lipids play an important role in the biosynthesis of 2-aminoethylphosphonate (AEP). In *Tetrahymena pyriformis* AEP occurs bound to protein and bound to lipid as well as in a free state[55]. The phosphonate lipid (I) of this organism is structurally analogous to phosphatidylethanolamine[56]. Interestingly, a major proportion of the

$$\begin{array}{c} \text{CH}_2\text{O}-\overset{O}{\underset{\|}{C}}-R \\ R'-\overset{O}{\underset{\|}{C}}-O\overset{|}{C}H \\ \overset{|}{CH_2}O-\underset{OH}{\overset{O}{\underset{\|}{P}}}-CH_2CH_2NH_2 \end{array}$$
(I)

fraction from *Tetrahymena* consists of the glyceryl ether analog rather than the diacyl form[57]. Liang and Rosenberg[56] demonstrated the ability of a *Tetrahymena* homogenate to form the complete lipid from diglyceride and cytidinemonophosphate–aminoethylphosphonate. Thus, apart from the synthesis of AEP itself, the biogenesis can proceed according to the classical pattern. However, in *Tetrahymena* the specific radioactivity of lipid-bound AEP increases much more rapidly than that of free AEP after the administration[55] of $^{32}P_i$, indicating that the cytidine nucleotide pathway might not be of physiological significance. Instead, it has been proposed[58] that AEP may be formed by the rearrangement of lipid-bound phosphoenolpyruvate, the most efficient precursor yet tested for the AEP carbon chain.

The biosynthesis of other phosphonate lipids, such as the phosphatidylserine analog or the ceramide aminoethylphosphonate described in the previous section, has not been examined in detail.

(g) Summary

While it may be more orderly to consider the biosynthesis of each phospholipid separately, this treatment renders it very difficult to impart a proper appreciation for the interrelationships among various species. The summary outline presented below (Fig. 2) was designed with an aim to illustrate some of the more important cross-currents of phospholipid metabolism.

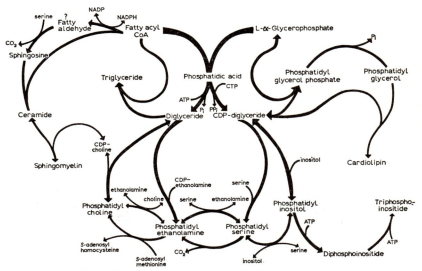

Fig. 2. Summary outline of phospholipid biosynthesis.

3. Catabolism

(a) Introduction

Much of our early knowledge concerning the pathways for phospholipid degradation was gained using enzymes of extracellular origin, *e.g.*, from snake venoms and bacterial toxins. A certain amount of faith was required to extrapolate the results of such work to the degradative processes occurring inside animal cells. However, in recent years, normal catabolic pathways in animal tissues have been examined in considerable detail. For the most part, the intracellular enzymes have been shown to function in a fashion not unlike that determined with venoms and toxins.

In this survey of phospholipid catabolism, emphasis will be placed on the enzymes known to function within animal cells during the normal physiological process of lipid turnover. A detailed account of many other aspects of the subject may be found in the monograph of Ansell and Hawthorne[59] and the review of Van Deenen and De Haas[60].

Substrate specificities of degradative enzymes have seldom been determined under conditions approaching those found within the cell, but our present knowledge suggests that most phospholipases will attack several types of phospholipid. For this reason, the discussion is organized so as to group information concerning individual enzymes rather than considering in sequence the stepwise degradation of specific lipids.

(b) Enzymes removing fatty acyl groups

(i) Phospholipase A

This enzyme, originally called lecithinase A and more recently termed phosphatide acyl-hydrolase (EC 3.1.1.4), is the well known phospholipase of snake venom. It catalyzes the reaction:

$$\begin{array}{c} \text{O} \\ \| \\ \text{R'}-\text{C}-\text{OCH} \\ | \\ \text{CH}_2\text{O}-\text{C}-\text{R} \\ | \\ \text{CH}_2\text{O}-\overset{\text{O}}{\underset{|}{\text{P}}}-\text{OCH}_2\text{CH}_2\text{N}^+(\text{CH}_3)_3 \\ \text{OH} \end{array} \longrightarrow \begin{array}{c} \text{O} \\ \| \\ \text{CH}_2\text{O}-\text{C}-\text{R} \\ | \\ \text{HOCH} \\ | \\ \text{CH}_2\text{O}-\overset{\text{O}}{\underset{|}{\text{P}}}-\text{OCH}_2\text{CH}_2\text{N}^+(\text{CH}_3)_3 \\ \text{OH} \end{array} + \text{R'COOH}$$

The enzyme specifically attacks at the 2 position of the glycerol moiety regardless of whether the fatty acid bound there is unsaturated or saturated.

Enzymes carrying out this reaction have been discovered in a number of animal tissues[61]; although the degree of activity varies widely, it is likely that the enzyme is almost ubiquitous. The activity is localized in mitochondria and requires Ca^{2+} for activity. De Haas et al.[62] have purified the porcine pancreatic enzyme and found it to have a molecular weight of about 14 000, less than half that estimated for the *Crotalus adamanteus* venom enzyme studied by Saito and Hanahan[63]. The enzyme is remarkably heat-stable; activity persists in the pancreatic protein even after boiling for 5 min at pH 4.

Of particular interest is the recent discovery of a phospholipase A zymogen in porcine pancreas[64]. The protein was isolated after it was observed that phospholipase A activity increases markedly upon storage of pancreas homogenates at room temperature. Trypsin was shown to activate the zymogen form by cleaving off a heptapeptide. Because it is possible that phospholipase A is secreted by the pancreas for eventual utilization in the intestinal digestive process, zymogen production may be peculiar to this tissue.

Although phospholipase A shows a rather pronounced specificity as to phospholipid type when studied with purified lipid substrates, there is evidence that the specificity is reduced in the presence of lipid mixtures. The affinity of a lipid for the enzyme would appear to depend upon the overall charge of the micelle in which it is contained[65].

The enzymes from different sources display different substrate specificities. Thus the snake-venom enzyme preferentially hydrolyzes lecithin while pancreatic phospholipase A rapidly attacks very acidic lipids, such as phosphatidic acid, phosphatidylglycerol, and cardiolipin, and hydrolyzes lecithin rather slowly[62]. Even among venom enzymes, differences can be noted. For example, phospholipase A of *Crotalus atrox* is much less active towards choline plasmalogen than towards phosphatidylcholine, but phospholipase A of *Naja naja* shows no strong preference for the diacyl phosphatide[66].

It is impossible at this time to interpret the *in vitro* specificity data in a way that would allow us to understand the *in vivo* reactivity. It is likely that all the common phosphoglycerides are hydrolyzed to some extent. A certain degree of specificity is probably retained *in vivo*, however, as evidenced by the preferential hydrolysis of mitochondrial phosphatidylethanolamine by endogenous phospholipase A[67].

Within the past few years, it has become apparent that some tissues contain a phospholipase activity differing from that of the classical phospholipase A in that it catalyzes the specific hydrolysis of fatty acids located at the C-1 position of phosphatides. This enzyme is currently designated as phospholipase A_1 (specific for the 1-acyl ester) as contrasted to phospholi-

References p. 197

pase A_2 (the original phospholipase A, specific for the 2-acyl ester). Waite and Van Deenen[68] have succeeded in separating the two enzymes from homogenates of rat liver. Phospholipase A_1 occurs mainly in the microsomal fraction while phospholipase A_2 is found in mitochondria. The former enzyme does not require a divalent cation and is less stable to heat than phospholipase A_2. It appears to be different from the enzyme pancreatic lipase, which has also been shown to attack the 1-acyl ester of phospholipids[69].

A phospholipase A_1 activity has also been isolated from brain[70] and from human plasma[71]. The enzyme from plasma is only found after the injection of heparin. Further purification of the plasma enzyme is necessary before it can be decided if the deacylation of phosphatidylcholine, phosphatidylethanolamine, and triglyceride is the work of a single protein.

(ii) Lysophospholipase

This enzyme, also termed lysolecithin acyl-hydrolase (EC 3.1.1.5) and sometimes phospholipase B, has been the subject of some confusion. The term phospholipase B originally signified an enzyme capable of hydrolyzing fatty acids from both the 1 and 2 positions of phospholipids, and this is still considered to be the reaction catalyzed by the *Penicillium notatum* enzyme[72]. However, the existence of such an activity in animal tissues has been questioned (see discussion by Ansell and Hawthorne[59], p. 161). The Enzyme Commission of the International Union of Biochemistry has listed phospholipase B as a synonym for lysolecithin acyl-hydrolase, an enzyme catalyzing the following reaction,

$$\text{Lysolecithin} + H_2O \rightarrow \text{glycerylphosphorylcholine} + \text{fatty acid}$$

No specificity is indicated with respect to the position which can be attacked, *i.e.*, the C-1 or C-2 position of the glyceryl moiety.

The hydrolysis of phosphatidylcholine to yield glycerylphosphorylcholine and fatty acids has been observed in many animal tissues[73,74]. Intestinal mucosa is a particularly rich source of activity. Typically, little or no lysophosphatide accumulates. In rat-liver microsomes endogenous phosphatidylethanolamine is deacylated at an even faster rate than endogenous phosphatidylcholine. It was concluded that the hydrolysis is due to combined action by phospholipase A and lysophospholipase, with the former enzyme being rate-limiting[75]. In purified mitochondria, on the other hand, lysophospholipase action is quite limited[67], and lysophosphatides constitute the major end-product of degradation.

Further information concerning the rat-liver system has come from the work of Waite and Van Deenen[68]. The deacylation of lysophosphatides was observed to be much more sensitive to temperature than was the phospholipase A activity. This was taken as evidence that the lysophospholipase(s) are distinct from the phospholipases. Furthermore, the phospholipase A activity was found to be localized in particulate fractions while the supernatant after sedimentation of microsomes at $100\,000 \times g$ for one hour contained most of the lysophospholipase(s). It has not yet been determined whether both the 1-acyl and the 2-acyl lysophosphatides are hydrolyzed by the same lysophospholipase.

(iii) Enzymes catalyzing the deacylation of sphingolipids

There is no evidence at present to suggest that animal tissues can carry out the cleavage of fatty acid from its amide linkage in sphingomyelin. The normal pathway of sphingomyelin catabolism would appear to involve the following reactions[76].

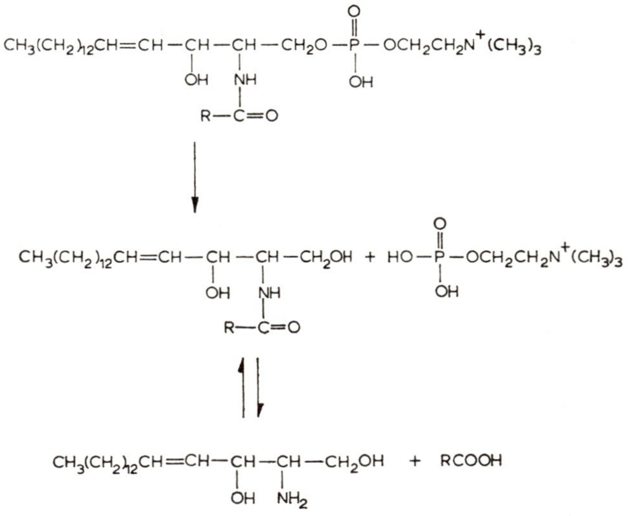

$$CH_3(CH_2)_{12}CH{=}CH{-}\underset{OH}{CH}{-}\underset{\underset{R-C=O}{NH}}{CH}{-}CH_2O{-}\underset{OH}{\overset{O}{\underset{\|}{P}}}{-}OCH_2CH_2N^+(CH_3)_3$$

$$\downarrow$$

$$CH_3(CH_2)_{12}CH{=}CH{-}\underset{OH}{CH}{-}\underset{\underset{R-C=O}{NH}}{CH}{-}CH_2OH \;+\; HO{-}\underset{OH}{\overset{O}{\underset{\|}{P}}}{-}OCH_2CH_2N^+(CH_3)_3$$

$$\updownarrow$$

$$CH_3(CH_2)_{12}CH{=}CH{-}\underset{OH}{CH}{-}\underset{NH_2}{CH}{-}CH_2OH \;+\; RCOOH$$

Little is known about the second step in which fatty acid is released from ceramide. The reaction has been observed in homogenates of rat brain, liver, and kidney by Gatt[47]. The purified brain enzyme has no metal ion requirement, and hydrolysis of ceramide is inhibited by addition of the reaction

References p. 197

products. The amide bonds of sphingomyelin or cerebrosides are not attacked. The reverse reaction, ceramide formation from sphingosine and fatty acid, is also catalyzed by the enzyme.

(c) Enzymes cleaving the phosphate–glycerol ester linkage

(i) Phospholipase C

Phospholipase C is the widely accepted trivial name for the enzyme phosphatidylcholine cholinephosphohydrolase (EC 3.1.4.3). As in the case with certain other lipolytic enzymes, many workers consider the trivial name more desirable than the systematic name, which does not indicate the activity of the enzyme for substrates other than lecithin.

The sources of the phospholipase C which have been extensively studied are the toxins of *Clostridium welchii, Bacillus cereus,* and other bacteria. Such preparations catalyze the reaction:

$$R'-\overset{O}{\overset{\|}{C}}-O\overset{CH_2O-\overset{O}{\overset{\|}{C}}-R}{\underset{CH_2O-\overset{}{\underset{OH}{P}}-OCH_2CH_2N^+(CH_3)_3}{\overset{|}{C}H}} \longrightarrow R'-\overset{O}{\overset{\|}{C}}-O\overset{CH_2O-\overset{O}{\overset{\|}{C}}-R}{\underset{CH_2OH}{\overset{|}{C}H}} + HO-\overset{O}{\underset{OH}{\overset{\|}{P}}}-OCH_2CH_2N^+(CH_3)_3$$

Under the proper conditions, the Ca^{2+}-activated enzyme can attack phosphatidylcholine, phosphatidylethanolamine, sphingomyelin, the ethanolamine and 2-aminoethylphosphonate analogs of sphingomyelin[76a], choline and ethanolamine plasmalogens[77], and the phosphonic analog of phosphatidylcholine[78].

Enzymes possessing this activity are not well known in animal tissues. Interpretations are often complicated by the concurrent action of other phospholipases in the systems under examination. Heller and Shapiro[79] isolated a preparation from rat-liver mitochondria (or lysosomes) which hydrolyzed sphingomyelin to ceramide and phosphorylcholine. The reaction proceeded more rapidly with the *erythro* isomer than with the *threo* form. Unlike the bacterial enzymes, this phospholipase was much more reactive towards sphingomyelin than towards phosphatidylcholine. Phosphatidic acid was also hydrolysed swiftly. The brain enzyme is further distinguished from those of bacterial origin by showing no dependence on Ca^{2+} for full activity.

Atherton and Hawthorne[80] have described the properties of a phospholipase C type enzyme occurring in intestinal mucosa and a number of other animal tissues. The enzyme exhibits a very pronounced specificity, attacking only phosphatidylinositol and di- and tri-phosphoinositides among all the more common glycerophosphatides examined. Sphingomyelin was not tested. The enzyme, termed phosphoinositide inositolphosphohydrolase, was localized in the $105\,000 \times g$ supernatant fraction and depended upon the presence of Ca^{2+} for activity.

(ii) Phosphatidate phosphohydrolase

Another enzyme falling into the general category under discussion is L-α-phosphatidate phosphohydrolase (EC 3.1.3.4), often called phosphatidic acid phosphatase. The reaction, which produces inorganic phosphate and diglyceride, is obviously a key step in the biosynthetic pathway of several important

$$\underset{\begin{array}{c}\\ \text{O}\\ \|\\ \text{R}'-\text{C}-\text{OCH}\\ |\\ \text{CH}_2\text{O}-\text{P}-\text{OH}\\ |\\ \text{OH}\end{array}}{\overset{\begin{array}{c}\text{O}\\ \|\\ \text{CH}_2\text{O}-\text{C}-\text{R}\\ |\\ \text{O}\\ \|\end{array}}{}} \longrightarrow \underset{\begin{array}{c}\\ \text{O}\\ \|\\ \text{R}'-\text{C}-\text{OCH}\\ |\\ \text{CH}_2\text{OH}\end{array}}{\overset{\begin{array}{c}\text{O}\\ \|\\ \text{CH}_2\text{O}-\text{C}-\text{R}\\ |\end{array}}{}} + \underset{\text{OH}}{\overset{\begin{array}{c}\text{O}\\ \|\end{array}}{\text{HO}-\text{P}-\text{OH}}}$$

phospholipids. In addition, as mentioned in an earlier section, the coupled action of the phosphohydrolase and diglyceride kinase creates the so-called phosphatidic acid cycle, which may be of importance in the process of active transport.

The possibility of at least two more or less independent functions for phosphatidate phosphohydrolase has been heightened by evidence that enzymatic activity occurs in various particulate fractions of the cell and that in these fractions certain properties of the enzyme differ significantly[81]. Some of the apparent differences in pH optima and kinetic parameters might be attributed to variations in the extent to which the enzyme is embedded within membranes.

Recent findings from the laboratories of Hübscher[82] and Johnston[83] demonstrate the existence of still another seemingly different phosphatidate phosphohydrolase in the high speed supernatant from liver and intestinal mucosa. Whether the intracellular location and properties of the enzyme determine its *in vivo* function is of considerable current interest.

References p. 197

(d) Enzymes cleaving the phosphate–base bond

The well-known enzyme phosphatidylcholine phosphatidohydrolase (EC 3.1.4.4), or phospholipase D, catalyzes the formation of phosphatidic acid from phosphatidylcholine or other phosphoglycerides. Thus far the enzyme has been found only in plants. For this reason details of the reaction

will not be considered here. Several references to recent work are included in a paper by Dawson and Hemington[84]. Although animal lipids are not the normal substrate for phospholipase D, they are susceptible to attack, and the enzyme is frequently used for the preparation of phosphatidic acid from animal phosphatides.

(e) Phosphoinositide phosphomonoesterases

Triphosphoinositides and diphosphoinositides occur in easily measurable levels only in the nervous system. It has been appreciated for several years that brain tissue contains polyphosphoinositide-specific phosphomonoesterases so potent that special precautions must be taken to avoid appreciable degradation of the lipids during lipid extraction by classical methods. Dawson and Thompson[85] succeeded in partially purifying the active triphosphoinositide phosphomonoesterase from brain acetone powders. The enzyme attacked triphosphoinositide in the presence of Mg^{2+} or Mn^{2+} ions, liberating inorganic phosphate and phosphatidylinositol. Only limited activity was exhibited towards free inositol triphosphate.

(f) *Enzymes catalyzing the cleavage of ether bonds*

(i) *Degradation of the vinylic ether linkage of plasmalogens*

We do not know what pathway is followed during the *in vivo* catabolism of plasmalogens. With reference to the cleavage of the ether bond proper, three routes have been detected *in vitro*. Attack may take place on the intact plasmalogen, the lysoplasmalogen, or the glyceryl vinylic ether remaining after phosphate and acyl ester hydrolysis. An outline of these alternate modes of ether-bond scission is given below. Further details of the reactions may be found in the review by Snyder[34].

Ansell and Spanner[86] showed in 1965 that rat brain contains an enzyme capable of cleaving the ether bond of ethanolamine plasmalogen or its lyso analog. The conversion of ethanolamine plasmalogen to 2-acylglycerylphosphorylethanolamine and fatty aldehyde required Mg^{2+} and was most active at approximately pH 7.4.

$$\begin{array}{c}
O \quad CH_2O-CH=CHR \\
\| \quad | \\
R'-C-OCH \\
| \quad O \\
CH_2O-P-OCH_2CH_2NH_2 \\
| \\
OH
\end{array} \xrightarrow{Mg^{2+}} \begin{array}{c}
O \quad CH_2OH \\
\| \quad | \\
R'-C-OCH \\
| \quad O \\
CH_2O-P-OCH_2CH_2NH_2 \\
| \\
OH
\end{array} + O=CHCH_2R$$

Warner and Lands[87] had previously described an enzyme in rat-liver microsomes which is capable of hydrolyzing the ether bond of lysocholine plasmalogen but not choline plasmalogen itself. The analogous ethanolamine-containing phosphatides were not attacked. This enzyme required no divalent cation.

Kapoulas *et al.*[88] observed that cell-free preparations of *Tetrahymena pyriformis* which were active in disrupting the ether bond of the structurally similar glyceryl ethers (see below), could also cleave the vinylic ether bond of free glyceryl vinylic ethers. Details of this reaction are unknown.

(ii) *Degradation of the ether linkage of glyceryl ether phospholipids*

There are no reported observations of the saturated ether bond of intact glyceryl ether phospholipids being cleaved. However, several tissues are capable of metabolizing free glyceryl ethers. The first report[89] of degradation in intact animals appeared in 1956. Chimyl alcohol (1-*O*-hexadecyl glycerol) fed to rats was rapidly absorbed with a significant amount of degradation. These findings have since been confirmed, and enzyme systems capable of *in vitro* ether cleavage have been described.

Tietz *et al.*[90] uncovered the principal details of the etherase reaction. By

References p. 197

administering radioactive glyceryl ethers to a preparation from rat-liver microsomes, they obtained data enabling them to formulate the following sequence of reactions:

$$\begin{array}{c} CH_2O-CH_2-R \\ | \\ HOCH \\ | \\ CH_2OH \end{array} + O_2 \xrightarrow[\text{PtH}_4]{\text{NADP} \quad \text{NADPH}} \left[\begin{array}{c} OH \\ | \\ CH_2O-CH-R \\ | \\ HOCH \\ | \\ CH_2OH \end{array} \right]$$

$$\begin{array}{c} CH_2OH \\ | \\ HOCH \\ | \\ CH_2OH \end{array} + \quad R-\overset{O}{\overset{\|}{C}}H \longleftarrow$$

$$\downarrow \begin{array}{c} \text{NAD} \\ \text{NADH} \end{array}$$

$$R-COOH$$

The points of special interest are the requirements for molecular oxygen and a tetrahydropteridine cofactor. The addition of a NADPH-generating system to the incubation mixture permitted regeneration of the fully reduced pteridine. The existence of a hemiacetal intermediate rather than an α,β-unsaturated glyceryl ether was postulated because chemical degradation of the reaction mixture revealed no aldehydogenic compounds, thus ruling out the presence of the α,β-unsaturated derivative in more than trace amounts. Although the proposed hemiacetal has never been demonstrated, the other intermediate, fatty aldehyde, was shown to accumulate when NAD was omitted from the test system.

Further details of the rat-liver system were obtained by Pfleger et al.[91]. who discovered that the active enzyme(s) could be easily dissociated from the microsomes. Under certain conditions fatty alcohols were detected in the system, apparently arising through NADPH-mediated reduction of fatty aldehydes.

Of a number of tissues assayed for the ether-cleaving enzyme, only liver and intestine possessed significant activity[91]. The role of the enzyme in liver is almost certainly the catabolism of dietary ethers, since liver lipids themselves are essentially devoid of ether analogs. Kapoulas et al.[88] have studied an etherase occurring in the glyceryl ether-rich ciliate *Tetrahymena pyriformis*. This enzyme, which supposedly participates in the normal turnover of cellular phospholipids, differs from the liver enzyme in certain respects, notably in lacking the pteridine cofactor requirement.

(g) Enzymes catalyzing sphingosine degradation

Several groups have reported the catabolism of sphingosine and its analogs to fatty acids. For example, Barenholz and Gatt[92] found that two-carbon fragments were split from radioactive sphingosine, dihydrosphingosine, and phytosphingosine injected intravenously into rats, yielding palmitic acid

$$CH_3(CH_2)_{14}-\underset{OH}{CH}-\underset{NH_2}{CH}CH_2OH \longrightarrow CH_3(CH_2)_{14}-\underset{O}{\overset{}{C}}-\underset{NH_2}{C}HCH_2OH$$

$$\downarrow$$

$$CH_3(CH_2)_{14}COOH + H_2NCH_2CH_2OH$$

from the first two compounds and pentadecanoic acid from phytosphingosine. Stoffel and Sticht[93] reported the ability of rats to convert [1-^{3}H, 3-^{14}C] *erythro*-DL-dihydrosphingosine to [1-^{14}C]palmitic acid and [1-^{3}H]ethanolamine. Sphingosine follows the same pathway. Because the same products resulting from sphingosine or dihydrosphingosine breakdown were produced at an even faster rate following the administration of [3-^{14}C]3-oxodihydrosphingosine, the following catabolic pathway was proposed. Thus the intermediate 3-oxodihydrosphingosine has been implicated in both the degradative and biosynthetic sequences.

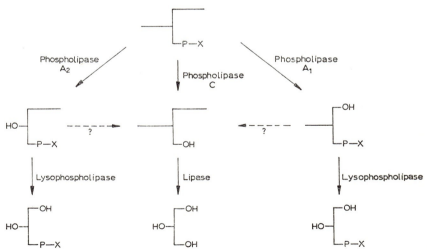

Fig. 3. Catabolism of phosphoglycerides. X represents choline, ethanolamine, serine, inositol, hydrogen, and perhaps other constituents, *e.g.*, glycerol.

References p. 197

(h) Summary

The complete degradation of phospholipids to fatty acids and water-soluble products requires action by more than one enzyme. It is therefore reasonable to ask by what sequence of reactions native phospholipids are degraded under physiological conditions. In most instances the answer to this question still eludes us.

A number of difficulties remain unsolved. Full information concerning the reactivity of known catabolic enzymes towards certain phosphatides is

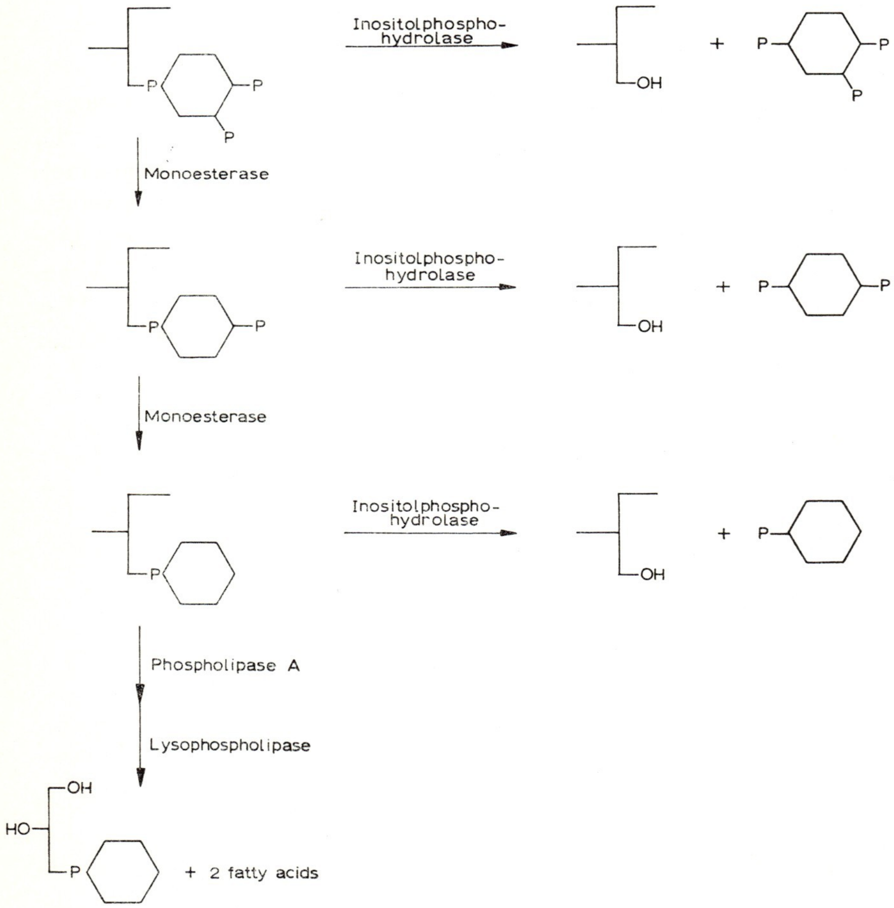

Fig. 4. Catabolism of phosphoinositides.

not available. The observed differences in rates of reaction with different substrates should serve as a warning that mixtures of lipids may be dealt with in diverse ways. Yet it is never quite certain that variations noted in *in vitro* systems are not artificial.

Ignoring the dangers of generalizing, I have outlined a series of summary schemes (Figs. 3, 4 and 5) portraying the probable catabolic fate of the more common phospholipids in animal tissues. The outline does not take into consideration the widespread reutilization of catabolic intermediates for synthetic purposes. This topic will be discussed in the following section.

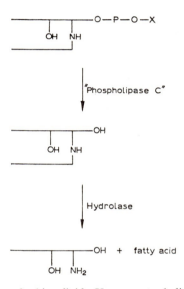

Fig. 5. Catabolism of sphingolipids. X represents choline or ethanolamine.

4. Exchange processes

(a) Mechanisms for the turnover of specific phospholipid components

(i) Acyl groups

As I implied earlier, catabolism of a phospholipid, once initiated, does not necessarily proceed to the complete dissolution of the molecule. Reutilization of partially degraded compounds is thought to be an important mechanism whereby the cell can conserve energy and, at the same time, achieve the subtle changes in lipid composition which may become necessary.

References p. 197

Enzymes capable of acylating lysophosphatides have been studied extensively by Lands and associates. The acyl-CoA:phospholipid acyltransferases appear to be a family of microsomal enzymes, each with its own substrate specificity. The general formulation of the reaction is as follows:

$$\begin{array}{c} \mathrm{CH_2O-\overset{O}{\overset{\|}{C}}-R} \\ \mathrm{HO-CH} \\ \mathrm{CH_2O-\underset{OH}{\overset{O}{\overset{\|}{P}}}-O-N\text{-base}} \end{array} + \mathrm{R'\overset{O}{\overset{\|}{C}}SCoA} \longrightarrow \begin{array}{c} \mathrm{CH_2O-\overset{O}{\overset{\|}{C}}-R} \\ \mathrm{R'-\overset{O}{\overset{\|}{C}}-OCH} \\ \mathrm{CH_2O-\underset{OH}{\overset{O}{\overset{\|}{P}}}-O-N\text{-base}} \end{array} + \mathrm{CoASH}$$

Most studies have involved the use of 1- or 2-lysolecithin or lysophosphatidylethanolamine. On the basis of inhibitor studies, the acyl-CoA:phospholipid acyltransferases reactive towards these substrates would seem to be different from the enzymes acylating glycerol 3-phosphate[8]. Even among the enzymes acylating phospholipids, different species act on the 2- and 1-lysophosphatides. Although it is not yet clear how many distinct enzymes are acting in concert, the overall effect in microsomal preparations is the rather specific esterification of saturated fatty acids at the α'-hydroxyl group and unsaturated acids at the β-hydroxyl group[94]. The selectivity of the enzymes depends upon the position of the hydroxyl group to be acylated and is not influenced by the nature of the acid attached at the other position of the lysophosphatide. The specificity of the liver microsomal enzyme(s) in transferring test acyl-CoA derivatives to 1-acyl-glycerylphosphorylcholine leads to the following preferences in transfer: oleate > linoleate > laurate > palmitate ⩾ stearate[95]. Acting on 2-acyl-glycerylphosphorylcholine, the same enzyme preparation showed this order of preference: stearate > palmitate > laurate ⩾ oleate > linoleate.

The selectivity of the transferase for certain fatty acids does not seem to be determined by some physical property related to the melting points. Over the temperature range of 15–40°, there is no change in the ability of the enzyme to distinguish between the various thiol esters[96]. In addition to its reactivity towards the substrates described above, the acyltransferase from erythrocytes is also active with lysocholine plasmalogen but, surprisingly, the liver enzyme does not acylate the latter derivative[97].

One of the most clearly demonstrated instances where acyltransferases are of quantitative significance is that of the erythrocyte membrane. Here drastic changes in diet alter the fatty acid composition of phospholipids more rapidly than can be accounted for by net phospholipid synthesis. However, the reaction rate and specificity of the membrane-bound acyltransferases was shown to be adequate to control the observed changes[98]. Thus the activity of these acyltransferases can account for the modification of membrane phospholipids in a way that might change the physical characteristics and therefore the physiological properties of the membranes.

Under physiological conditions the lysophosphatides acted upon by the erythrocyte acyltransferases would probably be derived from the surrounding plasma, since phospholipase A activity is slight in red blood cells[99]. Indeed, an enzyme which catalyzes the formation of lysolecithin is present in plasma[100]. The reaction mediated by this enzyme, designated plasma lecithin: cholesterol acyltransferase, is thought to be the major source of plasma esterified cholesterol in man.

The product of lysophosphatide utilization depends, of course, not only upon the nature of the fatty acid added but also upon the type of lysophosphatide available for reacylation. This is determined by the reactivity of the enzymes involved in its formation. Little is known about the *in vivo* selectivity of the phospholipase with regard to fatty acyl groups. We do recognize, as discussed by Lands[101], that there is *in vitro* a markedly reduced reactivity of phospholipases, including phospholipase A from some sources, on plasmalogens. With respect to the nitrogenous base involved we know that mitochondrial phospholipase A attacks endogenous phosphatidylethanolamine more rapidly than phosphatidylcholine[67]. But even this limited information was derived from *in vitro* incubations of purified mitochondria and may not be truly representative of events within the cell.

References p. 197

In addition to the formation of diacyl phospholipids from lysophosphatides and acyl-CoA, a second scheme for lysophosphatide acylation has been detected. By virtue of an exchange reaction, the fatty acid from one lysolecithin can be transferred to the free hydroxyl group of a second lysolecithin yielding lecithin and glycerylphosphorylcholine. The enzyme has been found in the high speed supernatant remaining after sedimentation of rat-liver[102,103]

$$2 \begin{array}{c} CH_2O-\overset{O}{\underset{\|}{C}}-R \\ HOCH \\ CH_2O-\overset{O}{\underset{\|}{P}}-OCH_2CH_2N^+(CH_3)_3 \\ OH \end{array} \longrightarrow \begin{array}{c} CH_2O-\overset{O}{\underset{\|}{C}}-R \\ R-\overset{O}{\underset{\|}{C}}-OCH \\ CH_2O-\overset{O}{\underset{\|}{P}}-OCH_2CH_2N^+(CH_3)_3 \\ OH \end{array}$$

$$+$$

$$\begin{array}{c} CH_2OH \\ HOCH \\ CH_2O-\overset{O}{\underset{\|}{P}}-OCH_2CH_2N^+(CH_3)_3 \\ OH \end{array}$$

and yeast[103] microsomes. By employing as substrate 1-[^{14}C]acylglycero-3-[^{32}P]phosphorylcholine, it was possible to assess the relative importance of the reaction as compared to the acyl-CoA requiring pathway previously discussed. In the absence of added ATP and CoA, the phosphatidylcholine isolated gave a ^{14}C/^{32}P ratio approximately twice that of the lysolecithin substrate, indicating that the fatty acid added at the 2 position of lysolecithin molecule was derived from a sister lysolecithin molecule. The transfer seems rather direct, since free fatty acids added to the incubation mixture did not reduce the ^{14}C/^{32}P ratio in the product. ATP and CoA were not stimulatory, although in the liver system, these cofactors effected a marked stimulation in the acyl-CoA:phospholipid acyltransferase activity. In the yeast supernatant, this latter enzyme seems to be absent.

We may conclude that in these cells and also in a number of other tissues investigated, lysolecithin is metabolized by three different reactions.

$$\text{2 Lysolecithin} \rightarrow \text{lecithin} + \text{glycerylphosphorylcholine} \quad (1)$$

$$\text{Lysolecithin} + \text{acyl-CoA} \rightarrow \text{lecithin} \quad (2)$$

$$\text{Lysolecithin} \rightarrow \text{glycerylphosphorylcholine} + \text{fatty acids} \quad (3)$$

Under normal conditions, where acyl-CoA is readily available, reaction 2 predominates. The physiological significance of reaction 1, which permits the transfer of an acyl group, usually saturated in nature, from the 1 to the 2 position, remains to be determined.

(ii) Nitrogen bases

Aside from the information presented in the section on biosynthesis, little can be added to shed light on the mechanism of *in vivo* base exchange. The fact that certain of the common bases involved in the Ca^{2+}-stimulated exchange inhibit each other competitively indicates that their relative abundance at the intracellular reaction site is precisely controlled.

(b) Exchange of intact molecules

We have learned much about lipid mobility by studying the exchange of phospholipids and cholesterol among erythrocytes and various plasma lipoproteins. The relatively more difficult task of isolating homogenous membrane fractions from other cell types has retarded the study of this phenomenon in more complex tissues.

Some of the more pertinent data have recently been published by Reed[104]. He used ^{32}P-labeled phospholipids to study the turnover of four major erythrocyte lipids, both *in vivo* and *in vitro*. By incubating labeled plasma lipoproteins with unlabeled erythrocytes, and *vice versa*, it was estimated that 60% of human erythrocyte phosphatidylcholine is exchangeable with phosphatidylcholine of plasma. The exchangeable pool of erythrocyte sphingomyelin amounted to only 30% of the total in man but 75% in the dog. No independent turnover of the phosphate or phosphorylcholine moieties could be detected. The finding that phosphatidylserine and phosphatidylethanolamine seemed not to exchange may be related to their low concentration in plasma. Although the turnover rate of red-cell phospholipids is slower than observed for cholesterol, it is sufficiently fast to be metabolically significant. As an example, 13% of the exchangeable lecithin pool of human erythrocytes turns over in 12 h. Although much of this exchange may be due to the transfer of intact phospholipids, some is doubtless accounted for by the expulsion of one lecithin molecule by a lysolecithin residue absorbed from the plasma and acylated by the acyl-transferases described above.

A few attempts have been made to measure any exchange of phospholipids that might take place between intracellular organelles. If most of the phospholipids of a cell are synthesized in the microsomal fraction (there is evidence[105] supporting the assumption, but the case is certainly not foolproof), it would seem logical that some transfer might occur. Wirtz and Zilversmit[106] have performed such experiments following the same general approach used by Reed in the erythrocyte study. Endogenous phospholipid labeled in the glycerol moiety with ^{14}C or in the phosphate group with ^{32}P was observed to exchange easily from labeled mitochondria to unlabeled microsomes and also in the reverse direction when the labeling was reversed. The microsomal supernatant fraction was required for most active exchange. Phosphatidylcholine was most actively transferred, followed by phosphatidylethanolamine. No exchange of cardiolipin took place. The experiments could not distinguish between the movement of intact molecules and the movement of lysophosphatides followed by reacylation. Thus while many confirming experiments remain to be done, it seems likely that under some conditions a rapid transfer of phospholipid molecules can take place from one region of the cell to another as well as into the cell from the outside environment.

(c) Summary and conclusions

The active participation of exchange processes complicates our efforts to understand phospholipid metabolism in the living organism. Not enough systems have been characterized as yet to permit the formulation of any general principles. Yet it would seem that some thread of conformity exists among the cases which have been examined.

In most tissues fatty acids, especially those in the β-position, can be exchanged readily. A particular acid can be replaced by an identical molecule or by a different species. The lysophosphatide intermediates in the exchange process and probably the intact phospholipids themselves can be exchanged from their position in the membrane. The replacement must apparently be of the same phospholipid class, although perhaps bearing different acyl groups.

Other exchange reactions, such as the replacement of one nitrogenous base by another, appear to be of less quantitative significance. Thus, while these transformations may be equally important in fixing the precise pattern of membrane organization, it is the former processes that account for the remarkably active intracellular mobility of the phospholipids.

5. Rates of lipid metabolism *in vivo*

(*a*) *Introduction*

In the previous sections of this chapter, I have outlined the pathways whereby animal phospholipids can be synthesized and degraded in nature. Emphasis has been placed on the enzymatic mechanisms and on the intermediates involved. Information concerning these points has usually come from *in vitro* studies in which the normal balance of cell metabolism was necessarily upset. It would now be well to consider briefly the evidence relating to the actual rates of phospholipid metabolism in intact normal cells. Such information is essential in establishing the relative importance of the several pathways under different environmental and growth conditions. Furthermore, it will hopefully aid in evaluating the role of exchange processes as opposed to *de novo* synthesis as a means of providing phospholipids of structure ideal for cellular needs.

(*b*) *Rates of lipid metabolism in non-growing tissues*

In mature animal tissues maintained under constant conditions one would expect to find little or no net synthesis of phospholipid. Under such conditions the pathways for synthesis, degradation and exchange operate only to renew preexisting cell structures and allow for the replacement of autolyzed cells. A number of studies have sought to measure the rates of lipid turnover under these circumstances.

Cuzner *et al.*[107] have followed the decline of radioactivity in brain fractions of young adult rats previously injected with ^3H_2O and [^{14}C]acetate. The overall rate of lipid metabolism of brain mitochondria was found to be approximately three times slower than that of liver mitochondria, in which the half-life is 10.6 days. Even so, the brain mitochondrial lipids are metabolized much more rapidly than are those of myelin. Cholesterol, cardiolipin, and phosphatidic acid appear to turn over considerably more slowly in brain mitochondria than do most of the other phospholipids.

Cardiolipin was also shown to be metabolized less actively than other phospholipids in liver mitochondria[108]. By labeling the structural proteins with [^{35}S]methionine, a half-life of 9 days was calculated, while at least one pool of $^{32}P_i$ or [^{14}C]acetate-labeled mitochondrial phospholipids demonstrated a half-life of less than two days. Other phospholipids were more stable.

References p. 197

A somewhat similar study was carried out on the endoplasmic reticulum membranes of rat hepatocytes[109]. In this cell, whose lifetime is 160–400 days, endoplasmic reticulum membrane proteins have a half-life of 3–5 days while the half-life of phospholipids is 10–30% shorter. The radioactive fatty acid components were lost much more slowly than the labeled glycerol moiety, indicating a recycling of the acids.

Gurr et al.[110] have examined the rates of liver-phospholipid metabolism using a slightly different approach, namely, by following the time course of $^{32}P_i$ uptake into various lipids. The results are in general agreement with those described above. Phosphatidylethanolamine and phosphatidylcholine became more highly labeled than the other phosphatides, and the specific radioactivity of cardiolipin was the lowest of all, averaging only about 20% of that for the most radioactive lipids.

(c) Rates of lipid metabolism in growing tissues

In dividing cells *de novo* synthesis of lipids must exceed degradation to produce the net lipid increase. Of course, superimposed upon these processes is the sort of turnover described above. Nevertheless, one might expect to find that the predominance of synthesis would have a damping effect upon the observed differences in turnover rates. On the contrary, Gurr et al.[110] found great similarity in the pattern of phospholipid labeling in normal and regenerating rat liver. Regrettably, it was impossible in this case to estimate just how much net synthesis had taken place during the experimental period. Tsao and Cornatzer[111] found marked differences in the specific radioactivity of various phospholipids isolated from $^{32}P_i$ treated HeLa or KB cells. In these experiments, incorporation into cardiolipin was relatively great as compared to that observed in rat liver.

Even in these instances of relatively rapid growth then, we find that the selective turnover of specific phosphatides is easily discernable above the generally increased biosynthetic rates for all lipids.

(d) Possible physiological significance of phospholipid turnover

The experiments briefly summarized above should not be accepted as representative of the general scheme of phospholipid metabolism in all animals. However, it seems firmly established that in at least some organisms the components of membranes are not all metabolized at the same rate.

It is puzzling as to why this should be. At one time it was thought to reflect differences in the degree of turnover of the several membrane-containing organelles within the cell. Evidence that both the phospholipid composition and the specific radioactivity patterns agree rather closely in the isolated subcellular fractions of non-growing[110] and growing[111] tissues suggests that this is not the cause of divergent turnover rates. Rather, phospholipids appear to be plucked from the membrane matrix at various rates, depending on their structure, and disassembled into their component parts, some of which are rapidly reused.

What advantage does this energy-requiring, continuous replacement of membrane lipids have for an organism? Could it be that phospholipids "wear out"? Is turnover necessary for the dynamic functioning of membranes? These and other questions have been discussed in the literature, but there is really only circumstantial evidence to support any answer.

Because of the scarcity of firm evidence, none of the postulations will be developed in detail here. However, it is not difficult to imagine how interesting and perhaps useful hypotheses have developed from observations such as those examples mentioned below.

It is known that under certain conditions polyunsaturated fatty acids are very susceptible to attack by oxygen, yielding lipid peroxides. Certainly in vitamin E deficient animals this would be considered the most unstable component of the phospholipid molecule. There are some indications that peroxidation might occur while the unsaturated fatty acid is still attached to the phospholipid[112], but further details are necessary before the significance of this reaction can be judged. In fact, to my knowledge there is no evidence concerning the capacity of an enzyme to transfer a peroxidized acyl sidechain. Nevertheless, it may be noteworthy that the polyunsaturated fatty acids are almost invariably found attached to the 2 position of phospholipid glycerol, and it is the acyl group at this location that turns over more rapidly than any other component.

Perhaps a more clearly significant advantage, at least for certain organisms, is the versatility which metabolic turnover affords for altering the physical properties of membranes. This is nowhere more dramatically demonstrated than in the adaptation of poikilotherms to change in environmental temperature. The brain phospholipids of adult goldfish, for example, contained significantly more unsaturation when the animals were acclimated to 5° than to 30°, although the amounts of the different phospholipid types remained constant[113]. The lower melting point of the more unsaturated lipids would at least partially offset the reduced temperature in maintaining a constant "fluidity" of the membranes.

References p. 197

This same ability of cells under stress to alter the fatty acid composition of their phospholipids has been observed in mammals as a result of drastic dietary changes, particularly as it involves alteration in the amounts of polyunsaturated fatty acids fed. In this case possible advantages to the organism are more obscure.

These are but two examples selected from much evidence suggesting the possible cellular benefits of lipid turnover. It should be reemphasized that the others are no less speculative than the ones mentioned here.

6. The role of lipid metabolism in the process of membrane fabrication

One of the most important questions facing biologists today concerns the mechanism whereby cellular membrane components, principally lipids and proteins, are assembled into various functional membranes. Our ignorance of these processes is almost complete. As regards the phospholipids in particular, it is not known whether they are transferred to the growing membrane in aggregates or unimolecularly, whether they achieve their specific distribution by self-assembly at a growing "boundary," intussusception into existing membranes, or binding to a template.

Because the different membrane-containing organelles are formed at varying rates during the course of cell development, there must be a mechanism for controlling phospholipid synthesis and directing the synthesized products to the appropriate site at the appropriate time. If many of the intracellular membranes are interconnected, as has been proposed by Robertson[114] and others, distribution of membrane material may occur by a "flow" of newly formed membrane through a series of functionally different structures towards an ultimate location. Similarities between the specific radioactivities in phospholipids of diverse organelles after relatively short-term labeling experiments argue against this scheme, but the concurrent action of exchange processes such as those observed by Wirtz and Zilversmit[106] might mask the pattern of net lipid movement.

If, on the other hand, the newly formed lipids and proteins are distributed on a unimolecular or small aggregate basis, a more complex control system may be needed to channel the material to the required sites. In any postulated mechanism, there must exist a complex interplay between enzymatic reactions and purely physical interactions.

These aspects of phospholipid metabolism are now being actively studied in many laboratories. They constitute a new frontier which is destined to contribute much towards our understanding of the cell.

REFERENCES

1. L. L. M. VAN DEENEN, *Progr. Chem. Fats Lipids*, 8 (1965) 1.
2. R. M. C. DAWSON, *Essays Biochem.*, 2 (1966) 69.
3. A. KORNBERG AND W. E. PRICER JR., *J. Biol. Chem.*, 204 (1953) 345.
4. G. P. AILHAUD AND P. R. VAGELOS, *J. Biol. Chem.*, 241 (1966) 3866.
5. H. GOLDFINE, *J. Biol. Chem.*, 241 (1966) 3864.
6. P. W. MAJERUS AND P. R. VAGELOS, *Advan. Lipid Res.*, 5 (1967) 1.
7. E. MARTENSSON AND J. KANFER, *J. Biol. Chem.*, 243 (1968) 497.
8. W. E. M. LANDS AND P. HART, *J. Biol. Chem.*, 240 (1965) 1905.
9. W. E. M. LANDS AND P. HART, *J. Lipid Res,.* 5 (1964) 81.
10. L. E. HOKIN AND M. R. HOKIN, *J. Histochem. Cytochem.*, 13 (1965) 113.
10a. A. K. HAJRA AND B. W. AGRANOFF, *J. Biol. Chem.*, 243 (1968) 3542.
11. D. J. HANAHAN AND H. BROCKERHOFF, in M. FLORKIN AND E. H. STOTZ (Eds.), *Comprehensive Biochemistry*, Vol. 6, Elsevier, Amsterdam, 1965, p. 83.
12. J. WITTENBERG AND A. KORNBERG, *J. Biol. Chem.*, 202 (1953) 431.
13. A. KORNBERG AND W. E. PRICER, *Federation Proc.*, 11 (1952) 242.
14. E. P. KENNEDY AND S. B. WEISS, *J. Biol. Chem.*, 222 (1956) 193.
15. P. BJØRNSTAD AND J. BREMER, *J. Lipid Res.*, 7 (1966) 38.
16. H. PAULUS AND E. P. KENNEDY, *J. Biol. Chem.*, 235 (1960) 1303.
17. J. R. CARTER AND E. P. KENNEDY, *J. Lipid Res.*, 7 (1966) 678.
18. B. W. AGRANOFF, R. M. BRADLEY AND R. O. BRADY, *J. Biol. Chem.*, 233 (1958) 1077.
19. J. N. HAWTHORNE AND P. KEMP, *Advan. Lipid Res.*, 2 (1964) 127.
19a. J. A. BENJAMINS AND B. W. AGRANOFF, *J. Neurochem.*, (1968) in the press.
20. R. H. MICHELL, J. L. HARWOOD, R. COLEMAN AND J. N. HAWTHORNE, *Biochim. Biophys. Acta*, 144 (1967) 649.
21. M. KAI AND J. N. HAWTHORNE, *Biochem. J.*, 102 (1967) 19P.
22. A. SHELTAWY AND R. M. C. DAWSON, *Biochem. J.*, 100 (1966) 12.
23. J. KANFER AND E. P. KENNEDY, *J. Biol. Chem.*, 239 (1964) 1720.
24. J. Y. KIYASU, R. A. PIERINGER, H. PAULUS AND E. P. KENNEDY, *J. Biol. Chem.*, 238 (1963) 2293.
25. Y. Y. CHANG AND E. P. KENNEDY, *J. Lipid Res.*, 8 (1967) 447.
26. N. Z. STANACEV, Y. Y. CHANG AND E. P. KENNEDY, *J. Biol. Chem.*, 242 (1967) 3018.
27. J. BREMER, P. H. FIGARD AND D. M. GREENBERG, *Biochem. Biophys. Acta*, 43 (1960) 477.
28. K. D. GIBSON, J. D. WILSON AND S. UDENFRIEND, *J. Biol. Chem.*, 236 (1961) 673.
29. D. REHBINDER AND D. M. GREENBERG, *Arch. Biochem. Biophys.*, 109 (1965) 110.
30. G. A. SCARBOROUGH AND J. F. NYC, *J. Biol. Chem.*, 242 (1967) 238.
31. J. A. BALINT, D. A. BEELER, D. H. TREBLE AND H. L. SPITZER, *J. Lipid Res.*, 8 (1967) 486.
32. G. HÜBSCHER, *Biochim. Biophys. Acta*, 57 (1962) 555.
33. H. D. CRONE, *Biochem. J.*, 104 (1967) 695.
34. F. SNYDER, *Progr. Chem. Fats. Lipids*, 10 (1968) 287.
35. J. Y. KIYASU AND E. P. KENNEDY, *J. Biol. Chem.*, 235 (1960) 2590.
36. G. A. THOMPSON JR. AND D. J. HANAHAN, *Arch. Biochem. Biophys.*, 96 (1962) 671.
37. G. A. THOMPSON JR., *J. Biol. Chem.*, 240 (1965) 1912.
38. J. ELLINGBOE AND M. L. KARNOVSKY, *J. Biol. Chem.*, 242 (1967) 5693.
38a. F. SNYDER, R. L. WYKLE AND B. MALONE, *Biochem. Biophys. Res. Commun.*, 34 (1969) 315.
39. G. A. THOMPSON JR., *Biochemistry*, 5 (1966) 1290.

40 G. A. THOMPSON JR., *Biochim. Biophys. Acta*, 152 (1968) 409.
41 R. O. BRADY AND G. J. KOVAL, *J. Biol. Chem.*, 233 (1958) 26.
42 R. O. BRADY, J. V. FORMICA AND G. J. KOVAL, *J. Biol. Chem.*, 233 (1958) 1072.
43 W. STOFFEL, D. KEKIN AND G. STICHT, *Z. Physiol. Chem.*, 349 (1968) 664.
43a R. N. BRADY, S. J. DIMARI AND AND E. E. SNELL, *J. Biol. Chem.*, 244 (1969) 491.
43b P. E. BRAUN AND E. E. SNELL, *Proc. Natl. Acad. Sci. (U.S.)*, 58 (1961) 298.
44 H. E. CARTER AND R. C. GAVER, *Biochem. Biophys. Res. Commun.*, 29 (1967) 886.
45 H. E. CARTER AND C. B. HIRSCHBERG, *Biochemistry*, 7 (1968) 2296.
46 M. SRIBNEY, *Biochem. Biophys. Acta*, 125 (1966) 542.
47 S. GATT, *J. Biol. Chem.*, 241 (1966) 3724.
48 M. SRIBNEY AND E. P. KENNEDY, *J. Biol. Chem.*, 233 (1958) 1315.
48a Y. FUJINO, MO. NAKANO, T. NEGISHI AND S. ITO, *J. Biol. Chem.*, 243 (1968) 4650.
49 R. O. BRADY, R. M. BRADLEY, O. M. YOUNG AND H. KALLER, *J. Biol. Chem.*, 240 (1965) PC 3693.
50 Y. FUJINO AND T. NEGISHI, *Biochim. Biophys. Acta*, 152 (1968) 428.
51 T. HORI, I. ARAKAWA AND M. SUGITA, *J. Biochem.*, 62 (1967) 67.
52 G. ROUSER, G. KRITCHEVSKY, D. HELLER AND E. LIEBER, *J. Am. Oil Chemists' Soc.*, 40 (1963) 425.
53 M. HORIGUCHI AND M. KANDATSU, *Nature*, 184 (1959) 901.
54 L. D. QUIN, *Topics Phosphorus Chem.*, 4 (1967) 23.
55 H. ROSENBERG, *Nature*, 203 (1964) 299.
56 C. R. LIANG AND H. ROSENBERG, *Biochim. Biophys. Acta*, 125 (1966) 548.
57 G. A. THOMPSON JR., *Biochemistry*, 6 (1967) 2015.
58 C. R. LIANG AND H. ROSENBERG, *Biochim. Biophys. Acta*, 156 (1968) 437.
59 G. B. ANSELL AND J. N. HAWTHORNE, *Phospholipids (BBA Library, Vol. 4)*, Elsevier, Amsterdam, 1964.
60 L. L. M. VAN DEENEN AND G. H. DE HAAS, *Ann. Rev. Biochem.*, 35 (1966) 157.
61 J. J. GALLAI-HATCHARD AND R. H. S. THOMPSON, *Biochim. Biophys. Acta*, 98 (1965) 128.
62 G. H. DE HAAS, N. M. POSTEMA, W. NIEUWENHUIZEN AND L. L. M. VAN DEENEN, *Biochim. Biophys. Acta*, 159 (1968) 103.
63 K. SAITO AND D. H. HANAHAN, *Biochemistry*, 1 (1962) 521.
64 G. H. DE HAAS, N. M. POSTEMA, W. NIEUWENHUIZEN AND L. L. M. VAN DEENEN, *Biochim. Biophys. Acta*, 159 (1968) 118.
65 A. D. BANGHAM, *Advan. Lipid Res.*, 1 (1963) 65.
66 G. COLACICCO AND M. M. RAPPORT, *J. Lipid Res.*, 7 (1966) 258.
67 P. BJØRNSTAD, *J. Lipid Res.*, 7 (1966) 612.
68 M. WAITE AND L. L. M. VAN DEENEN, *Biochim. Biophys. Acta*, 137 (1964) 498.
69 L. SARDA, M. F. MAYLIÉ, J. ROGER AND P. DESNUELLE, *Biochim. Biophys. Acta*, 89 (1964) 183.
70 S. GATT, *Biochim. Biophys. Acta*, 159 (1968) 304.
71 W. C. VOGEL AND E. L. BIERMAN, *J. Lipid Res.*, 8 (1967) 46.
72 R. M. C. DAWSON AND H. HAUSER, *Biochim. Biophys. Acta*, 137 (1967) 518.
73 A. F. ROBERTSON AND W. E. M. LANDS, *Biochemistry*, 1 (1962) 804.
74 A. OTTOLENGHI, *J. Lipid Res.*, 5 (1964) 532.
75 P. BJØRNSTAD, *Biochim. Biophys. Acta*, 116 (1966) 500.
76 P. B. SCHNEIDER AND E. P. KENNEDY, *J. Lipid Res.*, 9 (1968) 58.
76a I. ARAKAWA, M. SUGITA, AND T. HORI, *J. Japan. Biochem. Soc.*, 40 (1968) 154.
77 G. B. ANSELL AND S. SPANNER, *Biochem. J.*, 97 (1965) 375.
78 E. BAER AND N. Z. STANACEV, *Can. J. Biochem.*, 44 (1966) 893.
79 M. HELLER AND B. SHAPIRO, *Biochem. J.*, 98 (1966) 763.

80 R. S. ATHERTON AND J. N. HAWTHORNE, *Eur. J. Biochem.*, 4 (1968) 68.
81 B. SEDGWICK AND G. HÜBSCHER, *Biochim. Biophys. Acta*, 106 (1965) 63.
82 M. E. SMITH, B. SEDGWICK, D. N. BRINDLEY AND G. HÜBSCHER, *Eur. J. Biochem.*, 3 (1967) 70.
83 J. M. JOHNSTON, G. A. RAO, P. A. LOWE AND B. E. SHWARE, *Lipids*, 2 (1967) 14.
84 R. M. C. DAWSON AND N. HEMINGTON, *Biochem. J.*, 102 (1967) 76.
85 R. M. C. DAWSON AND W. THOMPSON, *Biochem. J.*, 91 (1964) 244.
86 G. B. ANSELL AND S. SPANNER, *Biochem. J.*, 94 (1965) 252.
87 H. R. WARNER AND W. E. M. LANDS, *J. Biol. Chem.*, 236 (1961) 2404.
88 V. M. KAPOULAS, G. A. THOMPSON JR. AND D. J. HANAHAN, *Biochim. Biophys. Acta*, 176 (1969) 237, 250.
89 S. BERGSTRÖM AND R. BLOMSTRAND, *Acta Physiol. Scand.*, 38 (1956) 166.
90 A. TIETZ, M. LINDBERG AND E. P. KENNEDY, *J. Biol. Chem.*, 239 (1964) 4081.
91 R. C. PFLEGER, C. PIANTADOSI AND F. SNYDER, *Biochim. Biophys. Acta*, 144 (1967) 633.
92 Y. BARENHOLZ AND S. GATT, *Biochemistry*, 7 (1968) 2603.
93 W. STOFFEL AND G. STICHT, *Z. Physiol. Chem.*, 348 (1967) 1345.
94 W. E. M. LANDS AND I. MERKL, *J. Biol. Chem.*, 238 (1963) 898.
95 H. VAN DEN BOSCH, L. M. G. VAN GOLDE, A. J. SLOTBOOM AND L. L. M. VAN DEENEN, *Biochim. Biophys. Acta*, 152 (1968) 694.
96 P. JEZYK AND W. E. M. LANDS, *J. Lipid Res.*, 9 (1968) 525.
97 K. WAKU AND W. E. M. LANDS, *J. Biol. Chem.*, 243 (1968) 2654.
98 K. WAKU AND W. E. M. LANDS, *J. Lipid Res.*, 9 (1968) 12.
99 E. MULDER AND L. L. M. VAN DEENEN, *Biochim. Biophys. Acta*, 106 (1965) 348.
100 J. A. GLOMSET, *J. Lipid Res.*, 9 (1968) 155.
101 W. E. M. LANDS, *Ann. Rev. Biochem.*, 34 (1965) 313.
102 J. F. ERBLAND AND G. V. MARINETTI, *Biochim. Biophys. Acta*, 106 (1965) 128.
103 H. VAN DEN BOSCH, H. A. BONTE AND L. L. M. VAN DEENEN, *Biochim. Biophys. Acta*, 98 (1965) 648.
104 C. F. REED, *J. Clin. Invest.*, 47 (1968) 749.
105 G. F. WILGRAM AND E. P. KENNEDY, *J. Biol. Chem.*, 238 (1963) 2615.
106 K. W. A. WIRTZ AND D. B. ZILVERSMIT, *J. Biol. Chem.*, 243 (1968) 3596.
107 M. L. CUZNER, A. N. DAVISON AND N. A. GREGSON, *Biochem. J.*, 101 (1966) 618.
108 E. BAILEY, C. B. TAYLOR AND W. BARTLEY, *Biochem. J.*, 104 (1967) 1026.
109 T. OMURA, P. SIEKEVITZ AND G. E. PALADE, *J. Biol. Chem.*, 242 (1967) 2389.
110 M. I. GURR, C. PROTTEY AND J. N. HAWTHORNE, *Biochim. Biophys. Acta*, 106 (1965) 357.
111 S. S. TSAO AND W. E. CORNATZER, *Lipids*, 2 (1967) 424.
112 H. E. MAY AND P. B. MCCAY, *J. Biol. Chem.*, 243 (1968) 2288.
113 B. I. ROOTS, *Comp. Biochem. Physiol.*, 25 (1968) 457.
114 J. D. ROBERTSON, *Biochem. Soc. Symp.*, 16 (1959) 3.

Chapter IV

Ganglioside Metabolism

LARS SVENNERHOLM

*Department of Neurochemistry, Psychiatric Research Centre,
University of Göteborg (Sweden)*

1. Introduction

(a) *Chemical structure and nomenclature*

Ganglioside is the generic term for a glycosphingolipid containing sialic acid. Brain tissue contains the largest concentrations of gangliosides and they were given their name because Klenk[1] assumed them to be localized in the ganglion cells. Gangliosides also occur outside the nervous system; they have been chromatographically identified in all organs studied. The gangliosides of various sources differ in the patterns of fatty acids and sphingosines and in the number of units in the carbohydrate chain. Pure gangliosides so far prepared have only been uniform in regard to their sugar moiety which therefore has been used for the characterization of the gangliosides. All the major brain gangliosides have in common a neutral carbohydrate moiety of four units: galactosyl-β-(1 → 3)N-acetylgalactosaminyl-β-(1 → 4)-galactosyl-β-(1 → 4)-glucosyl, which is attached to the terminal, primary hydroxyl of sphingosine. The sialic acid of mammalian brain gangliosides is N-acetylneuraminic acid (NAN) which is linked by a ketosidic bond to two main positions: C-3 of galactose and C-8 of another sialic acid. The visceral

---

The nomenclature and the abbreviations used in this article are as follows: Cer = ceramide = N-acylsphingosine where the acyl group is primarily stearic acid and sphingosine is the naturally-occurring C_{18} and C_{20} mixtures of these long-chain bases; sphingosine = [(2S:3R)-2-amino-trans-4-octadecene-1,3 diol]; all sugars are of the D configuration, and glycosides are pyranosides; Glc = glucose; Gal = galactose; GalNAc = N-acetylgalactosamine = 2-acetylamino-2-deoxy-D-galactose; NAN = N-acetylneuraminic acid; NGN = N-glycolylneuraminic acid; TLC = thin-layer chromatography.

References p. 226

TABLE I

GANGLIOSIDES OF MAMMALIAN BRAIN

Chemical structure	Generic term	Code system	
NAN(2 → 3)Gal(β,1 → 4)Glc(1 → 1)Cer	Monosialosyl-lactosylceramide	G_{M3}	G_{Lact}^1
NAN(2 → 8)NAN(2 → 3)Gal(β,1 → 4)Glc(1 → 1)Cer	Disialosyl-lactosylceramide	G_{D3}	G_{Lact}^2
GalNAc(β,1 → 4)Gal(β,1 →4)Glc(1 →1)Cer $\begin{pmatrix}3\\\leftarrow\\2\end{pmatrix}$ NAN	Monosialosyl-N-triglycosylceramide	G_{M2}	G_{NTrI}^1
GalNAc(β,1 → 4)Gal(β,1 → 4)Glc(1 → 1)Cer $\begin{pmatrix}3\\\leftarrow\\2\end{pmatrix}$ NAN(8 → 2)NAN	Disialosyl-N-triglycosylceramide	G_{D2}	G_{NTrII}^2
Gal(β,1 → 3)GalNAc(β,1 → 4)Gal(β,1 → 4)Glc(1 → 1)Cer $\begin{pmatrix}3\\\leftarrow\\2\end{pmatrix}$ NAN	Monosialosyl-N-tetraglycosylceramide	G_{M1}	G_{NT}^1
Gal(β,1 → 3)GalNAc(β,1 → 4)Gal(β,1 → 4)Glc(1 → 1)Cer $\begin{pmatrix}3\\\leftarrow\\2\end{pmatrix}$ NAN(8 ← 2)NAN	Disialosyl-N-tetraglycosylceramide	G_{D1a}	G_{NT}^{2a}

ERRATUM

Comprehensive Biochemistry, Vol. 18, Chapter IV, pp. 202 and 203, Table I.

The structural formulae for G_{D1a} and G_{D1b} have been interchanged.

The last chemical structure on p. 202 should read:

Gal(β,1 → 3)GalNAc(β,1 → 4)Gal(β,1 → 4)Glc(1 → 1)Cer   Disialosyl-*N*-tetraglycosyl-ceramide G_{D1a} G_{NT}^{2a}

(3 ← 2)

NAN   NAN

and the first structural formula on p. 203 should read:

Gal(β,1 → 3)GalNAc(β,1 → 4)Gal(β,1 → 4)Glc(1 → 1)Cer   Disialosyl-*N*-tetraglycosyl-ceramide G_{D1b} G_{NT}^{2b}

(3 ← 2)

NAN(8 ← 2)NAN

CHEMICAL STRUCTURE AND NOMENCLATURE

Structure	Name		
Gal(β,1 → 3)GalNAc(β,1 → 4)Gal(β,1 → 4)Glc(1 → 1)Cer $\quad$(3 ← 2) $\quad$NAN	Disialosyl-N-tetraglycosylceramide	G_{D1b}	G_{NT}^{2b}
[a]Gal(β,1 → 3)GalNAc(β,1 → 4)Gal(β,1 → 4)Glc(1 → 1)Cer $\quad$(3 ← 2) $\quad$NAN NAN(8 → 2)NAN	Trisialosyl-N-tetraglycosylceramide	G_{T1a}	(G_{NT}^{3})
Gal(β,1 → 3)GalNAc(β,1 → 4)Gal(β,1 → 4)Glc(1 → 1)Cer $\quad$(3 ← 2) $\quad$NAN(8 → 2)NAN	Trisialosyl-N-tetraglycosylceramide	G_{T1b}	G_{NT}^{3}
[a]Gal(β,1 → 3)GalNAc(β,1 → 4)Gal(β,1 → 4)Glc(1 → 1)Cer $\quad$(3 ← 2) $\quad$NAN(8 → 2)NAN NAN(8 → 2)NAN	Tetrasialosyl-N-tetraglycosylceramide	G_{Q1}	G_{NT}^{9}
Gal(1 → 1)Cer $\quad$(3 ← 2) $\quad$NAN	Monosialosylgalactosylceramide		G_{ga1}^{1}

[a] Assumed chemical structure

gangliosides show much larger variations in their carbohydrate chains but the chemical structures are still only known for a limited number. The major ganglioside of most visceral organs is, however, very simple: monosialosyllactosylceramide.

As in the field of carbohydrate chemistry there has been a great need for simple generic terms for the different brain gangliosides instead of the very long systematic designations. The terms mono-, di-, and trisialogangliosides[2] were early adopted but no common system of trivial names or symbols for the individual gangliosides were accepted. Most investigators preferred to introduce their own designations. Nevertheless, the code systems suggested by Kuhn and Wiegandt[3] and Svennerholm[4] are used today more than any of the others. Kuhn and Wiegandt based their original code system on the migratory rate, while Svennerholm used the composition of the carbohydrate chain. In the latter system G stands for gangliosides, subscript M, D, T and Q for the number of sialosyl groups and subscript 1 for the major neutral tetrasaccharide chain, 2 for the chain lacking the terminal galactose and 3 for the chain lacking galactosyl-N-acetylgalactosamine. Wiegandt[5,6] has suggested a new code system which is also based on the composition of the carbohydrate chain not only for the mammalian-brain gangliosides but also for gangliosides with other neutral oligoglycosyl chains. Many of these gangliosides have not yet been identified but they can be assumed to occur because corresponding neutral glycolipids or oligosaccharides have been found in many tissues or tissue fluids.

The new system of Wiegandt may seem convenient when written but it is too complicated for oral communication and the reader has to be familiar with the short designations for oligosaccharides used by Dr. R. Kuhn. In Table I the structure formulas of the brain gangliosides, their generic terms assigned according to the rules suggested by IUPAC–IUB[7], and the two code systems discussed are given. Short designations adopted by other investigators have not been tabulated because they are given in two recent reviews on gangliosides[8,9] to which the reader is referred. Because the knowledge of the metabolism of extraneural gangliosides is negligible their structure is not given in Table I.

(b) *Topographical distribution of gangliosides*

The highest concentrations of gangliosides in nerve tissue are found among mammals and birds while the concentration is lower in reptiles, amphibians

TABLE II

THE GANGLIOSIDE CONTENT IN BRAIN TISSUE OF VERTEBRATES

Material	Lipid-NAN (μg/g wet weight)	Distribution of NAN in %								Ref.
		G_{M3}	G_{M2}	G_{D2}	G_{M1}	G_{D1a}	G_{D1b}	G_{T1}	G_{Q1}	
Human:										
Cerebral cortex: newborn	400	1.0	3.6	1.1	14.6	71.6	1.8	7.3	—	10
adult	1002		1.3		11.3	22.4	28.3	29.9	5.9	11
White matter: newborn	400	1.0	6.9	1.4	19.1	57.8	2.1	3.4	—	10
adult	156		1.0		9.3	14.0	31.4	38.2	6.4	11
Rat	1047		1.2		13.0	32.1	20.3	27.3	6.0	12
Rabbit	497		3		15	38	16	23		5
Chicken	660									13
Turtle	220									14
Snake	320									14
Frog	220									14
Fish	370									14

References p. 226

and fishes (Table II). Studies of some invertebrates (octopus, lobster)[15] did not show detectable amounts of gangliosides in the nerves and ganglia but minute amounts have been isolated from the eye and eyestalk of crab[5]. The concentration of gangliosides is highest in the ganglion-cell-rich structures of brain. In human cerebral cortex and some of the larger nuclei (*e.g.* caudate nucleus) the concentration of gangliosides is about 2%, and in adult white matter 0.4% when calculated on a dry weight basis. This means that the gangliosides comprise about 6% of the lipids in grey matter but only about 0.6% in white matter. Before the initiation of myelination the ganglioside concentration is about the same in the presumptive white matter as in the cortex. The concentration of gangliosides is lower in spinal cord than in brain and it is still lower in peripheral nerve[16]. On subcellular fractionation the gangliosides are enriched in the synaptosome fraction[17,18] while only minute amounts or nil have been found in myelin[19,20].

The amount of brain gangliosides undergoes very rapid changes during early development. The ganglioside content of rat brain increased 50-fold from birth up to the end of the fifth week[21]. A still more rapid increase of the ganglioside content was observed in chicken embryos[13], a discovery which has been of utmost importance for the elucidation of the biosynthesis of gangliosides.

The ganglioside pattern also shows large variations with age and localization. In human and rat brain a rapid increase of the disialoganglioside G_{D1a} occurs during the maturation of the brain[6,22] and in frog brain[23] a rapid change from predominantly simple monosialoganglioside G_{M1} to tetrasialosyltetraglycosylceramide G_{Q1} occurred in a few weeks.

Spinal cord and peripheral nerve have a pattern which is characterized by lower amounts of the oligosialosylgangliosides but more of the monosialosyltri- and diglycosylceramides, G_{M2} and G_{M3}[24].

Gangliosides occur in rather high concentrations in some visceral organs such as adrenal medulla[25], spleen[26] and placenta[27], but their concentration is low in most other organs and tissues. The visceral gangliosides show some variations in composition among different species of mammals but in general mono- and disialosyllactosylceramides (G_{M3} and G_{D3}) predominate. Except for the brain-type of gangliosides there have also been found gangliosides with another neutral carbohydrate chain. In beef spleen and red cells the neutral tetraglycosyl moiety contained *N*-acetylglucosamine instead of *N*-acetylgalactosamine, and *N*-glycolylneuraminic acid was the major sialic acid[28].

2. Biosynthesis of gangliosides

The formation of gangliosides can proceed *via* essentially two different types of pathways: (*1*) Sugars are added to a lipid acceptor as oligosaccharide units[29]. (*2*) Sugars are added to a lipid acceptor as monosaccharide units in a step-wise manner[30].

There is still no evidence that the first route is used although free nucleotide oligosaccharides have been isolated from mammalian sources. Roseman and associates[30-32] have instead convincingly demonstrated the step-wise addition of "active" sugars to the appropriate acceptors. From their extensive studies the complete pathway for the biosynthesis of gangliosides from ceramide to disialosyltetraglycosylceramide (Fig. 1) can be outlined. The biosynthesis of the "active" sugars results from a sequence of reactions involving the conversion of glucose to the monosaccharides and their derivatives and "activation" of the monosaccharides. The "activated" monosaccharides are all nucleotide derivatives of the sugars. The pathways leading to their formation can be found in some comprehensive reviews[33-35].

The enzymes which catalyse the incorporation of the nucleotide sugars have a high or moderately high specificity. Their activities are at least 10-fold larger with the appropriate lipid acceptors than with other acceptors.

(a) Biosynthesis of ceramide

The enzymatic synthesis of ceramide has been achieved in three essentially different ways: Ceramide was formed when a brain homogenate of young rats was incubated with palmitoyl-CoA plus factors necessary for the formation of sphingosine[36] or with free sphingosine[37].

$$\text{Palmitoyl-CoA} + \text{serine} + (\text{NADP}) \xrightarrow[\text{nicotinamide}]{\text{Mg}^{2+}} \text{palmitoylsphingosine}$$

The reversal of the reaction by which ceramide is split into sphingosine and fatty acid was achieved with a particulate fraction of rat brain by Gatt[38]. Except for the addition of detergent and cholate the enzyme did not require any cofactor.

$$\text{Fatty acid} + \text{sphingosine} \rightleftharpoons \text{ceramide} + \text{H}_2\text{O}$$

The synthesis of ceramide might also be brought about indirectly by the hydrolytic removal of galactose from cerebrosides (galactosylceramides).

References p. 226

These lipids have been shown to be synthesized by the acylation of galactosylsphingosine (psychosine)[39]. Burton[40] has reported evidence for the biosynthesis from ceramide and UDP–galactose which has been confirmed in two recent reports[31,40a].

Sphingosine + UDP–galactose → galactosylsphingosine (+ UDP)
Galactosylsphingosine + fatty acyl-CoA → cerebroside
Cerebroside + H_2O → ceramide + galactose

There is no evidence that ceramide derived from cerebrosides has been utilized for the biosynthesis of gangliosides but Kopazyk and Radin[41] have demonstrated that ceramide derived from cerebrosides was utilized for the biosynthesis of sphingomyelin *in vivo*.

It is a characteristic feature of the brain gangliosides that they have an extremely homogenous fatty acid composition. Stearic acid comprises about 90% or more of the fatty acids while brain cerebrosides (galactosylceramides) mainly contain very long-chain fatty acids (C_{24}–C_{26})[42]. Because of this reason it might not seem likely that ceramides derived from galactosylceramide are the precursors for the gangliosides and in fact the largest increase of gangliosides occurs during the embryologic and early postnatal development when there are only minute amounts of galactosylceramide[10]. However, it does not seem to be any problem of chain-length "specificity" which gives the gangliosides the uniform fatty acid composition, because C_{24}-acylsphingosine was about as good a precursor as C_{18}-acylsphingosine for the formation of glycosylceramide, the first step in the biosynthesis of gangliosides[43]. The uniform fatty acid composition of gangliosides seems more to be a reflection of the available ceramide pool, because brain sphingomyelins with the same anatomical distribution as gangliosides have a similar fatty acid composition[44].

The ceramide of gangliosides shows another characteristic feature with development. At birth practical all sphingosine is C_{18}-sphingosine plus a few percent of C_{18}-dihydrosphingosine but with ageing the percentage of C_{20}-sphingosine increases and reaches in senile human brain about 70% of total sphingosine[21,45,46]. The sphingosines of sphingomyelin, isolated from the same brains as the gangliosides, did not show any similar changes in their composition.

(b) Biosynthesis of glucosylceramide and lactosylceramide

Several previous attempts to demonstrate a substantial incorporation of "activated" glucose or galactose into a glycosylceramide with ceramide as acceptor have failed. This was, however, achieved with a particulate fraction of chicken embryonic brain[31]. The enzyme was detected in homogenates of embryonic chicken brain ranging from 6 to 20 days of age. The enzymatic activity was completely sedimented in the fraction between 800 and 20 000 g and was further purified on a discontinuous sucrose gradient. All the glucosyltransferase activity was located in the junction between 1.0–1.2 M sucrose (a synaptosome fraction).

UDP-[^{14}C]glucose + ceramide → [^{14}C]glucosylceramide (+ UDP)

The enzyme required detergent for optimum activity while it had no requirement for any metal ion. It was extremely labile to heat; it was completely inactivated when incubated for 15 min at 37° in the absence of UDP–[^{14}C]glucose. This heat-lability combined with the ignorance of the importance of detergent seems to be the reason why the reaction has not been detected before. The transferase was most active at 32° and at pH 7.8 at which the approximate K_m values were $1.2 \cdot 10^{-4}$ M and $0.8 \cdot 10^{-4}$ M for UDP–glucose and ceramide, respectively. Substrate-specificity experiments showed ceramide to be the most active lipid acceptor while sphingosine, dihydrosphingosine and galactosylsphingosine had only 10–20% of the activity of ceramide. The importance of the chain length of the ceramide was not discussed—it can be assumed that C_{16}- and C_{24}-acids were the two major fatty acids of the beef-lung ceramide used in their study.

UDP–glucose could not be replaced by UDP–galactose with particulate preparations of young embryos but when the particulate fraction was prepared from the brain of 17–19 day-old embryos UDP–galactose gave a labeled product which migrated on chromatography as glycosylceramide. This may be an alternative pathway to that already described for the biosynthesis of galactosylceramide (cerebroside)[39].

Ceramide + UDP–galactose → galactosylceramide (+ UDP)

Another glycosyltransferase was also purified from the same homogenate of chicken embryonic brain that catalysed the reaction:

$$\text{Glc} \rightarrow \text{Cer} + \text{UDP-[}^{14}\text{C]Gal} \xrightarrow[\text{Mg}^{2+}]{\text{Mn}^{2+}} \text{[}^{14}\text{C]Gal} \rightarrow \text{Glc} \rightarrow \text{Cer} \; (+ \text{UDP})$$

The same reaction was reported shortly before with a homogenate of rat

References p. 226

spleen as enzyme source[47]. With rat spleen the reaction required detergent and Mn^{2+}. All other metal ions tested were inactive. The product was isolated by TLC and incompletely characterized.

This brain galactosyltransferase had a pH-optimum of 6.8 and it had an absolute requirement for detergent and metal ion. Mg^{2+} and Mn^{2+} were equally active while of all other tested metal ions only Ca^{2+} had a slight activity. The different behaviour to Mg^{2+} seems to differentiate between the brain and spleen glucosyltransferases. The glycosyl donor UDP–galactose could not be replaced by any other nucleotide sugar while the glycosyl acceptor was partially replaced by glucosylsphingosine but not by sphingosines, ceramide and galactosylceramide. The product has only been partially characterized. It migrated as lactosylceramide, and after hydrolysis only one labeled spot migrating as galactose was identified.

(c) CMP–NAN: ganglioside sialosyltransferases

The sialosyltransferases are a family of enzymes which catalyse the incorporation of CMP–N-acetylneuraminic acid (CMP–NAN) or CMP–N-glycolylneuraminic acid (CMP–NGN) into polymers. They can be distinguished from each other on the basis of their specificities towards the

$$Gal \rightarrow Glc \rightarrow Cer + CMP\text{-}[^{14}C]NAN \longrightarrow Gal \rightarrow Glc \rightarrow Cer \; (+ CMP) \quad (A)$$
$$\uparrow$$
$$[^{14}C]NAN$$

$$Gal \rightarrow GalNAc \rightarrow Gal \rightarrow Glc \rightarrow Cer + CMP\text{-}[^{14}C]NAN \longrightarrow$$
$$\uparrow$$
$$NAN$$

(B)

$$Gal$$
$$\uparrow$$
$$[^{14}C]NAN \rightarrow GalNAc \rightarrow Gal \rightarrow Glc \rightarrow Cer \; (+ CMP)$$
$$\uparrow$$
$$NAN$$

$$Gal \rightarrow Glc \rightarrow Cer + CMP\text{-}[^{14}C]NAN \longrightarrow Gal \rightarrow Glc \rightarrow Cer \; (+ CMP) \quad (C)$$
$$\uparrow \qquad\qquad\qquad\qquad\qquad\qquad\qquad \uparrow$$
$$NGN \qquad\qquad\qquad [^{14}C]NAN \rightarrow NGN$$

acceptor molecules or on the basis of the chemical structure of the product[48]. The acceptor can be an oligosaccharide, a glycoprotein or a glycolipid. The sialyltransferases engaged in the biosynthesis of gangliosides have been studied systematically only by Roseman and associates[30,32]. The chicken embryonic brain, used for the isolation of the glycosyltransferases has been the enzymic source. The sialyltransferases were primarily located in the synaptosome fractions like the glycosyltransferases discussed above. Three different ganglioside sialyltransferases have been detected which catalyse the three reactions A, B and C.

Reaction A results in the formation of monosialosyllactosylceramide (G_{M3}) the basic precursor of the brain gangliosides[30]. All other neutral glycolipids depicted in Fig. 1 and galactosylceramide were also used as substrates. N-Tetraglycosylceramide (Gal→GalNAc→Gal→Glc→Cer) was the only active lipid acceptor except for lactosylceramide. Partial heat inactivation and substrate competition indicated two different enzymes. The lactosylceramide transferase was comparatively heat-stable and had no requirements for metal ions while the other was more labile and required Mg^{2+} for optimum activity. The lowest K_m value was obtained for N-tetraglycosylceramide and in adult tissue much larger incorporation was obtained with this glycolipid than with lactosylceramide.

The product, obtained from lactosylceramide, behaved as authentic G_{M3} on chromatography and had the theoretical composition, while the ganglioside formed from N-tetraglycosylceramide was incompletely characterized. It was assumed to have the same structure as G_{M1}, but the sialic acid was neuraminidase-labile which excludes binding to internal galactose. It migrated on TLC slightly slower than G_{M1} which suggests that NAN was bound to the terminal galactose.

Reaction B was catalysed by a sialyltransferase which showed similarities with the transferase that had N-tetraglycosylceramide as acceptor. The only reported difference is a lack of metal-ion requirement for G_{M1} as acceptor. It is, however, unfortunate that the neutral glycolipid was not tested with the same enzyme preparation and that no kinetic studies were performed. It is tempting to speculate that the sialyltransferase B is specific for the terminal galactose in a tetraglycosylceramide, neutral or containing sialic acid linked to the internal galactose. In further studies of sialyltransferase B it would be important to include the disialoganglioside G_{D1b} which is also a potential acceptor for a sialic acid.

The characterization of the product in reaction B is incompletely described

References p. 226

and the question will not be answered if any other ganglioside than G_{D1a}, e.g. G_{T1a}, was formed.

The reaction C differed from the other two in several respects. Product formation was not directly proportional to enzyme concentration, but the sigmoidal curve could be changed to a linear one by the addition of ethanolamine phosphoglyceride. The reaction rate was stimulated by histone and the optimum pH was low, pH 5.6. Sialosyltransferase C was further differentiated from transferase A by kinetic studies. The product of reaction C was isolated and it was shown by periodate oxidation that the labeled sialic acid was probably attached to the other sialic acid in a 2→8 linkage, since the internal sialic acid was resistant to the periodate.

The first indication for the incorporation of CMP–NAN into a glycolipid was given by Kanfer et al.[49]. Rat-kidney homogenate was used as the enzymatic source and N-triglycosylceramide served as acceptor. The reaction mixture also contained detergent and UTP, UDP–Glc, and UDP–Gal. It was found that more than 20% of the labeled CMP–NAN was incorporated in 3 h, and 80% of the labeling was recovered in monosialosyl-N-tetraglycosylceramide (G_{M1}). These results are not consistent with those obtained by Kaufman et al.[30]. When the experiment was repeated in our laboratory[24] two gangliosides were formed:

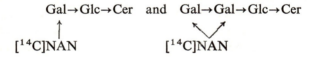

but no G_{M1}. It is noteworthy that the formation of these two labeled gangliosides was increased when N-triglycosylceramide was added to the incubation mixture.

All attempts of Roseman and associates[30,32] to solubilize the sialosyltransferases from the synaptosome fraction have failed. A solubilisation of sialosyltransferase from the microsome fraction of rat brain was reported by Acre et al.[50]. The microsomes were disrupted by sodium deoxycholate and the soluble enzyme was tested with different acceptors. The lowest K_m value was found for lactosylceramide while the value was higher for lactosylsphingosine and "asialogangliosides", the latter obtained by acidic hydrolysis of gangliosides. The products were only tentatively identified.

(d) *Elongation of the carbohydrate chain of ganglioside*

(i) *UDP–N-acetylgalactosamine: monosialosyllactosylceramide N-acetylgalactosaminosyltransferase*

The particulate fraction of chicken embryonic brain has been further shown to contain two enzymes which participate in the biosynthesis of the major brain gangliosides[30]. The least studied of the enzymes is that which transfers N-acetylgalactosamine from UDP–N-acetylgalactosamine to monosialosyllactosylceramide (G_{M3}). Several other glycolipids were tested as acceptors, of which only lactosylceramide and monosialosyl N-triglycosylceramide (G_{M2}) were slightly active.

$$\text{Gal}\to\text{Glc}\to\text{Cer} + \text{UDP–Gal-[}^{14}\text{C]NAc} \xrightarrow{Mn^{2+}}$$
$$\uparrow$$
$$\text{NAN}$$

$$\text{Gal-[}^{14}\text{C]NAc}\to\text{Gal}\to\text{Glc}\to\text{Cer}$$
$$\uparrow$$
$$\text{NAN}$$

$$\text{Gal}\to\text{Glc}\to\text{Cer} + \text{UDP–Gal-[}^{14}\text{C]NAc} \xrightarrow{Mn^{2+}}$$

$$\text{Gal-[}^{14}\text{C]NAc}\to\text{Gal}\to\text{Glc}\to\text{Cer}$$

The relative inactivity of lactosylceramide as acceptor is consistent with the suggestion that sialic acid is first introduced into lactosylceramide followed by N-acetylgalactosamine. The low but still existing incorporation of N-acetylgalactosamine in lactosylceramide supports the assumption that the neutral N-tri- and N-tetraglycosylceramides which normally occur in brain tissue have been formed by a synthetic route (Fig. 1). It has been proposed that these glycolipids were formed by degradation of the corresponding monosialosylgangliosides G_{M2} and G_{M1}[51] by the action of brain neuraminidase but all actual experiments have shown the sialic acid of these gangliosides not to be liberated by any neuraminidase.

(ii) *UDP–galactose: monosialosyl–N-triglycosylceramide galactosyltransferase*

This is the last of the 7 glycosyltransferases from the particulate fraction of chicken brain which has been shown to participate in the biosynthesis of the

References p. 226

brain gangliosides[30]. It had as the other brain glycolipid galactosyl-transferases an absolute requirement for a metal ion. Mn^{2+} gave optimum activity, but Mg^{2+} could not replace Mn^{2+} as was possible when glycosylceramide was the acceptor. Competition experiments with glucosylceramide and monosialosyl-N-triglycosylceramide (G_{M2}) at saturating concentrations showed that they did not compete, which also indicates that the chicken brain contains two specific galactosyltransferases.

The product obtained when G_{M2} was acceptor was homogenous on TLC and migrated with the same rate as authentic G_{M1}. Chemical analysis showed the expected composition of a G_{M1}-ganglioside.

The activity of the ganglioside galactosyltransferase showed great variation with age. There was a sharp increase in activity between the 7th–9th days and then the level was constant during the rest of foetal development, followed by a sharp drop in the adult chicken. A high activity of the same transferase was also found in fetal pig. After maturation of the brain a strong diminution of the ganglioside galactosyltransferase has been found for the chicken, pig, rat, calf, sheep, guinea pig[30] and human[53]; only the frog has been shown to be an exception to this rule[52], and to have a high adult level.

(e) *Incorporation of labeled precursors in vivo*

The formation of a particular glycolipid from an immediate precursor *in vitro* demonstrates that the synthesis can occur, but it provides limited quantitative information concerning the metabolic pathways involved. Only *in vivo* studies can give an unequivocal answer to the physiological importance of a reaction sequence shown *in vitro*. The *in vivo* incorporation studies with labeled precursors have given, however, very little information on the biosynthetic pathways for gangliosides. All the *in vivo* studies reported were done on rats. Burton *et al.*[53] found that all the sugar moieties of gangliosides—glucose, galactose, galactosamine and sialic acid—acquired the same specific activity on the combined use of labeled galactose, glucose and glucosamine. When the gangliosides were separated they were only able to detect incorporation into G_{D1a} and G_{D1b}, while Suzuki and Korey[54] found essentially the same activity in all the four major brain gangliosides, G_{M1}, G_{D1a}, G_{D1b} and G_{T1b}. In a more recent study Suzuki[55] found the relative rate of formation of the four gangliosides parallelled the changes in the percentage distribution. The rapid increase in the relative concentration of G_{D1a} during the early post-partum period was reflected by a higher labeling of this ganglioside. The study also showed that the pool of any major ganglioside could not be the precursor of any other. In short-term experiments with 5–10 day old rats, Holm and Svennerholm[56] found that the major route followed the monosialoganglioside G_{M1} pathway after intracerebral injection.

(f) Metabolic pathways for the biosynthesis

The *in vivo* incorporation studies with labeled precursors have failed to give any substantial information about the biosynthesis of brain gangliosides. Therefore, the pathways outlined in Fig. 1, have been based on the normal pattern of brain gangliosides and particularly on the extensive and brilliant studies by Roseman and associates[30-32]. Two important assumptions were made[30]: The neutral glycolipids and gangliosides formed are identical to those of the naturally occurring compounds although the linkages and anomeric configurations of the sugars have not been determined, and the observed substrate specificities represent the relative specificities of the different transferases towards these glycolipids. The conditions were, however, chosen to be optimal for the most active and the assumed glycolipid acceptor in each system. Different conditions particularly in the concentration and type of detergent and addition of other lipids to the incubation mixture might change the micellar size and structure of the acceptors and give other results with glycolipids that showed low activity. These circumstances are particularly important when the charged gangliosides are compared with the neutral glycolipids as acceptors. Thus, a direct comparison of K_m values for different substrates cannot be used as a single proof for their relative affinities towards the enzyme.

The available data with the glycosyltransferase systems strongly favour the conclusion that disialoganglioside G_{D1a} is formed from ceramide by the pathway depicted in Fig. 1. The evidence also strongly supports the proposal that there may be a second pathway for gangliosides with a disialosyl residue at the internal galactose. This proposal is corroborated by the finding of large increases of G_{D3} and G_{D2} in pathological conditions in which the levels of the normal major disialogangliosides with a disialosyl residue, G_{D1b} and G_{T1b}, are greatly reduced[16].

The discovery of three different sialosyltransferases, each one with a specific glycolipid acceptor affords some comments and raises several questions. Sialyltransferase A seems to have a high specificity for lactosyl-

References p. 226

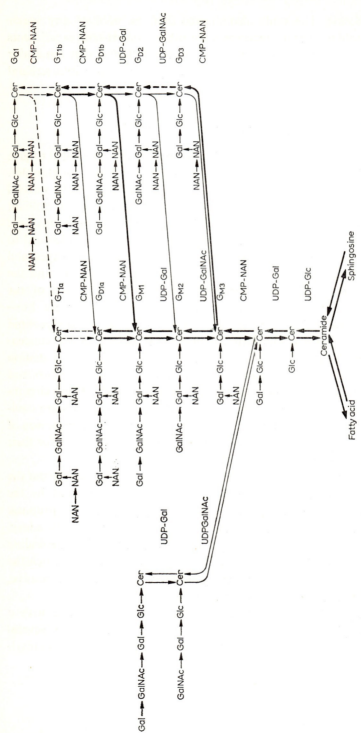

Fig. 1. Metabolic pathways for brain gangliosides; unbroken lines = pathway demonstrated; dashed lines = pathway assumed. Thick lines = major pathways; thin lines = minor pathways.

ceramide and might be designated as a CMP–NAN:lactosylceramide sialosyltransferase. The preferred substrate for sialyltransferase C was monosialosyllactosylceramide (G_{M3}, hematoside). Is it specific for this particular ganglioside or is it specific for a specific binding site, a sialosyl residue of a ganglioside? Will the same enzyme catalyze the incorporation of a second sialic acid of the terminal galactose, forming a trisialoganglioside (G_{T1a})?

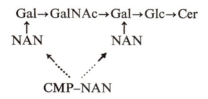

Sialosyltransferase C

Is there any steric hindrance which prevents the incorporation of a second sialic acid at the internal galactose?

Before it was known that chicken embryonic brain contains at least three CMP–NAN:glycolipid sialosyltransferases, a lower K_m value for N-tetraglycosylceramide (Gal→GalNAc→Gal→Glc→Cer) than for lactosylceramide was difficult to interpret[30]. The recent results[32] make it likely that the incorporation of sialic acid was catalyzed by the sialyltransferase B. If this assumption is correct, sialyltransferase B might be specific for the terminal galactose in a tetraglycosylceramide and catalyse the incorporation of sialic acid into the neutral glycolipid and into G_{M1} and G_{D1b} gangliosides.

In normal brain tissue there is no indication for a monosialosyl-N-tetraglycosylceramide with a sialic acid on the terminal galactose, although the corresponding glycolipid exists and has the same ceramide composition as the ganglioside[4]. This discrepancy may only be apparent. In the biosynthesis of gangliosides the product of each transferase reaction serves as the preferred substrate for the next transferase in the sequence[31]. Systems of this type were recently designated *multiglycosyltransferase systems* by Roseman[57]. The scanty results of the *in vivo* studies and the large number of studies of normal brain ganglioside composition also support the suggestion that the gangliosides are formed on a chain of enzymes and the biosynthesis is continued until the final step is reached. After the branching point (Fig. 1) of the common chain there are no connections likely.

References p. 226

3. Biodegradation

The results from the *in vivo* studies of the degradation of brain gangliosides are difficult to interpret[40]. The half-life for ganglioside turnover in young rats had been determined to be 10–25 days[55,58]. The turnover of the gangliosides rapidly diminished with increasing age of the animals[55]. Because a comprehensive review of the *in vivo* studies has recently been published, the reader is referred to this article for further details[40].

(a) *Neuraminidases*

The first studies of mammalian brain neuraminidase (or sialidase, *N*-acyl-neuraminosylhydrolase, E.C. 3.2.1.18) were performed with total brain homogenates. Neuraminidase activity of nervous tissue was determined by Carubelli *et al.*[59] in rat brain and by Morgan and Laurell[60] in human, guinea-pig and bovine brain. The latter determined liberated sialic acid after incubation and attributed the figure for free sialic acid to hydrolysis of endogenous ganglioside. In later studies gangliosides were added to homogenates, and the products were isolated and identified by chromatography. In this manner Korey and Stein[61] found a complete disappearence of some of the added ganglioside and introduced the term "gangliosidase system". In better controlled studies the stepwise degradation of trisialoganglioside G_{T1b} to G_{M1} has been shown with homogenates of human[60] and pig brain[62].

There are two serious disadvantages of the work with a total brain homogenate. It contains all the other glycosidases, which participate in the degradation of the gangliosides, and a large amount of neuraminidase-labile sialic acid, derived from gangliosides, glycopeptides and glycoproteins. No stoichiometric figures for the action of the neuraminidase can thus be obtained.

A substantial purification of the enzyme from calf brain was achieved by Leibovitz and Gatt[63]. An acetone powder of brain was extracted twice with cholate and the enzyme was then extracted with an anionic detergent. The enzyme was still particle-bound but kept in colloidal suspension by the detergent. All attempts to solubilize the enzyme led to losses of its activity. The optimal pH for hydrolysis was low, using acetate or citrate buffer. Detergent was a necessary requirement for activity. The enzyme preparation was specific for gangliosides and did not hydrolyse sialosyllactose or sialoglycoproteins. The highest activity was obtained with G_{T1} and G_{D1a} as substrates while considerably lower figures were obtained with G_{M3} and G_{D1b} (Table III).

The results reported for calf-brain neuraminidase are very similar to those found by us for the neuraminidase prepared with the same method from human brain[64]. The enzymatic activity and the purification factor was slightly better with the human-brain enzyme than with the calf enzyme, but it still had a high hexosaminidase activity and it contained a considerable amount of endogenous ganglioside. All efforts to solubilize the human-brain neuraminidase failed, but the neuraminidase-labile sialic acid could be released by incubation of the enzyme at 37° for 90 min. The enzyme obtained gave stoichiometric ratios for the degradation of ganglioside and yielded free sialic acid with a linear response for 2 h and a 10-fold change in protein content. It required detergent for optimum activity. The neuraminidase was specific for gangliosides and the highest hydrolysis rate was obtained with G_{T1b} and G_{D1a}. This preferential action on the sialic acid bound to the terminal galactose was not a detergent effect, since the same results were obtained with the endogenous gangliosides of the enzyme preparation. By competition experiments it was shown that only one enzyme was involved.

Tettamanti and Zambotti[65] have recently succeeded in preparing a soluble neuraminidase with considerable activity. The enzyme was isolated from pig brain by extraction with isotonic KCl. It was purified by conventional protein

TABLE III

HYDROLYSIS OF GANGLIOSIDES BY CALF- AND PIG-BRAIN NEURAMINIDASE

Incubation was performed at pH 4.4 for 2 h with the calf-brain neuraminidase and at pH 4.9 for 30 min with the pig-brain neuraminidase. The sialic acid released was determined.

Substrate	Calf enzyme[63]	Pig enzyme[65]
	Sialic acid released ($m\mu moles/mg\ protein/h$)	
Mixed brain gangliosides	65	N^a
Tetrasialoganglioside G_{Q1}	N^a	219
Trisialoganglioside G_{T1b}	77	190
Disialoganglioside G_{D1a}	71	230
Disialoganglioside G_{D1b}	33	64
Monosialoganglioside G_{M1}	0	0
Monosialoganglioside G_{M2}	0	0
Monosialoganglioside G_{M3}	50	75
Sialosyllactose	0	71
Ovine submaxillary mucin	0	44

^{a} N = not determined.

References p. 226

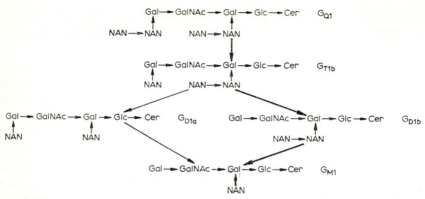

Fig. 2. Biodegradation of major brain gangliosides by mammalian brain neuraminidase (EC 3.2.1.18). Thick lines = major pathways; thin lines = minor pathways.

isolation methods: ammonium sulfate precipitation, heat denaturation, and chromatography on hydroxyl apatite and Sephadex. Optimum pH was 4.9 for gangliosides using citrate–phosphate buffer. The enzyme had no requirement for detergents. The relative rates of hydrolysis of different gangliosides are given in Table III.

The pathway for the degradation of the brain gangliosides can thus be outlined as in Fig. 2. G_{T1b} is mainly degraded to G_{D1b} because the sialic acid bound to the terminal galactose is the preferred action site for the neuraminidase. The ratio between G_{D1b} and G_{D1a} formed from G_{T1b} was about 10 in our studies[64] while Tettamanti and Zambotti[65] could not detect any G_{D1a} at all. Our data for the action of brain neuraminidase are very similar to those earlier obtained with neuraminidase from *Vibrio cholera*[4].

The soluble pig-brain neuraminidase also had rather high activity towards sialosyllactose and ovine submaxillary mucin. A soluble brain neuraminidase was also found in rat brain, while only 5% of the rabbit-brain neuraminidase was soluble[65].

It is important that different species be studied, as demonstrated in the paper of Tettamanti and Zambotti. For kinetic studies of the enzymes involved in the metabolism of gangliosides, there is a necessity to have soluble enzymes. It would be very interesting to know whether the other ganglioside glycosidases also are soluble in pig brain.

In view of the studies discussed it is evident that brain has different neuraminidases: one soluble enzyme which catalyses the hydrolysis of glycosidic-linked sialic acid in gangliosides, oligosaccharides and glycoproteins, and one particle-bound enzyme which is specific for gangliosides. It seems rather likely that it is the same enzyme. When the particle-bound neuraminidase is removed from its membrane attachment it will lose part of its specificity and much of its activity. As is evident from Table III the activity of the particle-bound enzyme is slightly lower than that of the soluble enzyme when expressed in activity/mg protein, but if expressed per initial brain weight the particle-bound will be many fold more active.

(b) Glycosidases

Brain neuraminidase degrades all the major brain gangliosides to G_{M1}:

$$Gal \rightarrow GalNAc \rightarrow Gal \rightarrow Glc \rightarrow Cer$$
$$\uparrow$$
$$NAN$$

5 enzymes except for neuraminidase would be required for complete hydrolysis of this ganglioside: (a) two β-galactosidases, (b) β-N-acetylgalactosaminidase, (c) β-glucosidase and (d) "ceramidase". The sialic acid bound to the internal galactose was not attacked by the neuraminidase, because of the steric hindrance from N-acetylgalactosamine[3], but will be it as soon as hexosamine is split off. As mentioned above a complete hydrolysis of the gangliosides was obtained with a brain homogenate[61], and the aim for the purification of the catabolic ganglioside enzymes was to bring the enzyme to a state of purity in which it was practically free of the enzyme catalyzing the following hydrolytic step in the ganglioside molecule. Previous methods for the isolation of tissue glycosidases did not give a separation of them and it was a general assumption that the same glycosidase could have both β-galactosidase and β-glucosidase activities[66].

A general purification method was finally elaborated by Gatt and associates[67] in which the different brain ganglioside glycosidases were separated. A crude mitochondrial fraction of beef, calf or rat brain was used as starting material[68]. The particles were subjected to sonic disintegration and then extracted with cholate. The cholate extract had both β-glucosidase and β-galactosidase activities. Acidification of the extract to pH 5 precipitated the β-glucosidase activity, while the supernatant contained the β-galactosidase

References p. 226

activity practically free from glucosidase activity. The β-glucosidase activity was recovered from the precipitate with cholate extraction. The purification of the glycosidases compared to the homogenate was 10–20 fold.

(i) β-Galactosidases

The brain β-galactosidase (β-D-galactoside galactohydrolase, EC 3.2.1.23) was assayed with several different substrates with a β-galactosidic linkage[68,69]. Galactose was liberated from the following glycolipids tested: monosialosyl-N-tetraglycosylceramide (G_{M1}), N-tetraglycosylceramide (obtained by acid hydrolysis of G_{M1}), triglycosylceramide [Gal(1→4)Gal-(1→4)Glc→Cer], lactosylceramide and digalactosylceramide [Gal-(1→4)Gal→Cer]. The enzyme did not liberate galactose from galactosylceramide (cerebroside). The enzyme required cholate for activity; optimum activity was obtained when a nonionic detergent was also added to the reaction mixture. Optimum pH was 5.0 and the reaction was inhibited by galactose and γ-galactonolactone, but also by ceramide, fatty acid and sphingosine. The inhibition by the latter compounds was likely not specific but might have changed the micellar state of the substrate.

The β-galactosidase seems to require for activity a β-galactosidic linkage in which galactose is bound to another sugar, such as glucose, galactose or N-acetylgalactosamine. The enzyme was not specific for glycosphingolipids since it also hydrolysed lactose and o- and p-nitrophenyl-β-D-galactopyranoside. The K_m value was, however, 10-fold larger for p-nitrophenylgalactoside than for the glycolipids, indicating a relatively high specificity for the latter substrates.

The β-galactosidase isolated by Gatt will thus release both the terminal and internal galactose of the ganglioside. Because no competition studies were performed with the different glycolipid substrates it is not possible to decide if the enzyme preparation contained one or several β-galactosidases with different substrate specificities. There is some evidence which supports the latter assumption. Brady et al.[70] purified from the intestinal tract an enzyme which specifically catalyzed the hydrolysis of the terminal galactose of triglycosylceramide [Gal(1→4)Gal(1→4)Glc→Cer]. Thus, this enzyme did not catalyse the hydrolysis of lactosylceramide. The enzyme was reported to occur in several tissues, and next to intestine, the highest activity was found in brain. In one of the inherited diseases of ganglioside metabolism, G_{M1}-gangliosidosis or generalized gangliosidosis, a large accumulation of the G_{M1}-ganglioside and its corresponding neutral glycolipid, N-

tetraglycosylceramide, occurs in brain and several visceral organs. Okada and O'Brien[71] demonstrated a nearly complete lack of a β-galactosidase which hydrolyses the terminal galactose of the ganglioside and the neutral glycolipid. Unfortunately, they did not test lactosylceramide as substrate, but since this lipid was not increased in the tissues it seems likely that the β-galactosidase activity towards this substrate was not seriously diminished.

There is also known another inherited ganglioside disorder in which ganglioside G_{M3} and lactosylceramide are strongly increased but in which the level of G_{M1} is low[72]. All evidence supports the assumption that this disease is caused by a lack of a β-galactosidase which is specific towards lactosylceramide.

Thus, the knowledge on the galactosidases which are engaged in the degradation of gangliosides is still very incomplete.

(ii) β-Glucosidase

The β-glucosidase (β-D-glucoside glucohydrolase, EC 3.2.1.21) of beef brain[68] hydrolysed glucosylceramide to ceramide and glucose[73]. It required for optimum activity cholate and a nonionic detergent and had an optimum pH at 5.0 with acetate buffer. It hydrolysed also p-nitrophenyl-β-D-glucopyranoside but not the β-glucosidic linkage of several oligosaccharides or methyl-β-glucoside. A glucosylceramide-cleaving enzyme has also been isolated from spleen tissue[74]. This enzyme was isolated from the 100 000 × g supernatant fraction and had a somewhat higher pH optimum. It was inactive towards galactosylceramide and the activity towards other compounds with a β-glucosidic linkage was not reported. A decreased activity of this soluble β-glucosidase was subsequently reported by Brady *et al.* in Gaucher disease[75]. In the infantile form of this inherited disease an increase of glucosylceramide was demonstrated in brain. It could be shown that this cerebroside had the same composition of the ceramide as gangliosides and its sphingosine composition showed it to be derived from the degradation of gangliosides[16].

(iii) β-N-Acetylhexosaminidase

β-N-Acetylhexosaminidases (β-2-acetylamino-2-deoxy-D-glycoside acetylaminodeoxyglycohydrolases, EC 3.2.1.30) have been demonstrated in brain and many other tissues. Three different β-N-acetylhexosaminidases were isolated by Frohwein and Gatt[76] from calf brain. In a fraction sedimenting between 1000 × g and 25 000 × g a particulate enzyme was isolated which hydrolysed the p-nitrophenyl derivatives of both β-N-acetylglucosamine and β-N-acetylgalactosamine. All attempts to separate the two enzymatic activities were unsuccessful. Hydrolyses of different substrates under various conditions and in the presence of inhibitors indicated two separate active sites of the enzyme preparation. The two other enzymes were isolated from the $100\,000 \times g$ supernatant and they could be separated into a β-N-acetylglucosaminidase and a β-N-acetylgalactosaminidase.

Three glycosphingolipids, having a terminal N-acetylgalactosamine unit, globoside (N-tetraglycosylceramide [GalNAc,β(1→3)Gal,β(1→4)Gal, β(1→4)Glc,β,N-acylsphingosine]), ganglioside G_{M2}, and its corresponding sialic acid free N-triglycosylceramide were used as substrates for the three enzyme preparations[77]. The two soluble enzymes were inactive towards all the three glycolipids. The N-acetylhexosaminidase had a pH optimum of 3.8 with the glycolipid substrates. The hydrolysis was inhibited by several compounds which presumably changed the micellar structure of the substrate but also by two sulfhydryl inhibitors. The hydrolysis rate was similar for the two neutral glycolipids but much lower for the ganglioside. The ganglioside also inhibited strongly the hydrolysis of the neutral glycolipids. The enzymatic hydrolysis of ganglioside G_{M2} is treated in a very brief manner in the report, and it would be necessary to repeat the studies with various detergents and at different concentrations to determine whether the slow hydrolysis and the inhibition will not depend on a change of the micellar structure by the acidic ganglioside.

4. Concluding remarks

The biosynthesis of gangliosides proceeds in a stepwise manner by a particulate "*multiglycosyltransferase system*" which is primarily localized in the synaptosome fraction. Each enzyme has a high or moderately high specificity for the preferred substrate which is the product of the previous transferase in the sequence.

The degradation of gangliosides is also catalyzed by a series of enzymes which degrade the gangliosides in a stepwise manner to sphingosine and fatty acid. The hydrolases are all primarily localized in the crude mitochondrial fraction, closely allied to the lysosomes. They have also a rather high degree of specificity for the glycolipid substrates, although this specificity may be falsely increased by the strong detergents of the incubation mixtures. The preferred substrate seems in general to be the product from the previous hydrolase although kinetic studies are necessary for confirmation of the preliminary observations. One enzyme seems to be an exception to this common rule: β-N-acetylhexosaminidase. It had a very low activity towards ganglioside G_{M2}, much lower than any other of the catabolic enzymes of the ganglioside chain. This finding is extremely interesting and it is in accordance with the observation of an increase of ganglioside G_{M2} in a number of inherited and exogenous diseases of the nervous system. If the β-N-acetylgalactosaminidase activity is the rate-limiting factor for the degradation of brain gangliosides, an increased catabolism of gangliosides or a generalized diminution of the enzymatic activities of the ganglioside hydrolases in a brain disease will lead to an accumulation of ganglioside G_{M2}.

References p. 226

REFERENCES

1 E. KLENK, *Z. Physiol. Chem.*, 273 (1942) 76.
2 L. SVENNERHOLM AND A. RAAL, *Biochim. Biophys. Acta*, 53 (1961) 422.
3 R. KUHN AND H. WIEGANDT, *Chem. Ber.*, 96 (1963) 866.
4 L. SVENNERHOLM, *J. Neurochem.*, 10 (1963) 613.
5 H. WIEGANDT, *Ergeb. Physiol. Biol. Chem. Exptl. Pharmakol.*, 58 (1966) 190.
6 H. WIEGANDT, *Angew. Chem.*, 80 (1968) 89.
7 *The Nomenclature of Lipids*, IUPAC–IUB Commission on Biochemical Nomenclature, April 1967 (also published in *Biochim. Biophys. Acta*, 152 (1968)1).
8 R. H. MCCLUER AND R. J. PENICK, in S. M. ARONSON AND B. VOLK (Eds.), *Inborn Disorders of Sphingolipid Metabolism*, Pergamon, Oxford, 1967, p. 241.
9 R. LEEDEN, *J. Am. Oil Chemist's Soc.*, 43 (1966) 57.
10 L. SVENNERHOLM, *J. Neurochem.*, 11 (1964) 839.
11 K. SUZUKI, *J. Neurochem.*, 12 (1965) 969.
12 K. SUZUKI, *J. Neurochem.*, 12 (1965) 629.
13 O. W. CARRIGAN AND E. CHARGAFF, *Biochim. Biophys. Acta*, 70 (1963) 452.
14 H. H. HESS AND E. ROLDE, *J. Biol. Chem.*, 239 (1964) 3215.
15 R. NORÉN AND L. SVENNERHOLM, The gangliosides of fish, crab, lobster and octopus (in manuscript).
16 L. SVENNERHOLM, in S. M. ARONSON AND B. W. VOLK (Eds.), *Inborn Disorders of Sphingolipid Metabolism*, Pergamon, Oxford, 1967, p. 169.
17 L. S. WOLFE, *Biochem. J.*, 79 (1961) 348.
18 H. WIEGANDT, *J. Neurochem.*, 14 (1967) 671.
19 K. SUZUKI, S. E. PODUSLO AND W. T. NORTON, *Biochim. Biophys. Acta*, 144 (1967) 375.
20 E. J. THOMPSON, H. GOODWIN AND J. N. CUMINGS, *Nature*, 215 (1967) 168.
21 A. ROSENBERG AND N. STERN, *J. Lipid Res.*, 7 (1966) 122.
22 K. SUZUKI, *Life Sci.*, 3 (1964) 1227.
23 J. DAIN, personal communication.
24 L. SVENNERHOLM, unpublished results.
25 R. LEDEEN, K. SALSMAN AND M. CABRERA, *Biochemistry*, 7 (1968) 2287.
26 L. SVENNERHOLM, *Acta Chem. Scand.*, 17 (1963) 860.
27 L. SVENNERHOLM, *Acta Chem. Scand.*, 19 (1965) 1506.
28 R. KUHN AND H. WIEGANDT, *Z. Naturforsch.*, 19b (1964) 80.
29 J. N. KANFER AND R. O. BRADY, in S. M. ARONSON AND B. VOLK (Eds.), *Inborn Disorders of Sphingolipid Metabolism*, Pergamon, Oxford, 1967, p. 187.
30 B. KAUFMAN, S. BASU AND S. ROSEMAN, in S. M. ARONSON AND B. VOLK (Eds.), *Inborn Disorders of Sphingolipid Metabolism*, Pergamon, Oxford, 1967, p. 193.
31 S. BASU, B. KAUFMAN AND S. ROSEMAN, *J. Biol. Chem.*, 243 (1968) 5802.
32 B. KAUFMAN, S. BASU AND S. ROSEMAN, *J. Biol. Chem.*, 243 (1968) 5804.
33 S. ROSEMAN, *Ann. Rev. Biochem.*, 28 (1959) 545.
34 G. W. JOURDIAN AND S. ROSEMAN, *Ann. N.Y. Acad. Sci.*, 106 (1963) 202.
35 L. WARREN, in S. M. ARONSON AND B. VOLK (Eds.), *Inborn Disorders of Sphingolipid Metabolism*, Pergamon, Oxford, 1967, p. 251.
36 I. ZABIN, *J. Am. Chem. Soc.*, 79 (1957) 5834.
37 M. SRIBNEY, *Federation Proc.*, 21 (1962) 280.
38 S. GATT, *J. Biol. Chem.*, 241 (1966) 3724.
39 R. O. BRADY, *J. Biol. Chem.*, 237 (1962) PC2416.
40 R. M. BURTON, in G. SCHETTLER (Ed.), *Lipid and Lipidoses*, Springer, Berlin, 1967, p. 122.

REFERENCES

40a P. MORELL AND N. S. RADIN, *Biochemistry*, 8 (1969) 506.
41 K. C. KOPACYK AND N. S. RADIN, *J. Lipid Res.*, 6 (1965) 140.
42 L. SVENNERHOLM, *J. Lipid Res.*, 5 (1964) 145.
43 S. BASU, personal communication.
44 S. STÄLLBERG-STENHAGEN AND L. SVENNERHOLM, *J. Lipid Res.*, 6 (1965) 146.
45 A. ROSENBERG, in S. M. ARONSON AND B. VOLK (Eds.), *Inborn Disorders of Sphingolipid Metabolism*, Pergamon, Oxford, 1967, p. 267.
46 L. SVENNERHOLM, *Biochem. J.*, 98 (1966) 20P.
47 G. HAUSER, *Biochem. Biophys. Res. Communs.*, 28 (1967) 502.
48 S. ROSEMAN, *Birth Defects, Original Article Series*, 2 (1966) 25.
49 J. N. KANFER, R. S. BLACKLOW, L. WARREN AND R. BRADY, *Biochem. Biophys. Res. Communs.*, 14 (1964) 287.
50 A. ACRE, H. F. MACCIONI AND R. CAPUTTO, *Arch. Biochem. Biophys.*, 116 (1966) 52.
51 H. JATZKEWITZ, H. PILZ AND K. SANDHOFF, *J. Neurochem.*, 12 (1965) 135.
52 J. A. YIAMOUYIANNIS AND J. A. DAIN, *Lipids*, 3 (1968) 378.
53 R. M. BURTON, L. GARCIA-BUNUEL, M. GOLDEN AND Y. BALFOUR, *Biochemistry*, 2 (1963) 580.
54 K. SUZUKI AND S. R. KOREY, *J. Neurochem.*, 11 (1964) 647.
55 K. SUZUKI, *J. Neurochem.*, 14 (1967) 917.
56 M. HOLM AND L. SVENNERHOLM, in preparation.
57 S. ROSEMAN, in E. ROSSI AND E. STOLL (Eds.), *Biochemistry of Glycoproteins and Related Substances, Proc. 4th Intern. Conf. on Cystic Fibrosis of the Pancreas*, Part II, Karger, Basel, 1968, p. 244.
58 R. M. BURTON Y. M. BALFOUR AND J. M. GIBBONS, *Fed. Proc.*, 23 (1964) 230.
59 R. CARUBELLI, R. E. TRUCCO AND R. CAPUTTO, *Biochim. Biophys. Acta*, 60 (1962) 196.
60 E. H. MORGAN AND C.-B. LAURELL, *Nature*, 197 (1963) 921.
61 S. R. KOREY AND A. STEIN, *J. Neuropathol. Exptl. Neurol.*, 22 (1963) 67.
62 V. ZAMBOTTI, G. TETTAMANTI AND B. BERRA, *Proc. Fed. Europ. Biochem. Soc., Vienna*, A236, 1965.
63 Z. LEIBOVITZ AND S. GATT, *Biochim. Biophys. Acta*, 152 (1968) 136.
64 R. ÖHMAN, A. ROSENBERG AND L. SVENNERHOLM, in preparation.
65 G. TETTAMANTI AND V. ZAMBOTTI, *Enzymologia*, 35 (1968) 61.
66 J. CONCHIE, J. FINDLAY AND G. A. LEVVY, *Biochem. J.*, 71 (1959) 318.
67 S. GATT, in S. ARONSON AND B. W. VOLK (Eds.), *Inborn Disorders of Sphingolipid Metabolism*, Pergamon, Oxford, 1967, p. 261.
68 S. GATT AND M. M. RAPPORT, *Biochim. Biophys. Acta*, 113 (1966) 567.
69 S. GATT, *Biochim. Biophys. Acta*, 137 (1967) 192.
70 R. O. BRADY, A. E. GAL, R. M. BRADLEY AND E. MARTENSSON, *J. Biol. Chem.*, 242 (1967) 1021.
71 S. OKADA AND J. S. O'BRIEN *Science*, 160 (1968) 1002.
72 H. PILZ, K. SANDHOFF AND H. JATZKEWITZ, *J. Neurochem.*, 13 (1966) 1282.
73 S. GATT, *Biochem. J.*, 101 (1966) 687.
74 R. O. BRADY, J. KANFER AND D. SHAPIRO, *J. Biol. Chem.*, 240 (1965) 39.
75 R. O. BRADY, J. N. KANFER AND D. SHAPIRO, *Biochem. Biophys. Res. Communs.*, 18 (1965) 221.
76 Y. Z. FROHWEIN AND S. GATT, *Biochemistry*, 6 (1967) 2775.
77 Y. Z. FROHWEIN AND S. GATT, *Biochemistry*, 6 (1967) 2783.

Chapter V

Bacterial Lipid Metabolism

WILLIAM M. O'LEARY

Cornell University Medical College, Department of Microbiology, New York, N.Y. (U.S.A.)

1. Introduction

It has become almost traditional to preface compilations of information on the chemistry and metabolism of microbial lipids with some comment on the prodigious increase of knowledge in this field in recent years. The expansion of this area of science is such a striking phenomenon that—at some risk of repeating both myself and others—I nevertheless cannot refrain from making a few observations of the way in which this once quiet backwater has broken into a rushing torrent.

For a great many years, few biologists or biochemists either knew or cared about microbial lipids. What little information was available (and much of this later proved to be erroneous) seemed to suggest that bacteria and other microorganisms contained only minor amounts of unremarkable lipoidal substances, and it was generally assumed that these substances were similar to and metabolized like the lipids of higher plants and animals. These concepts linked with the fact that lipids then were damnably hard to work with caused the field to be notable for its inactivity.

Then, about 15 years ago, a variety of seemingly unrelated events associated with microbiological assays, studies of biotin metabolism, and so on, which have been detailed elsewhere[1,2] combined to quicken interest in microbial lipids, and there began to accumulate numerous indications that microbial lipids might be considerably more unusual and distinctive than had been suspected. The first findings resulting from this *nouvelle vague* of investigation were so unexpected and so provocative that they understandably

References p. 263

stimulated more work by more people, and so the ripples spread. In an astonishingly well-timed manner, a number of valuable techniques including gas–liquid and thin-layer chromatography became available which greatly facilitated work in this field which had so long been so frustratingly difficult from an experimental standpoint. This happy concatenation of circumstances led to what now seems like an exponential increase in the study and understanding of microbial lipids in recent years. This simultaneous stirring of interest and the development of productive experimental methods is a prime example, I think, of a field "whose time had come".

We have gone through three overlapping and still coexisting stages. The first was the determination of the exact chemical nature of microbial lipids and the recognition that many of these compounds were unique to microorganisms. The second was the realization that unique compounds meant unique biosynthetic pathways, and the search for an elucidation of such pathways. The third was the realization that hitherto unrecognized lipids might have hitherto unrecognized functions. All three of these stages are now well advanced and are understood to an extent roughly in proportion to the order in which they are named.

Now we are in a sort of band-wagon phase in which there are literally more reviews each year than there used to be papers. By my own and undoubtedly incomplete count, in 1966 there were almost 600 papers published in scientific journals dealing with or germane to microbial lipids. 15 years ago, a lively graduate student could recite in a short period of time the entire literature of the field! So complex has this subject become that it is no longer possible in anything short of a monograph of book length to cover all of its aspects—that is, the chemistry and metabolism of all kinds of lipids in all kinds of microorganisms. There are available a number of books, monographs, and reviews which alone or in combination do give nearly comprehensive coverages as of approximately a year preceding their publication dates[2-7].

The purpose of this present writing is more limited, namely to survey what we now know—or believe—regarding the metabolism of lipids by the bacteria which will arbitrarily be considered to include the *Eubacteriales* and the *Pseudomonadales* with some reference to the mycobacteria. Even in so limited a coverage as *just* metabolism and *just* bacteria it is not possible, and probably not even desirable, to be all inclusive. Therefore, the following is explicitly intended to a synopsis of current understanding rather than an omnibus of attribution.

2. Unique aspects of bacterial lipids

The lipids of bacteria exhibit many unusual and distinctive characteristics with respect to their chemical nature, their metabolism, and their cellular functions. Indeed, bacterial lipids differ so markedly from those of higher plants and animals that they have become increasingly interesting to comparative biochemists, students of evolution, and even taxonomists.

In higher forms of life, one sees relatively few types of lipids metabolized by relatively few pathways, and serving relatively few functions. In contrast, the bacterial lipids seem very complex and diversified. To speak somewhat teleologically, it seems as if bacteria are "trying" many different lipid compounds and pathways before "settling down", as higher forms have, to relatively few compounds and pathways that are optimal for their purposes. Thus, bacteria contain numerous lipids not seen in higher forms, lack compounds that are found in plants and animals, use the same lipids for different purposes and different lipids for the same purposes as more advanced forms. Finally, bacteria produce some lipids that have no purpose —or, more probably, for which no use has yet been discovered.

A large number of such peculiarities are now known, many of which will be discussed in some detail later. For the present it will suffice to cite only a few of the more notable examples:

(1) Bacteria contain many fatty acids not found or not common in higher forms. This is particularly true of the specific isomers of monoenoic acids, branched-methyl acids, hydroxy acids and fatty acids containing cyclopropane rings. Also, while higher plants and animals usually contain only a few fatty acids, bacteria often have 20 or more such compounds.

(2) Bacteria possess various biosynthetic pathways for fatty acids not seen in higher forms. This is necessarily true for those acids found only in bacteria, but in some cases even the acids or types of acids found in higher forms are synthesized differently in bacteria.

(3) With rare exceptions, the true bacteria neither contain nor require polyunsaturated acids.

(4) In most bacteria, lipids are concentrated in the outer layers of the cell, especially the cytoplasmic membrane, rather than in the cytoplasm.

(5) Lipid inclusions in bacterial cytoplasm often prove to consist of fatty acid polymers instead of triglycerides.

(6) In many if not most bacteria, lipids are not used for energy storage. On the contrary, bacterial lipid content may actually increase with age or stress.

References p. 263

(7) Bacterial phospholipids are usually phosphatidyl serines and phosphatidyl ethanolamines. Less commonly, one sees phosphatidic acids and the mono- and di-methyl derivatives of phosphatidyl ethanolamines. Phosphatidyl cholines, or lecithins, while they do occur in bacteria are much less common than they are in higher forms.

(8) As is the case with bacterial fatty acid biosynthesis, the formation of phospholipids in bacteria often is accomplished by reactions not found or not usual in higher forms.

(9) Steroid contents of true bacteria are negligible to nil.

Such a list, while it could be extended, suffices to exemplify the many notable aspects of bacterial lipids. There are various reasons why these aspects have attracted as much interest as they have. Of course, there is the obvious consideration that they are a major part of the overall nature of bacteria as we see them today, and must be understood in order to understand bacteria in their entirety just as contemporary life forms. There is the interest that these substances and their metabolism have for students of phylogeny. Certainly the unique features of bacterial lipids serve as valuable signposts to and along the pathways of metabolism. The discovery of unsuspected lipoidal compounds has frequently led to unsuspected biosynthetic mechanisms for producing those compounds. Sometimes these mechanisms were limited to bacteria, sometimes once they were known to exist they were then sought for and found in other forms of life. Now that we are beginning to have a reasonably clear concept of the chemistry and biosynthesis of bacterial lipids, we naturally are becoming more and more interested in the functions and catabolism of these compounds. As yet we know relatively little of bacterial lipid function and catabolism, but findings in these areas cannot fail to be at least as interesting as what we have learned so far.

One last consideration in which the nature of bacterial lipids may be important to man is their possible role in the action of antibiotics. Ideally antibiotics are of use because they inhibit the growth of microorganisms while not affecting host cells; and ideally this selectivity depends on metabolic activities that are non-existent or non-essential in host cells. Right now there is a great deal of empiricism in the development and use of antibiotics, and we have only limited understanding of how most of them actually work. If we had clearer understanding of just what metabolic activities are unique to microorganisms, we might be able to do a much more organized and logical job of developing antibiotics, both natural and synthetic. As the preceding paragraphs have shown, one area in which bacterial cells differ most markedly from mammalian cells is the nature of their lipids, and it may be that there are many unique and essential features of bacterial lipids that are susceptible to specific inhibition by substances that would, by definition, be antibiotic. Indeed, it may well be that some antibiotics already known suppress bacterial growth because they interfere in some presently unsuspected way with the formation or function of lipids peculiar to these microorganisms. Thus increasing knowledge in this field gives promise of being of considerably more than just—accursed phrase!—academic interest.

3. Biosynthesis of fatty acids

(a) Saturated straight-chain fatty acids

It has by now become well known that the biosynthesis of saturated straight-chain fatty acids, particularly those with even numbers of carbon atoms, is accomplished by means of the malonyl-coenzyme A pathway. The details of this pathway have been contributed by many workers among whom Vagelos[8] and Wakil[9] have been particularly prominent. Much of the work was originally done in animal systems, but there is extensive evidence that the malonyl pathway occurs widely in bacteria[6,7,10].

Since it was long believed that fatty acid biosynthesis was probably merely the reversal of the oxidative pathway, it might be useful to list the major points of difference between the degradation and biosynthesis of fatty acids.

(1) In biosynthesis, the basic "adding unit" is malonyl-CoA, not acetyl-CoA.

(2) Biosynthesis employs NADPH whereas NAD is involved in oxidation.

(3) In biosynthesis, the hydroxy acids involved are the D(−)-β isomers rather than the L(+)-β isomers that occur in oxidation.

(4) Acyl moieties are connected to protein (acyl-carrier protein or ACP) rather than to coenzyme A during the biosynthetic reactions.

The essential steps involved in this pathway are as follows:

$$HCO_3^- + ATP + \text{acetyl-CoA carboxylase} \underset{}{\overset{Mn^{2+}}{\rightleftharpoons}} CO_2\text{-acetyl-CoA carboxylase} + ADP + P_i$$
$$\text{(biotin-containing)}$$

References p. 263

CO_2-acetyl-CoA carboxylase + CH_3CO-S-CoA $\rightleftharpoons$
(acetyl-CoA)

acetyl-CoA carboxylase + $HOOCCH_2CO$-S-CoA
(malonyl-CoA)

Malonyl-CoA so generated and acetyl-CoA are then used to produce the corresponding acyl-carrier protein (ACP) complexes. Thereafter, all reactions involve intermediates in which acyl groups are covalently bound to ACP.

CH_3CO-S-CoA + ACP-SH $\xrightleftharpoons{\text{Enz. 1}}$ CH_3CO-S-ACP + CoA-SH
(acetyl-CoA) (acetyl–ACP)

$HOOCCH_2CO$-S-CoA + ACP-SH $\xrightleftharpoons{\text{Enz. 2}}$ $HOOCCH_2CO$-S-ACP + CoA-SH
(malonyl-CoA) (malonyl–ACP)

CH_3CO-S-ACP + $HOOCCH_2CO$-S-ACP $\xrightleftharpoons{\text{Enz. 3}}$
CH_3COCH_2CO-S-ACP + ACP-SH + CO_2
(acetoacetyl–ACP)

CH_3COCH_2CO-S-ACP + NADPH + H^+ $\xrightleftharpoons{\text{Enz. 4}}$
D(−)-$CH_3CHOHCH_2CO$-S-ACP + $NADP^+$
(hydroxybutyryl–ACP)

D(−)-$CH_3CHOHCH_2CO$-S-ACP $\xrightleftharpoons{\text{Enz. 5}}$ $CH_3CH=CHCO$-S-ACP + H_2O
(crotonyl–ACP)

$CH_3CH=CHCO$-S-ACP + NADPH + H^+ $\xrightleftharpoons{\text{Enz. 6}}$
$CH_3CH_2CH_2CO$-S-ACP + $NADP^+$
(butyryl–ACP)

The enzymes involved in this sequence of reactions are: Enzyme 1, acetyl transacylase; Enzyme 2, malonyl transacylase; Enzyme 3, β-ketoacyl–ACP synthetase; Enzyme 4, β-ketoacyl–ACP reductase; Enzyme 5, enoyl–ACP hydrase; Enzyme 6, crotonyl–ACP reductase.

At the conclusion of the events just described, a 4-carbon acyl fragment has been generated. The butyryl–ACP can then be "recycled" and condensed with another malonyl–ACP to give a 6-carbon fragment, the 6-carbon fragment recycled to give an 8-carbon fragment, and so on until the long-chain fatty acids found in bacterial cells are formed. In such organisms, acids ranging from 8 to 24 carbon atoms in chain length have been reported; however, the principal acids in by far the majority of bacteria are palmitic, myristic and stearic acids in approximately that order of abundance. A few organisms, notably the corynebacteria and the mycobacteria contain very-long-chain acids with 32 or even more carbon atoms. It is not yet clear whether these very large acids are also formed by the malonyl pathway. Once suitable chain lengths are attained, presently available data suggest that the long-chain fatty acid–ACP complexes are used in the biosynthesis of the various cellular lipids. Only small amounts of fatty acids seem to be released as free acids by hydrolysis of the acyl–ACP complexes.

The mechanism that regulates the termination of the malonyl pathway and that thereby controls the relative amounts of different saturated fatty acids in bacterial cells is as yet unknown. There is some degree of species specificity in this regulatory mechanism since under carefully controlled conditions one can expect to detect, within some limits of variation, certain ratios of fatty acids in given species of bacteria. However, this regulation must not be too stringent or critical because it is easy to alter considerably the proportions of different straight-chain fatty acids in a given bacterial species merely by minor modifications of growth conditions. Indeed, the ease with which one can alter, wittingly or unwittingly, the percentage composition of the total fatty acid content of bacterial cells is a major problem in the study of such substances.

A particularly striking aspect of this biosynthetic pathway is the acyl-carrier protein. This substance, which we are now coming to recognize as being of great importance in various areas of lipid biosynthesis, was completely unknown only a few years ago. It is a very stable protein being unaffected by 0.1 N HCl at 25° and even by boiling water[6]. Its minimum molecular weight based on amino acid analysis is 9488, and calculations based on a sedimentation constant of 1.44 give a molecular weight[6] of 9750. As pointed out be Lennarz[6] and others, the functional group of ACP

References p. 263

is very similar to the structure of coenzyme A, as shown below.

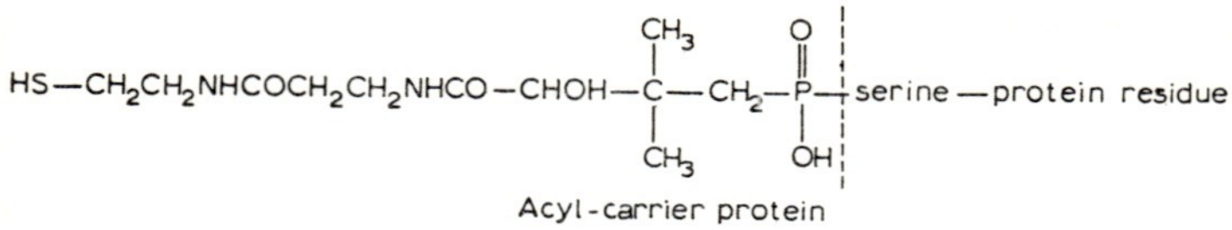

$$HS-CH_2CH_2NHCOCH_2CH_2NHCO-CHOH-\underset{\underset{CH_3}{|}}{\overset{\overset{CH_3}{|}}{C}}-CH_2-\underset{\underset{OH}{|}}{\overset{\overset{O}{\|}}{P}}-serine-protein\ residue$$

Acyl-carrier protein

$$HS-CH_2CH_2NHCOCH_2CH_2NHCO-CHOH-\underset{\underset{CH_3}{|}}{\overset{\overset{CH_3}{|}}{C}}-CH_2-\overset{\overset{O}{\|}}{\underset{\underset{OH}{|}}{P}}-O-\overset{\overset{O}{\|}}{\underset{\underset{OH}{|}}{P}}-O-CH_2-\ldots adenine$$

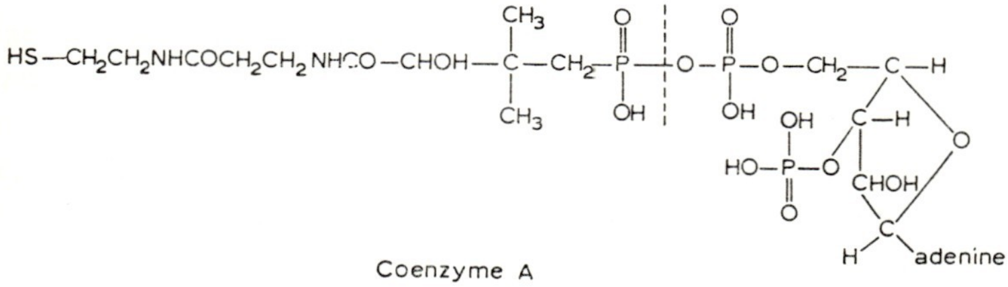

Coenzyme A

Not only are the acyl–ACP complexes utilized as intermediates in saturated fatty acid biosynthesis in bacteria, but the acyl–CoA esters are not utilized in such biosynthesis other than in the early steps where acetyl- and malonyl-CoA complexes are utilized.

One other means of producing saturated fatty acids in bacteria should be mentioned before leaving this section, and that is the ability of a few organisms to biohydrogenate unsaturated acids. This activity has been discovered and studied in populations of rumen bacteria[11,12]. The main fatty acid converted by biohydrogenation to saturated fatty acid is linolenic acid which is introduced into the rumen in the natural pasture diet of ruminants. This is converted by rumen bacteria in a stepwise fashion to dienoic, monoenoic, and finally stearic acid. *In vitro* studies have shown that linoleic and oleic acids can also serve as substrates, albeit poorer ones, for this reaction. Many details regarding the specific isomeric transitions of the substrate have been elucidated and are set forth at length in the references cited above, but little is yet known of the enzymes, cofactors, or organisms involved. It is recognized that there is a soluble, heat-stable accessory factor in cell-free rumen liquor that greatly accelerates the reaction when added to washed cells of rumen bacteria, but this has not been identified. While mixed populations of rumen bacteria actually obtained from fistulated ruminants are very active in such biohydrogenations, experiments with pure cultures of selected alimentary microorganisms have given poor results. However, our knowledge of rumen flora and biochemistry is exceedingly sketchy, and it could well be that we simply have not yet investigated—or even identified—

the species primarily responsible for this reduction of unsaturated fatty acids. Also, it could be that this is a synergistic phenomenon and that maximal activity is observed only in the presence of the correct proportions of the correct organisms.

(b) *Unsaturated fatty acids*

The formation of unsaturated fatty acids in bacteria is an interesting and complex matter and has been treated in considerable detail elsewhere[2,13]. Basically, there are two modes of synthesis for these compounds—elongation of already unsaturated short-chain intermediates, and dehydrogenation of long-chain saturated fatty acids. It was long assumed that the latter method was the sole means by which bacteria formed their monoenoic acids simply because this was the operation known to occur in higher plants and animals. However, extensive investigations at the University of Pittsburgh in the 1950's showed that few true bacteria could desaturate long-chain saturated acids, and that they instead made successive 2-carbon additions to short-chain monoenoic acids. Subsequent investigations at other laboratories, particularly Harvard, verified and extended these observations. In the Pittsburgh studies, the first identifiable compound in this series of reactions was *cis*-3,4-decenoic acid. Eventually it was found that the source of the decenoic acid was the β,γ-dehydration of 3-hydroxydecanoic acid as shown below.

$$CH_3(CH_2)_6-\underset{\underset{H}{|}}{\overset{\overset{OH}{|}}{C}}-CH_2COOH \xrightarrow{\beta,\gamma\text{-dehydration}} CH_3(CH_2)_5CH=CHCH_2COOH$$

cis-3,4-decenoic acid

3-hydroxydecanoic acid

$$\quad\quad\quad\quad\quad\quad\quad\quad\quad\quad\quad\quad\quad\quad\quad\quad\quad\quad\quad\quad\quad\quad\quad\quad\quad +3\ C_2\ \downarrow$$

$$CH_3(CH_2)_5CH=CH(CH_2)_9COOH \xleftarrow{+1\ C_2} CH_3(CH_2)_5CH=CH(CH_2)_7COOH$$

cis-11,12-octadecenoic acid   *cis*-9,10-hexadecenoic acid
  (*cis*-vaccenic acid)   (palmitoleic acid)

It will be noted that the product of the reaction sequence just described will be *cis*-vaccenic acid (*i.e.*, Δ^{11}) rather than oleic acid (Δ^9) which is more

References p. 263

common in higher forms. If the same series of reactions is begun with 3-hydroxydodecanoic acid, the end-product will be oleic acid. Indeed, using this basic pathway and a variety of initial substrates, it has been possible to construct a generalized concept for the formation of monoenoic acids by bacteria[13-15] which is shown in the following diagram. It can be seen that many of the compounds "predicted" by this concept have actually been found in one bacterial species or another.

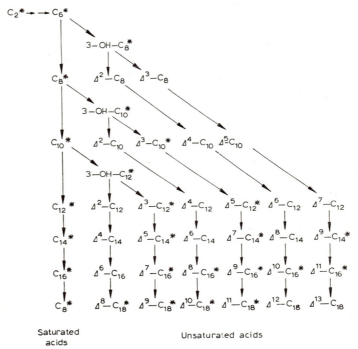

Fig. 1. Bacterial biosynthesis of monoenoic acids by desaturation and elongation. (* = Compounds already detected in bacteria.)

The discovery of *cis*-vaccenic acid in bacteria was somewhat surprising since biologists had become accustomed to think of oleic acid as *the* octadecenoic acid. We now know that many if not most bacteria contain *cis*-vaccenic acid and its precursors instead of oleic acid and its precursors, although some contain both types of acids and some contain only oleic acid and its precursors. These two groups of acids differ basically in the structure of the methyl ends of their chains, and on this basis it has proven useful to speak of the *cis*-vaccenic series and the oleic series which are shown in Table I.

TABLE I

Fatty acid	Structure
The cis-vaccenic series	
cis-3-decenoic	$CH_3\text{-}(CH_2)_5\text{-}\underset{H}{C}=\underset{H}{C}\text{-}CH_2\text{-}COOH$
cis-5-dodecenoic	$CH_3\text{-}(CH_2)_5\text{-}\underset{H}{C}=\underset{H}{C}\text{-}(CH_2)_3\text{-}COOH$
cis-7-tetradecenoic	$CH_3\text{-}(CH_2)_5\text{-}\underset{H}{C}=\underset{H}{C}\text{-}(CH_2)_5\text{-}COOH$
cis-9-hexadecenoic	$CH_3\text{-}(CH_2)_5\text{-}\underset{H}{C}=\underset{H}{C}\text{-}(CH_2)_7\text{-}COOH$
cis-11-octadecenoic	$CH_3\text{-}(CH_2)_5\text{-}\underset{H}{C}=\underset{H}{C}\text{-}(CH_2)_9\text{-}COOH$
The oleic series	
cis-3-dodecenoic	$CH_3\text{-}(CH_2)_7\text{-}\underset{H}{C}=\underset{H}{C}\text{-}CH_2\text{-}COOH$
cis-5-tetradecenoic	$CH_3\text{-}(CH_2)_7\text{-}\underset{H}{C}=\underset{H}{C}\text{-}(CH_2)_3\text{-}COOH$
cis-7-hexadecenoic	$CH_3\text{-}(CH_2)_7\text{-}\underset{H}{C}=\underset{H}{C}\text{-}(CH_2)_5\text{-}COOH$
cis-9-octadecenoic	$CH_3\text{-}(CH_2)_7\text{-}\underset{H}{C}=\underset{H}{C}(CH_2)_7\text{-}COOH$

Interestingly, there are some discrepancies in our present concept of the elongation mode of unsaturated fatty acid biosynthesis. For one, the rate of dehydration of 3-hydroxyacyl-CoA complexes is slower than the *de novo* synthesis of monoenoic acids from malonyl- and acetyl-CoA complexes. Also, decenoyl-CoA will not serve as a precursor of long-chain monoenoic acids. These observations were once regarded with some uneasiness since they seemed at variance with the overall concept of the elongation mode of unsaturated acid synthesis. However, recent evidence indicates that this is merely one more function of the seemingly ubiquitous acyl-carrier protein because 3-hydroxyacyl and β,γ-unsaturated acyl–ACP derivatives do serve as efficient precursors of long-chain unsaturated fatty acids in bacteria[16,17].

The elongation of already unsaturated short-chain acyl derivatives to form long-chain unsaturated acids has frequently been referred to as the

References p. 263

"anaerobic pathway" because it does not involve the direct utilization of molecular oxygen as does the dehydrogenation mechanism found in higher life forms. The dehydrogenation pathway, as I have already mentioned, was the original, unevidenced, "by analogy", and almost intuitive guess regarding the way in which microorganisms formed unsaturated acids. Then there was the flurry of activity in which the elongation or anaerobic pathway was elucidated and for a time regarded as the sole route to such acids in bacteria. Now we have come full circle, and it is clear that at least some microorganisms do, after all, dehydrogenate saturated long-chain fatty acids to form the analogous long-chain monoenoic acids. As this pathway does involve the utilization of molecular oxygen (and NADPH), it has been called the aerobic pathway*.

Bloch and his associates[18] first convincingly demonstrated this reaction in the yeast *Saccharomyces cerevisiae*, and in this reaction it is the long-chain acyl-CoA derivatives that serve as substrates and intermediates. A typical reaction of this type is shown below. It is noteworthy that the products derived from this type of reaction are predominantly of the oleic series, so

$$CH_3(CH_2)_{16}CO\text{-}SCoA \xrightarrow[O_2]{NADPH} CH_3(CH_2)_7CH=CH(CH_2)_7CO\text{-}SCoA$$
stearyl-CoA $\qquad\qquad\qquad\qquad\qquad$ oleyl-CoA

that we see a general association of the oleic series compounds with the dehydrogenation mechanism, and the *cis*-vaccenic series compounds and the elongation mechanism with, of course, some exceptions to both generalizations.

A great deal of information has now been accumulated on the frequency of each mode of monoenoic fatty acid biosynthesis among different microorganisms, particularly by Erwin and Bloch[14]. What we now see is that among the true bacteria the most common mechanism is the elongation procedure whereas the dehydrogenation procedure is more common among the higher microorganisms including the mycobacteria and other actinomycetes, yeast, fungi, algae, protozoa and metazoa. As always, there are exceptions to this generalization, and we do find that some true bacteria

---

* I should remark parenthetically that I am reluctant to employ this aerobic/anaerobic terminology with respect to monoenoic acid biosynthesis since these words have so many shadings of meaning in today's biosciences. For example, aerobic microorganisms (with respect to oxygen requirements during cultivation) may conduct "anaerobic" biochemical reactions, and so on. However, what's done is done, and I bow to current usage.

produce unsaturated fatty acids by the desaturation mechanism but these are very few as compared with those that elongate. Thus we might make the following general comparisons:

EUBACTERIA	HIGHER FORMS AND A FEW EUBACTERIA
cis-vaccenic series unsaturated acids	oleic series unsaturated acids
elongation mechanism	dehydrogenation mechanism

The only true bacteria so far shown to utilize the dehydrogenation mechanism are *Micrococcus lysodeikticus*, *Corynebacterium diphtheriae* (which accumulating evidence suggests is incorrectly classified as a true bacterium, being much more similar to the mycobacteria and other actinomycetes), and a strain of *Bacillus megaterium*.

While it does anticipate a subsequent section of this discussion, it should be mentioned here that a major function of long-chain unsaturated fatty acids in many bacteria—not *all*, but many— is to serve as the immediate precursors of cyclopropane fatty acids. The latter compounds are quite unique substances in which a 3-carbon ring is located at the same site as the double bond in monoenoic acids. Two carbons of this ring are those on each end of the double bond; the third methylene group forming this ring is derived from the methyl group of methionine by a process somewhat resembling transmethylation[19]. This will be discussed in detail in a later portion of this chapter.

(c) *Cyclopropane fatty acids*

The cyclopropane fatty acids are among the most distinctive features of bacterial lipids, their occurrence elsewhere being limited—as far as we now know—to only a few plant oils. In contrast, in the true bacteria these acids are very common although not universal constituents. The general nature of their structure is shown below (in all examples so far encountered these acids have the *cis* configuration). Several such acids have been reported to

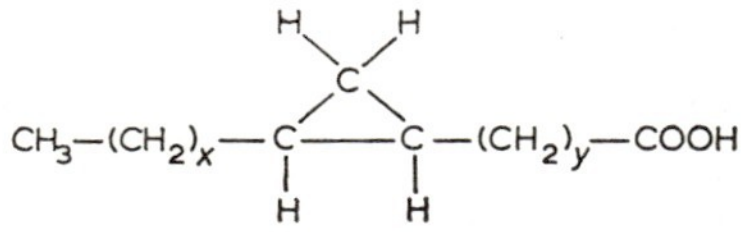

$$CH_3-(CH_2)_x-\overset{H}{\underset{H}{C}}\overset{\diagdown\,C\,\diagup}{\underset{}{}}\overset{H}{\underset{H}{C}}-(CH_2)_y-COOH$$

occur in various bacterial species ranging in carbon content from C_{11} to C_{19}, but the two most common compounds seen are *cis*-11,12-methylene octadec-

References p. 263

anoic acid ($x=5$, $y=9$) and cis-9,10-methylene hexadecanoic acid ($x=5$, $y=7$). The 19-carbon acid was the first discovered and was given the common name of lactobacillic acid for the organisms in which it first was detected. Both of these are clearly related to the *cis*-vaccenic series, and the reader would think it likely that the C_{17} cyclopropane acid is derived from palmitoleic acid while the C_{19} cyclopropane acid is derived from *cis*-vaccenic acid, and such is exactly the case. Radioisotope studies have shown that the chains of cyclopropane acids are entirely derived from analogous monoenoic acids; *i.e.*, cyclopropane acids consist of unsaturated acids with a methylene bridge built over the double bond[20]. Thus, all but one carbon atom in such acids can be accounted for by the immediate precursor which is a monounsaturated acid. What then of the last or ring carbon? In the case of lactobacillic acid, this was shown to be derived from the methyl carbon of methionine in the form of S-adenosylmethionine[21,22]. Subsequently this was shown to be a general mechanism, that cyclopropane acids were formed by something like transmethylation on the double bond of monounsaturated acids. While this was unequivocally so on the basis of many studies, there still remained at least one problem which was this:

$$-\underset{\underset{H}{|}}{C}=\underset{\underset{H}{|}}{C}- \quad + \quad H-\underset{\underset{H}{|}}{\overset{\overset{H}{|}}{C}}- \quad \longrightarrow \quad [?] \quad \longrightarrow \quad -\underset{\underset{H}{|}}{\overset{\overset{H}{\diagup}\overset{C}{\diagdown}\overset{H}{}}{C}}\!\!-\!\!\underset{\underset{H}{|}}{C}- \quad + \quad H(?)$$

Monoenoic   Methyl   Cyclopropane
acid   group of   acid
            methionine

This discrepancy of one hydrogen atom was the reason for the general uneasiness: was the "extra" hydrogen discarded from the monoenoic acid or from the methyl group of methionine (*i.e.*, S-adenosyl-methionine)? This inability to account for all the hydrogen atoms in a single reaction seemed to suggest the existence of more than one reaction and intermediate. Much work was expended on this problem by various laboratories[1,2,7,19] without, frankly, much progress. Finally, a series of studies by Law and his associates[23,24] showed that the chain hydrogens of the cyclopropane ring were those of the monoenoic precursor and that the methylene hydrogens (*i.e.*, those on the third ring carbon) were derived from the methionine methyl group. It would seem, then, that the "lost" hydrogen is one of those on the methionine methyl group, although just what becomes of it is still not known. In spite of this seemingly advanced understanding of this biosynthesis, there are some nagging details that need to be worked out (see, e.g., ref. 2).

The cyclopropane acids are singularly uncommon substances: they are, to start with, compounds virtually unique to bacteria; they are formed by transmethylation-like reactions not seen in other forms of life; they have, as yet, no discernible function in bacteria and yet they appear to be indispensible to those organisms that contain them[1-3]. Finally, they are biosynthesized in a way that has no parallel in the biological world. It appears that bacteria can only synthesize cyclopropane acids if the precursor monoenoic acids are already incorporated into a phospholipid in the manner illustrated below. This procedure is markedly affected by the physical state

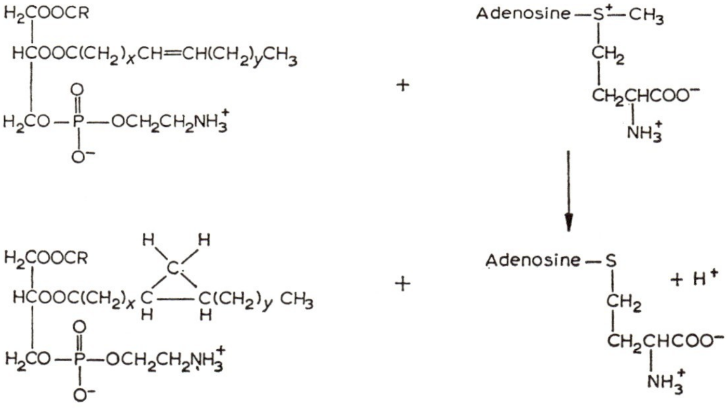

$$
\begin{array}{c}
H_2COOCR \\
HCOOC(CH_2)_xCH=CH(CH_2)_yCH_3 \\
H_2CO-\overset{O}{\underset{O^-}{P}}-OCH_2CH_2NH_3^+
\end{array}
\quad + \quad
\begin{array}{c}
Adenosine-S^+-CH_3 \\
CH_2 \\
CH_2CHCOO^- \\
NH_3^+
\end{array}
$$

$$\downarrow$$

$$
\begin{array}{c}
H_2COOCR \\
HCOOC(CH_2)_x\overset{H}{\underset{H}{C}}\!\!\!-\!\!\!\overset{H}{\underset{H}{C}}(CH_2)_yCH_3 \\
H_2CO-\overset{O}{\underset{O^-}{P}}-OCH_2CH_2NH_3^+
\end{array}
\quad + \quad
\begin{array}{c}
Adenosine-S \\
CH_2 \\
CH_2CHCOO^- \\
NH_3^+
\end{array}
\quad + H^+
$$

of the phospholipid. It is necessary for the substrate to be in micellar suspension, and anionic surfactants greatly stimulate the rate of formation of cyclopropane acid (although cationic and neutral surfactants are actually inhibitory). Also, temperature changes strongly affect this biosynthesis possibly because of changes in the physical states of the phospholipids. To further complicate this picture, Chung and Law[25] noted that purification of crude cyclopropane synthetase preparations seemed to cause a decrease in enzyme activity. This anomaly was resolved when it was found that the S-adenosylhomocysteine generated during this reaction is a potent inhibitor of the methylene addition step. In the crude preparations there was a hydrolase that destroyed this inhibitor as it formed, but during purification of the synthetase, this hydrolase was removed permitting S-adenosylhomocysteine to accumulate and interfere with the biosynthesis of cyclopropane acids.

References p. 263

Cyclopropane synthetase exhibits considerable specificity with respect to phospholipid substrate, location of the fatty acid in the phospholipid, and the nature of the cyclopropane product. In most bacteria, the phospholipid involved in this reaction is phosphatidyl ethanolamine. Phosphatidyl glycerol and phosphatidic acid can also function in this capacity, but phosphatidyl serine and phosphatidyl choline function only poorly. Within the phospholipid substrate, it has been found that in most organisms the unsaturated and cyclopropane acids are located at the β-position[26]. The cyclopropane acids formed by this mechanism are very predominantly the C_{17} and C_{19} compounds. C_{11}, C_{13} and C_{15} cyclopropane acids have also been reported, but these are believed to occur in only a few species and in quite small concentrations.

(d) Branched-chain fatty acids

Both iso and anteiso types of branched methyl fatty acids are found in bacteria and are particularly common in such Gram-positive organisms as the members of the genera *Sarcina*, *Micrococcus*, and *Bacillus*; in the *Pseudomonadales*; and in various rumen bacteria.

In many organisms it has been shown that iso and anteiso acids are formed by the elongation of short-chain branched acids. Thus, isovalerate (derived from leucine) is elongated to form iso-C_{15} and -C_{17} acids, and anteiso-C_{15} and -C_{17} acids can be formed by elongation of 2-methylbutyrate derived from isoleucine[7].

$$\begin{array}{c} CH_3 \\ | \\ CH_3CHCH_2CO\text{-}SCoA \\ \text{isovalerate} \end{array} \begin{array}{c} +5\ C_2 \\ \nearrow \\ \searrow \\ +6\ C_2 \end{array} \begin{array}{c} CH_3 \\ | \\ CH_3CH(CH_2)_{11}CO\text{-}SCoA \\ \\ CH_3 \\ | \\ CH_3CH(CH_2)_{13}CO\text{-}SCoA \end{array}$$

$$\begin{array}{c} CH_3 \\ | \\ CH_3CH_2CHCO\text{-}SCoA \\ \text{2-methylbutyrate} \end{array} \begin{array}{c} +5\ C_2 \\ \nearrow \\ \searrow \\ +6\ C_2 \end{array} \begin{array}{c} CH_3 \\ | \\ CH_3CH_2CH(CH_2)_{10}CO\text{-}SCoA \\ \\ CH_3 \\ | \\ CH_3CH_2CH(CH_2)_{12}CO\text{-}SCoA \end{array}$$

Iso-C_{14} and -C_{16} acids have been produced from isobutyrate by some species. The elongation mechanism involved in these pathways is believed to be connected with the malonate scheme discussed above. If this is so, it is likely that the various intermediates in the branched-chain fatty acid biosyntheses are actually ACP esters rather than CoA esters as have so far been reported by students of these reactions.

The mycobacteria produce their varied branched acids by entirely different means. One of these is typified by the biosynthesis of tuberculostearic acid (10-methylstearic) by *M. phlei* and other species. These organisms dehydrogenate stearic acid to oleic, and then transfer a C_1 group from the methyl group of methionine in a manner reminiscent of the way in which cyclopropane acids are formed. Similarly, if palmitic acid is used as the initial

$$CH_3(CH_2)_{16}COOH \xrightarrow{-2H} CH_3(CH_2)_7CH=CH(CH_2)_7COOH$$
stearic oleic

$+ C_1$ from methionine

$$CH_3(CH_2)_7\overset{\overset{\displaystyle CH_3}{|}}{C}HCH_2(CH_2)_7COOH$$
tuberculostearic
(10-methylstearic)

substrate, the product is methylpalmitate. It would seem that this should be a straightforward transfer of the methionine methyl group to the unsaturated fatty acid, and this was believed for some time. However, recently some disquieting information has come to light which suggests that there may be more to this than was originally thought. Isotope studies conducted by Jauréguiberry and his associates[27] indicate that in the methyl group of the branched fatty acid, only two of the three hydrogens are derived from the methionine methyl group; that somehow one methionine methyl hydrogen is lost and replaced by one from another source. How or why this happens is not yet known.

Still another method for making branched methyl fatty acids has been found in the mycobacteria. Tubercle bacilli contain an assortment of fatty acids with multiple-branched methyl groups called mycocerosic acids. The most prominent of these is a 32-carbon compound, 2,4,6,8-tetramethyl-

References p. 263

octacosanoic acid. Etémadi and Lederer[28] have suggested that this is formed by the successive addition of 4 molecules of propionic acid to a straight-chain saturated C_{20} acid.

$$CH_3(CH_2)_{18}COOH + 4\ \underset{\underset{CH_3}{|}}{CH_2COOH} \quad \Bigg\downarrow$$

$$CH_3(CH_2)_{19}\text{-}\underset{\underset{CH_3}{|}}{CH}CH_2\text{-}\underset{\underset{CH_3}{|}}{CH}CH_2\text{-}\underset{\underset{CH_3}{|}}{CH}CH_2\text{-}\underset{\underset{CH_3}{|}}{CH}COOH$$

Phthienoic acids which are similar but also β,γ-unsaturated acids are believed to be formed in a similar way.

(e) Hydroxy fatty acids

Hydroxy acids, particularly the D-(−)-3-hydroxy compounds are widely distributed in bacteria. They are frequently associated with surface and extracellular lipids such as the cell-envelope lipopolysaccharides of Gram-negative bacilli and the extracellular rhamnolipid of *Pseudomonas aeruginosa*. While similar hydroxy acids do occur in the malonyl pathway to saturated acids, they seem to remain bound to carrier protein and therefore not accumulate. Presumably the hydroxy acids that do accumulate in bacterial cells are produced by some other mechanism, but there is virtually no information on such biosyntheses in the *Eubacteriales*. Kates[7] has recently summarized what information is available on such disparate forms as yeasts, mycobacteria, and slime molds.

(f) Fatty acids peculiar to mycobacteria and corynebacteria

Various lines of evidence have begun to suggest that the corynebacteria are more closely related to the mycobacteria than our present taxonomic schemes indicate. It may be that the corynebacteria should be removed from their present classification and placed in the *Actinomycetales* along with the mycobacteria. One of the areas in which the similarity between these two genera is most evident is that of lipid composition. Both these genera are characterized by a profusion of different and complex fatty acids not seen in other organisms. So bizarre and complicated are these substances that their study could well constitute a subspecialty in itself. Fortunately, this subject has been recently and thoroughly reviewed by Asselineau[4,5]. While this is far too large a subject to cover extensively here, and not strictly germane, some mention should be made at least of the major types of acids.

The mycocerosic acids were discussed above in connection with branched methyl fatty acids, as were the phthienoic acids. Another major category is that of the mycolic acids which contain both branching and hydroxyl groups, and are found in both corynebacteria and mycobacteria. The acid found in the corynebacteria, called corynomycolic acid is formed by the condensation of 2 molecules of palmitic acid as shown below. Similarly, if one palmitic molecule is replaced by one of palmitoleate, the product obtained would be the unsaturated corynomycolenic acid shown below.

$$CH_3(CH_2)_{14}COOH + CH_2COOH \rightarrow CH_3(CH_2)_{14}CHOHCHCOOH$$
$$\underset{\underset{CH_3}{|}}{\underset{|}{(CH_2)_{13}}} \qquad \underset{\underset{CH_3}{|}}{\underset{|}{(CH_2)_{13}}}$$

corynomycolic acid

$$CH_3(CH_2)_5CH=CH(CH_2)_9CHOHCHCOOH$$
$$\underset{\underset{CH_3}{|}}{\underset{|}{(CH_2)_{13}}}$$

corynomycolenic acid

A related fatty acid is the so-called α-smegmamycolic acid produced by *Mycobacterium smegmatis*. Actually there are several such acids (C_{76} to C_{79}), and this is the structure of one.

$$CH_3(CH_2)_{17}CH=CH(CH_2)_{13}CH=CHCH(CH_2)_{17}CHOHCHCOOH$$
$$\underset{CH_3}{|} \qquad\qquad \underset{C_{22}H_{45}}{|}$$

α-smegmamycolic acid

It has been found that the biosynthesis of this compound involves the utili-

References p. 263

zation of tetracosanoic acid as such for the 24 carbons at the carboxyl end of the chain, and that the methyl group is derived from the methyl carbon of methionine[28]. The source of the rest of the molecule is as yet obscure.

4. Biosynthesis of complex lipids

(a) Phospholipids

While our understanding of the chemical nature and biosynthesis of bacterial fatty acids is now fairly extensive if not quite complete, our knowledge of the lipid complexes into which these acids are assembled is much more limited. We have accumulated considerable information regarding the general types of lipids that are to be found in different bacterial species, and this information is well summarized by both Kates[13] and Ikawa[29].

Phospholipids are major components of most bacteria, often constituting 70–90% of the total cellular lipids, and a variety of such substances have been recognized in one microbial species or another including phosphatidic acid, phosphatidylcholine, phosphatidylethanolamine, phosphatidylserine, phosphatidylinositol, phosphatidylglycerol, and diphosphatidylglycerol. Of considerable interest in this connection is the fact that phosphatidylcholine which is the most common phospholipid in higher plants and animals is uncommon in bacteria with the exception of some *Pseudomonadales* and *Rhizobiaceae*. Phosphatidyl-N-mono- and -N,N-dimethylethanolamines are also found in a few species. While the occurrence and relative abundance of these compounds have been determined in a large number of organisms, much less is known of the details of their biosynthesis.

Pieringer and his associates[30,31] have investigated the formation of the "fundamental" phospholipid, phosphatidic acid, in *E. coli* and found two different reactions by which this type of lipid can be formed. One is the reaction of ATP and α,β-diglyceride catalyzed by a particulate enzyme to give phosphatidic acid. Similarly, the same enzyme preparation will convert monoglycerides to lysophosphatidic acids. The other pathway to phosphatidic acids involves the esterification of α-glycerol phosphate by acyl–CoA complexes or, as recent evidence suggests, by acyl–ACP complexes.

Phosphatidic acids formed by either mechanism then serve as the precursors of other bacterial phospholipids, at least in *E. coli*. As Kanfer and Kennedy[32] have shown, phosphatidic acid is converted to a lipid–nucleotide complex by reacting with cytidine triphosphate. The cytidine diphosphate diglyceride so formed can then be converted into a variety of phospholipids as shown in the following diagram.

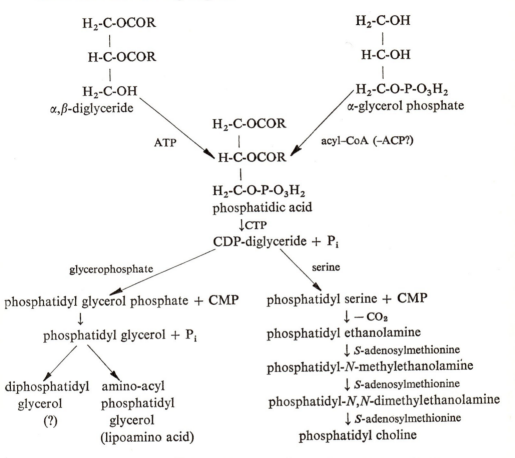

Chang and Kennedy[42] found that *E. coli* contains a diglyceride kinase that can convert diglyceride to phosphatidic acid, but saw no evidence of a metabolically active pool of diglyceride in these cells. They feel this suggests that glycerol phosphate is an obligate intermediate in the conversion of glycerol to lipid by this organism.

It should be noted that, as other authors have pointed out[6,33], there is no one bacterial species in which all these reactions or compounds have been observed. The sequential methylations of phosphatidyl ethanolamine are particularly uncommon which accounts for the fact that the intermediates

References p. 263

and final product, phosphatidyl choline, are found in only a few species. Conversely, while many organisms are capable of forming phosphatidyl serine, numerous species contain relatively little of this substance because they rapidly decarboxylate it to phosphatidyl ethanolamine.

The aminoacyl phosphatidyl glycerols (or "lipoamino acids") are currently the subject of considerable interest, investigation, and speculation. Their biosynthesis appears to involve two steps: (a) in the presence of ATP and Mg^{2+} a soluble enzyme catalyzes the combination of amino acid and sRNA forming aminoacyl–sRNA; then (b) a particulate enzyme facilitates the transfer of the aminoacyl fragment from the sRNA to phosphatidyl glycerol to yield an O-acyl ester[6]. The sRNA does not appear to be species-specific in this reaction since aminoacyl–sRNA preparations from, e.g., *E. coli* or even *Neurospora* can function in preparations produced from *S. aureus* cells. At the present time, the functions of aminoacyl phosphatidyl glycerols are both obscure and disputed. It has been suggested that they may be structural components of the cytoplasmic membrane; that they may be involved in cell wall formation; that they play some role in protein synthesis; that they help maintain membrane potential; or that they aid in membrane transport. Such observations as the low concentrations that are found in actively growing cells and the relatively few amino acids so far found to be incorporated in these substances further complicate hypotheses regarding their metabolic activities.

(b) Glycolipids

The evidence accumulated by several laboratories in recent years indicates that glycolipids are much more common constituents of bacterial cells than was formerly suspected. The types of compounds most frequently encountered are the mono- and diglycosyl diglycerides, and the carbohydrates involved are usually glucose, galactose, and mannose. Such substances are now known to occur in a wide variety of bacteria including species of *Micrococcus*, *Diplococcus*, *Streptococcus*, *Staphylococcus*, *Clostridium*, numerous Gram-negative bacteria, *Mycoplasma* and photosynthetic bacteria[7,34,35].

Lennarz[36,37] has investigated the biosynthesis of the dimannosyl diglyceride found in *Micrococcus lysodeikticus*. The first step in this synthesis involves the transfer of the carbohydrate moiety from GDP–mannose to α,β-diglyceride in the presence of Mg^{2+} to form O-α-D-mannosyl-$(1 \to \alpha')$

diglyceride. Studies using a particulate enzyme preparation rather than whole cells require the presence of a cationic detergent, the most effective one tested being the sodium salt of the C_{15} branched-chain fatty acid found in this organism. The mannosyl diglyceride once formed by the above process can then be converted to O-α-D-mannosyl-(1→3)-O-α-D-mannosyl-(1→α') diglyceride by a soluble enzyme requiring, again, GDP–mannose and Mg^{2+}.

[Reaction scheme: Diglyceride + GDP-mannose/Mg^{2+} → Mannosyl diglyceride + GDP-mannose/Mg^{2+} → Mannosylmannosyl diglyceride]

Studies of the mono- and diglycosyl diglycerides found in *Pneumococci*[38,39] indicate that they also are formed by a similar transfer of glycosyl moieties from nucleotides to diglycerides, but with certain differences. For one, *Pneumococci* make use of UDP complexes rather than the GDP derivatives utilized by *M. lysodeikticus*. Also, the carbohydrates found in pneumococcal glycolipids are glucose and galactose instead of mannose. Thus, following the general scheme described above, *Pneumococci* first form glucosyl diglyceride from UDP–glucose and diglyceride. Then galactose is transferred from UDP–galactose to the glucosyl diglyceride to give galactosylglucosyl diglyceride.

Many other interesting glycolipids have been detected and identified in bacteria. For example, in *Halobacterium cutirubrum* Kates and his associates[7] have reported the occurrence of a dihydrophytyl diether analogue of a glycosylsulfate diglyceride. The carbohydrate portion of this molecule is a trisaccharide containing galactose, glucose and mannose. The sulfate is

References p. 263

esterified to the C-3 of galactose. ^{14}C from labeled acetate and mevalonate appears primarily in the dihydrophytyl groups; ^{14}C from labeled glycerol appears in the glycerol and the carbohydrates as well as the long chains; and ^{35}S -labelled sulfate appears as such in the glycolipid sulfate.

A different type of glycolipid is the so-called rhamnolipid that is produced by *Pseudomonas aeruginosa*[2]. This substance is now known to be 2-*O*-α-L-rhamnopyranosyl-α-L-rhamnopyranosyl-D-3-hydroxydecanoyl-D-3-hydroxydecanoate[40]. Studies with partially purified enzymes[41] have shown that this substance is formed by three successive reactions.

(1) 2 D-3-hydroxydecanoyl-CoA →
    D-3-hydroxydecanoyl-D-3-hydroxydecanoate (D-3-HD-D-3-HD)

(2) D-3-HD-D-3-HD + TDP-L-rhamnose →
    L-rhamnosyl-D-3-hydroxydecanoyl-D-3-hydroxydecanoate

(3) L-rhamnosyl-D-3-HD-D-3-HD + TDP-L-rhamnose →
    L-rhamnosyl-L-rhamnosyl-D-3-hydroxydecanoyl-D-3-hydroxydecanoate

An interesting aspect of this compound is that its rate of excretion into the culture medium is almost as great as its rate of synthesis so that relatively little rhamnolipid accumulates in the cell.

(c) *Lipopolysaccharides*

Gram-negative enteric bacteria possess a complex and multi-layered cell envelope consisting of protein, phospholipid, a glycopeptide responsible for the rigidity of bacterial cell walls, and a lipopolysaccharide which is responsible for immunological specificity. The lipopolysaccharide has also been thought to be associated with or the cause of the injurious effects of these bacteria. It is true that lipopolysaccharide extracts from various enteric organisms have certain toxic pharmacological effects, but these effects are generally similar regardless of the species from which the lipopolysaccharide is extracted and even regardless of whether the source organism is pathogenic or harmless. Consequently, these lipopolysaccharides—or "endotoxins"—can be at best only contributory in the pathogenesis of enteric diseases which differ considerably in their clinical manifestations.

These lipopolysaccharides are actually a type of glycolipid, but much more complex ones than those discussed in the preceding section; indeed, it has been pointed out that they exhibit a degree of complexity that may be unique among polysaccharide derivatives[43]. While details of their structure are still lacking, it is known that they are high molecular weight substances with four main components each of which in turn can be quite large and complex; these are the immunologically specific O-antigen side-chain, the outer core, the backbone, and the so-called lipid A. As an example, the nature of the lipopolysaccharide of *Salmonella typhimurium* is shown diagrammatically[43]. The identities of the compounds in each fraction of this

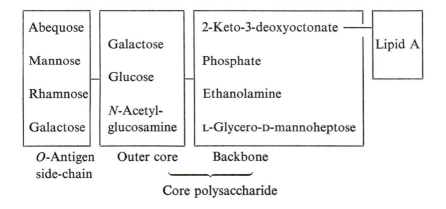

and various other lipopolysaccharides are known, but their abundance and arrangements are still under study.

The fraction of particular interest in our present context is the lipid A moiety (which is the portion of the lipopolysaccharide that is principally responsible for the toxic effects produced by such complexes and which include pyrexia, leucocytosis, tissue damage, vascular disturbances and a shock-like syndrome). Much research has been done on the nature of this fraction as well as that of the complex as a whole (see, *e.g.*, ref. 44), but relatively little "hard" information is yet available.

The lipid A of *E. coli* strain 0111 has been isolated and purified[45] and found to have a molecular weight of about 1700, a reducing end group, and to contain phosphate, glucosamine, acetyl and long-chain acyl groups in the ratio of 1:2:3–4:5. There is evidence for the linking of the two glucosamines by a glycosidic bond. A major amount of the fatty acid in this lipid proves to be β-hydroxymyristic acid, and β-hydroxy fatty acids of varying chain lengths have also been found in the lipopolysaccharides of other Gram-negative bacilli[6]. An interesting aspect of the fatty acid composition of the

References p. 263

lipopolysaccharides is the content of cyclopropane acids. When these unique acids were first being investigated, it was seen that they are (*a*) uncommon or non-existent in higher forms, (*b*) common in *Eubacteria*, (*c*) especially prominent in Gram-negative bacilli. As a consequence, there was a transient period where it was postulated that the toxic effects of Gram-negative lipopolysaccharides might be due at least in part to the presence of cyclopropane acids. However, like so many initially promising hypotheses, this one proved to be hampered by the fact that it wasn't true. Indeed, lipopolysaccharides—and lipid A fractions—are notably low in or even free of cyclopropane acids regardless of the total cellular content of such substances[6].

Unfortunately, relatively little is known of the details of lipid A structure or biosynthesis. A soluble enzyme has been extracted from *E. coli* cells that promotes the transfer of 2-keto-3-deoxyoctonoate (KDO) from cytidine monophosphate–KDO to a lipid acceptor containing at least glucosamine, phosphate and β-hydroxymyristate, and possibly being the lipid A moiety[6,46]. As pointed out above, KDO appears to be the site of attachment of lipid A to the core polysaccharide of the lipopolysaccharides.

In addition to being major constituents of bacterial lipopolysaccharides, lipids also appear to play important functional roles in the synthesis of the carbohydrate moieties of such complexes[6]. Various aspects of this involvement have been examined in one organism or another, but our present information and concepts are obviously fragmentary with respect to the total picture. At least two different roles of lipid in such syntheses have been detected in the *Salmonella*. One entails the combination of incomplete lipopolysaccharide with phosphatidyl-ethanolamine followed by glycosylation involving either glucose or galactose nucleotides. There is evidence that an intermediate complex of lipopolysaccharide/phosphatidyl-ethanolamine/glycosyl-transferase is formed[47,48].

The other known mode of involvement of lipid in *Salmonella* polysaccharide biosynthesis is in the formation of the repeating unit of the core polysaccharide of the lipopolysaccharide which is (D-mannosyl–L-rhamnosyl–D-galactosyl)$_n$. Di- and trisaccharide intermediates have been isolated and shown to be rhamnosyl–galactosyl–1-phosphate–X and mannosyl–rhamnosyl–galactosyl–1-phosphate–X. The "X" in these complexes appears to be a lipid of as yet unknown nature[49].

Lipid-bound intermediates have also been observed in the synthesis of Gram-positive cell walls[50]. Both *N*-acetylmuramyl–pentapeptide–lipid and

N-acetylglucosaminyl–N-acetylmuramyl–pentapeptide–lipid complexes have been observed. The latter is incorporated into cell wall glycopeptide at which point the carrier lipid is released.

As can be gathered from the above, our present understanding of the chemical nature, biosynthesis, function in microorganisms and role in pathogenesis of the bacterial lipopolysaccharides is far from complete. This is not due to lack of interest or of investigation; much work has been done and a mass of information has been collected as is evidenced by the hundreds, possibly thousands, of papers, reviews, symposia, annals, monographs and what not which have appeared on the subject. Unhappily, so far this means only that we know much about these substances while understanding little; we have almost excrutiating detail on many aspects but are still unclear as to what they are, what they are for, or what they can cause.

5. Cellular distribution of lipids

The localization of lipid within the bacterial cell is a matter of considerable interest not only in itself, but also because it may aid in fathoming the functions of these compounds. Actually, there seems to be little to say with respect to lipid location beyond the general statement that in most cells, most lipids are concentrated at the cell surface with relatively little in the cytoplasm. Following that generalization, some qualifications and details are in order.

In most organisms, the principal site of lipid accumulation is the cytoplasmic membrane which is the semi-permeable barrier of the microbial cell. This membrane is generally characterized as a lipoprotein unit membrane in which lipids constitute approximately 15–30% of the dry weight. Most of what we know regarding cell membranes has been derived from studies of Gram-positive bacteria which have discrete membranes readily obtained from the cells by physical and enzymatic procedures[51]. In such organisms, membranes constitute about 10% of the dry cell weight while containing 90–95% of the total cellular lipid. There are exceptions to this average, of course, and some cells do have significant amounts of lipid in the cytoplasm. It is noteworthy that in Gram-positive organisms where cell wall can be clearly separated from cytoplasmic membrane, there is little and usually no lipid in the cell wall.

It is more difficult to make clear distinctions between walls and membranes in Gram-negative bacteria because in these organisms one does not see two

References p. 263

discrete structures, but rather a complex multilayered cell "envelope" in which it is not presently possible to say what is cell wall and what is cytoplasmic membrane. Consequently, it has been necessary to analyse this envelope as one structure whether it is actually one or several in the cell. Here again, however, one finds almost all of the cellular lipid[2,6].

Microscopic studies have long suggested the presence in the cytoplasm of certain bacteria of visible lipid deposits as evidenced by staining with lipophilic dyes. Such accumulations do indeed occur in some organisms; in some species they consist of globules of liquid lipid, in others of granules of poly-β-hydroxybutyrate[2,52]. However, appearances can be deceptive. Some organisms that, on the basis of staining and microscopic examination, appear to contain cytoplasmic lipid inclusions can be shown by chemical analyses to contain little or no detectable cytoplasmic lipid. It has been suggested that in many cases what appear to be lipid inclusions are actually accretions of non-lipid substances covered with a thin veneer of lipid (which tends to accumulate at interfaces), and that it is the lipoidal veneer which is stained by lipophilic dyes[53].

Similarly, electron micrographs of bacterial spores often show numerous vacuoles which have been tentatively identified as the sites of lipid inclusions. However, chemical analyses of such spores usually show little or no lipid[53]. It is not always easy to translate microscopic observations into chemical certitude.

In discussing the location of lipids within bacterial cells, some mention might also be made of those located outside the cell, that is, the extracellular lipids. A few bacteria synthesize lipid and then secrete it into the environment. The best known of these is *Pseudomonas aeruginosa* which produces an extracellular rhamnolipid (described in an earlier section) in profusion. Whether this curious activity is a biochemical "error", the consequence of an incomplete pathway, or an activity serving some positive purpose for the cell is as yet veiled.

6. The functions of bacterial lipids

The ultimate goal of research on bacterial lipids is an understanding of the functional roles these substances play in microbial cells. At the present, however, "ultimate" in the sense of "the most remote" is an apt word for this goal. As has already been described, most of the work done to date has centered on the necessary and difficult labors of firmly establishing the chemical natures and modes of biosynthesis of bacterial lipids, and even this work is far from complete. Consequently, we have barely begun to study, let alone understand, the functions of such compounds.

Considering the observations that have been made in this area, it seems that there are four general ways in which bacterial lipids could function: energy storage, structural components, active transport, and involvement in biosynthetic activities.

Energy storage is a function that might reasonably be expected in view of what we know of lipid functions in higher forms of life. In some bacteria, particularly those that contain cytoplasmic granules of poly-β-hydroxybutyrate, lipid does seem to serve as a storage form of energy. However, for the majority of bacteria that have been studied, lipids are not the energy reserves; instead this function is performed by amino acids, peptides, and carbohydrates such as glycogen[54]. The triglycerides which are major reserve material in higher forms are minor constituents of bacteria and absent in many species. Also, it has been observed that in many bacteria, lipid concentrations actually increase during conditions of stress, depletion of nutrients in the environment, or during ageing of the cells. This is, of course, the reverse of what would be expected if these lipids were serving as energy reserves.

We have already noted that lipids are prominent structural components of bacteria, particularly the cytoplasmic membranes of Gram-positive bacteria and the wall/membrane "envelope" of the Gram-negative bacteria. Membranes that have been isolated from Gram-positive bacteria have the appearance of unit membranes; the envelopes of Gram-negative bacteria are more complex and multilayered so that it is more difficult to distinguish between wall and membrane, but are nevertheless rich in lipids[55]. The fact that bacterial lipids form much of the structure of membranes and envelopes does not, however, mean that these lipids do not have other, more kinetic functions. Indeed, their locations and consequent structural roles may be both necessary to and secondary to their more dynamic roles. Some mention might be made here of the mesosomal structures seen in some bacteria which are convoluted membrane structures having high lipid content. Present evidence indicates that these lamellar intracellular structures are really invaginations of the cytoplasmic membrane and so of similar composition.

The involvement of lipids in active transport is an attractive concept for those contemplating possible functions for bacterial lipids, however, there have not yet been any unequivocal demonstrations of this hypothesis. Most

References p. 263

bacterial lipids are phospholipids, most phospholipids are concentrated in membranes and envelopes, these structures are primarily responsible for controlling the passage of compounds in and out of the cell; therefore, it seems most plausible that lipids do function in transport and its regulation. There has been some evidence presented in the last few years that lipids are involved in the transport of amino acids and in β-galactoside permease activity[6,25], but as yet we have no definitive information.

Some investigations currently underway in my laboratory strongly suggest that there is some relationship between bacterial surface lipids and sensitivity to a variety of antibiotics. As yet we have made only a cursory examination of the tip of what may be a very large iceberg, but what we have seen is very exciting. There are definite qualitative and quantitative correlations between antibiotic sensitivity and bacterial lipid composition. What evidence we now have leads us to suspect that specific surface lipid complexes serve as receptor sites for specific antibiotics, and that when these sites are absent, the organism is then insensitive to that specific antibiotic. We have been interested so far in the tetracyclines and polymyxins, but we have some data suggesting that other types of antibiotics may also be involved. This is a very promising line of investigation, but also in a highly preliminary stage.

The last of the four ways in which we currently believe that bacterial lipids may be functional is that of biosynthesis. It is in this area that we have more definite information than in the other three. It seems that lipid may play two types of roles in biosynthesis, either that of a co-factor or that of a carrier or portion of a covalently bound intermediate.

In the discussion of lipopolysaccharides above, it was pointed out that lipids could function as cofactors in polysaccharide formation (see also refs. 6, 33, 47). Recently it has also been reported that lipid plays cofactor roles in the malate oxidizing systems of mycobacteria, and species of *Pseudomonas*[6].

Instances of the participation of lipid-bound intermediates in the biosynthesis of lipopolysaccharides and cell-wall components have also been described[33,49,50].

Law has described the need for a phospholipid intermediate in the biosynthesis of cyclopropane acids[25,26]. Although the immediate precursors of cyclopropane acids are the comparable monoenoic acids, these unsaturated acids must be incorporated into phosphatidyl ethanolamine or a similar compound before one-carbon addition and ring closure can occur to form the cyclopropane acids.

All of these observations of functional roles for bacterial lipids are quite recent—most of them have been made in the last five years. Therefore, it is virtually certain that the next few years will see many additions to and revisions of what we think we understand at present.

7. The effects of exogenous lipids on bacterial growth

Throughout this discussion we have been dealing with the formation and fate of lipids native to the bacterial cell. In this section some attention will be given to the interaction of bacterial cells with exogenous lipids in the environment.

Many organisms elaborate lipases (*i.e.*, triglyceridases) which split environmental fats into free fatty acids and glycerol (and/or mono- and diglycerides). The hydrolytic products are then utilized by the cells for energy and biosynthesis. The phenomenon of rancidity is a consequence of microbial lipolysis where the free fatty acids are short-chain compounds like butyric, caproic, caprylic and capric acids. Various other organisms produce a variety of phospholipases (*e.g.*, the α-toxin of *Clostridium perfringens* is actually phospholipase C). Still other organisms produce lipoproteinases (*e.g.* pathogenic strains of *Staphylococcus aureus*). So, there is widespread and diverse ability in the bacterial world to attack and utilize environmental lipids of varied types, and for many organisms lipids can serve as important sources of nutrition.

While shorter chain fatty acids, whether they occur as such in the environment or are the products of extracellular lipolysis, are readily tolerated and metabolized by bacteria, the picture is somewhat more complex regarding the long-chain fatty acids. In general they are not as readily utilized nor are they tolerated over as wide a concentration range as are the short-chain acids. In low concentrations long-chain fatty acids can be innocuous, stimulatory or even required while in slightly higher concentrations they become highly inhibitory to bacterial growth[56].

Stimulatory activity is principally referable to the long-chain unsaturated and cyclopropane acids which are capable of replacing biotin in the nutrition of many microorganisms. Biotin is involved in the biosynthesis of these two types of fatty acids which are essential for growth in many organisms, and if the fatty acids are supplied preformed, the need for biotin is decreased or eliminated[1,3]. However, as always, there can be too much of a good thing, and when optimal concentrations of either unsaturated or cyclopropane acids are exceeded, bacterial growth decreases and finally stops

References p. 263

altogether. Interestingly, these effects can be both enhanced and inhibited by the simultaneous presence of relatively small amounts of saturated long-chain acids. When minute amounts of stearic, palmitic or similar acids are added along with optimal concentrations of unsaturated or cyclopropane acids, bacterial growth is even more enhanced[3]. But if the concentration of saturated acid is increased only slightly, the growth stimulation of the unsaturated and cyclopropane acids is powerfully suppressed even to the point where all bacterial growth is stopped[57].

The exact effects that any given acid has vary considerably depending on the organism in question, and *vice versa*. However, for any given acid and any given bacterium the effects are remarkably constant, so much so that highly sensitive assays can be designed around such phenomena.

As can be seen, the seemingly simple matter of the response of bacteria to exogenous lipids is actually quite complex and, as a matter of fact, still not well understood in spite of some 20 years of study.

As a final complication, we might give some attention to lipid and lipid-related substances that are not part of or required by bacteria, but that can affect or be affected by bacteria. One area is that of microbial transformations or conversion oxidations where such compounds as sterols and related substances are supplied to microorganisms that don't require them but will partially metabolize them, producing altered compounds useful to man that can't be otherwise produced[58]. Often such transformations are produced by fungi, but species of *Eubacteriales* and *Pseudomonadales* are also capable of such activities. Such procedures are today principal sources of substances such as the corticosteroids and steroidal hormones.

Similarly, the bile acids while not present in, required by, or utilized by bacteria, do have powerful effects on them. This is related to the ability of bile acids and their salts to inhibit many bacteria (or even dissolve them if the concentration of bile salt is high enough) and yet at the same time to have no effect on other organisms. It has been assumed that this dissolution effect of bile salts on certain bacteria is due to their ability to disrupt lipoidal membranes, but this is not a sufficiently detailed concept to explain the differential effect on various species. As the microbiology-oriented observer will recognize, this phenomenon has various practical applications including the bile-solubility test to differentiate pneumococci from streptococci, the regulation of intestinal flora by the gall bladder, and the rationale behind the addition of bile salts to media intended to select enteric pathogens from mixed natural microflora. While we don't yet understand why these nonrequired, non-utilized substances make such great and discriminating differences to bacteria, we nevertheless see their role in important processes and can make important practical use of their effects.

8. Coda

The preceding pages have related how a long fallow field, fertilized with the proper mixture of new techniques and rising interest, has suddenly flowered into lush if somewhat tangled young growth. This field has become sprinkled with increasing numbers of enthusiasts, some of whom have been darting to and fro exclaiming over one blossom and another, and others who have been doggedly digging at some interesting root whose origin and significance are yet obscure. The halcyon days. However, passing from here to a bountiful and useful harvest is likely to be a long hard way, requiring the ordering of the field, expanding what is good and sound, and maybe weeding out what is feeble or dubious.

We have, indeed, come a long way in a relatively short time. The subject of microbial lipids and their metabolism which was so long regarded as a sort of scientific backwater, as a high effort/low yield field, has now proven to be an area of considerable interest and importance, and one that is now reasonably amenable to experimental treatment. It now begins to be evident that this is one of the areas, if not *the* area, in which bacteria differ most from other forms of life. The discovery of unique compounds and pathways of lipid metabolism encourages one to suspect that here is an area wherein at least some and perhaps many unique properties of bacteria may be explained.

We have struggled to a point now where we are reasonably clear about the nature and metabolism of bacterial fatty acids; we have made approaches to similar knowledge regarding the lipid classes; now we have hopes of progress on understanding the functions of these substances in the bacterial cell. As yet we can only speculate about what insights this work will give us. Will lipid researches clarify membrane-transport phenomena? Reveal the basis of antibiotic resistance? Explain the mode of action of certain antibiotics? Be useful in understanding microbial pathogenesis at the biochemical level? Probably all of these areas, and more not presently suspected, will benefit from our constantly increasing fund of information on the bacterial lipids.

It might be noted that the present discussion has dealt with what we have learned about bacterial lipids only; viral, rickettsial, and, for that matter,

References p. 263

fungal lipids still await extensive study by the more modern and discriminating methods. What these studies may reveal is veiled but enticing.

It is my belief that the next few years will see numerous and significant advances in the chemistry and physiology of microbial lipids, and that these will prove invaluable in understanding many phenomena which are now obscure. The prospect is exciting.

REFERENCES

1 W. M. O'Leary, *Bacteriol. Rev.*, 26 (1962) 421.
2 W. M. O'Leary, *The Chemistry and Metabolism of Microbial Lipids*, World, Cleveland, 1967.
3 K. Hofmann, *Fatty Acid Metabolism in Microorganisms*, Wiley, New York, 1963.
4 J. Asselineau, *Les lipides bactériens*, Hermann, Paris, 1962.
5 J. Asselineau, *The Bacterial Lipids*, Rev. ed., Holden-Day, San Francisco, 1966.
6 W. J. Lennarz, *Advan. Lipid Res.*, 4 (1966) 175.
7 M. Kates, *Ann. Rev. Microbiol.*, 20 (1966) 13.
8 P. R. Vagelos, *Ann. Rev. Biochem.*, 33 (1964) 139.
9 S. J. Wakil, *Ann. Rev. Biochem.*, 31 (1962) 369.
10 J. A. Olson, *Ann. Rev. Biochem.*, 35 (1966) 559.
11 C. E. Polan, J. J. McNeill and S. B. Tove, *J. Bacteriol.*, 88 (1964) 1056.
12 P. F. Wilde and R. M. C. Dawson, *Biochem. J.*, 98 (1966) 469.
13 M. Kates, *Advan. Lipid Res.* 2 (1964) 17.
14 J. Erwin and K. Bloch, *Science*, 143 (1964) 1006.
15 G. Scheuerbrandt and K. Bloch, *J. Biol. Chem.*, 237 (1962) 2064.
16 E. L. Pugh, F. Sauer, M. Waite, R. E. Toomey and S. J. Wakil, *J. Biol. Chem.*, 241 (1966) 2634.
17 D. J. H. Brock, L. R. Kass and K. Bloch, *Federation Proc.*, 25 (1966) 340.
18 D. K. Bloomfield and K. Bloch, *J. Biol. Chem.*, 235 (1960) 337.
19 W. M. O'Leary, in S. Shapiro and F. Schlenk (Eds.), *Transmethylation and Methionine Biosynthesis*, Univ. of Chicago Press, Chicago, 1965, p. 94.
20 W. M. O'Leary, *J. Bacteriol.*, 77 (1959) 367.
21 W. M. O'Leary, *J. Bacteriol.*, 78 (1959) 709.
22 W. M. O'Leary, *J. Bacteriol.*, 84 (1962) 967.
23 S. Pohl, J. H. Law and R. Ryhage, *Biochim. Biophys. Acta*, 70 (1963) 583.
24 J. W. Polachek, B. E. Tropp, J. H. Law and J. A. McCloskey, *J. Biol. Chem.*, 241 (1966) 3362.
25 A. E. Chung and J. H. Law, *Biochemistry*, 3 (1964) 1989.
26 J. G. Hildebrand and J. H. Law, *Biochemistry*, 3 (1964) 1304.
27 G. Jauréguiberry, J. H. Law, J. A. McCloskey and E. Lederer, *Biochemistry*, 4 (1965) 347.
28 A. H. Etémadi and E. Lederer, *Biochim. Biophys. Acta*, 98 (1965) 160.
29 M. Ikawa, *Bacteriol. Rev.*, 31 (1967) 54.
30 R. A. Pieringer, *Federation Proc.*, 24 (1965) 476.
31 R. A. Pieringer and R. S. Kunnes, *J. Biol. Chem.*, 240 (1965) 2833.
32 J. Kanfer and E. P. Kennedy, *J. Biol. Chem.*, 239 (1964) 1720.
33 J. H. Law, in B. D. Davis and L. Warren (Eds.), *The Specificity of Cell Surfaces*, Prentice Hall, Englewood Cliffs, N.J., 1967, p. 87.
34 M. Cohen and C. Panos, *Biochemistry*, 5 (1966) 2385.
35 P. F. Smith and W. L. Koostra, *J. Bacteriol.*, 93 (1967) 1853.
36 W. J. Lennarz, *J. Biol. Chem.*, 239 (1964) PC 3110.
37 W. J. Lennarz and B. Talamo, *J. Biol. Chem.*, 241 (1966) 2707.
38 J. Distler and S. Roseman, *Proc. Natl. Acad. Sci. (U.S.)*, 51 (1964) 897.
39 B. Kaufman, F. D. Kundig, J. Distler and S. Roseman, *Biochem. Biophys. Res. Commun.*, 18 (1965) 312.
40 J. R. Edwards and J. A. Hayashi, *Arch. Biochem. Biophys.*, 111 (1965) 415.
41 M. M. Burger, L. Glaser and R. M. Burton, *J. Biol. Chem.*, 238 (1963) 2595.
42 Y. Chang and E. P. Kennedy, *J. Biol. Chem.*, 242 (1967) 516.

43 M. J. OSBORN AND I. M. WEINER, *Federation Proc.*, 26 (1967) 70.
44 A. NOWOTNY (Ed.), *Ann. N.Y. Acad. Sci.*, 133 (1966) 277.
45 A. J. BURTON AND H. E. CARTER, *Biochemistry*, 3 (1964) 411.
46 R. M. MAYER, R. D. EDSTROM AND E. C. HEATH, *Federation Proc.*, 24 (1965) 479.
47 L. ROTHFIELD AND B. L. HORECKER, *Proc. Natl. Acad. Sci. (U.S.)*, 52 (1964) 939.
48 L. ROTHFIELD AND M. PEARLMAN, *J. Biol. Chem.*, 241 (1966) 1386.
49 I. M. WEINER, T. HIGUCHI, L. ROTHFIELD, M. J. OSBORN AND B. L. HORECKER, *Proc. Natl. Acad. Sci. (U.S.)*, 54 (1965) 228.
50 J. S. ANDERSON, M. MATSUHASHI, M. A. HASKIN AND J. L. STROMINGER, *Proc. Natl. Acad. Sci. (U.S.)*, 53 (1965) 881.
51 M. R. J. SALTON, *The Bacterial Cell Wall*, Elsevier, Amsterdam, 1964.
52 I. C. GUNSALUS AND R. Y. STANIER (Eds.), *The Bacteria*, Vol. 1, Academic Press, New York, 1960.
53 C. LAMANNA AND M. F. MALLETTE, *Basic Bacteriology, Its Biological and Chemical Background*, 3rd ed., Williams and Wilkins, Baltimore, Md., 1965.
54 E. A. DAWES AND D. W. RIBBONS, *Biochem. J.*, 95 (1965) 332.
55 M. R. J. SALTON, in B. D. DAVIS AND L. WARREN (Eds.), *The Specificity of Cell Surfaces*, Prentice Hall, Englewood Cliffs, N.J., 1967, p. 71.
56 C. NIEMAN, *Bacteriol. Rev.*, 18 (1954) 147.
57 K. HOFMANN, W. M. O'LEARY, C. W. YOHO AND T. Y. LIU, *J. Biol. Chem.*, 234 (1959) 1672.
58 W. CHARNEY AND H. L. HERZOG, *Microbial Transformations of Steroids*, Academic Press, New York, 1967.

Chapter VI

Fatty Acid Metabolism in Plant Tissues

P. K. STUMPF

Department of Biochemistry and Biophysics, University of California, Davis, Calif. (U.S.A.)

1. Introduction

In general, higher plants contain a complement of fatty acids and complex lipids which are quite similar in compositional patterns. Superimposed on this normal pattern may be lipids characteristically associated with specialised tissues of that species. For example, in the species *Ricinus communis*, the endosperm of the developing seed synthesizes large quantities of ricinoleic acid which is stored as triglycerides in discrete oil droplets in the cytoplasm of the cell. The mitochondria of the endosperm, however, contain phospholipids and galactolipids with fatty acid compositions devoid of ricinoleic acid but very similar to the mitochondrial lipids of other higher plants. Furthermore, the leaf lipids of *Ricinus* not only are free of ricinoleic acid but also contain fatty acids, namely, linoleic and α-linolenic acids, very similar to those found in leaves of other plants.

This constancy of fatty acid composition in important plant organelles must relate to membrane structures that have an architecture carefully designed for optimal functional capacity. Thus, their lipoidal composition cannot be varied according to the lipids characteristic of a given species. For example, to insert ricinoleic acid into the fatty acid moiety of either a phospholipid or a galactolipid in the mitochondrial membranes of cells of *Ricinus communis* would lead probably to serious changes in the physical structure of the membranes with a concomitant alteration in functional effectiveness. Evidently, higher plants allocate to the neutral lipids, namely, the triglycerides, the depository for bizarre fatty acids. In general, seed lipids contain fatty acids characteristic of a given plant species (see Wolff[1], and Table I).

References p. 291

TABLE I

FATTY ACID COMPOSITION OF SOME SEED LIPIDS

Species	Fatty acid composition (%)						Unusual fatty acid
	14:0	16:0	18:0	18:1	18:2	18:3	
Carthamus tinctorius L. (var. N-10)	—	7	1	16	76	—	—
Linum usitatissimum	—	6	4	22	16	52	—
Hydnocarpus wightiana	—	2	—	6	—	—	hydnocarpic 49 chaulmoogric 27
Petroselinum sativum	—	3	—	15	6	—	gorlic 12 petroselenic 76
Ricinus communis	—	—	2	7	3	—	ricinoleic 87
Picramnia lendiniana	22	33	3	22	—	—	tarisic 20
Veronónia anthelimenties	—	3	1.5	2	9	—	epoxyoleic 78

2. Oxidative systems

A number of metabolic systems have been described in higher plants by which a cleavage of a carbon–carbon bond leads to the breakdown of fatty acids to compounds containing fewer carbon atoms. These include (*1*) lipoxidation, (*2*) α-oxidation, (*3*) β-oxidation, and (*4*) modified β-oxidation. The oxygen-requiring systems responsible for hydroxylations of fatty acids and desaturation will be described in Section 4 of this Chapter (p. 286).

(*a*) *Lipoxidase* (EC 1.13.1.13)

The lipoxidases have been quite intensively studied for a number of years. Thus, soybean lipoxidase has been crystallized and its chemical and physical properties determined (Tappel[2]). The enzyme catalyzes the oxidation by molecular oxygen of specific fatty acids containing a *cis,cis*-1,4-pentadiene system to form a conjugated *cis,trans*-diene hydroperoxide.

$$RCH \stackrel{c}{=} CH-CH_2-CH \stackrel{c}{=} CH-R + O_2 \rightarrow RCH \stackrel{c}{=} CH-CH \stackrel{t}{=} CH-\underset{OOH}{CH}-R \quad (1)$$

The enzyme is widely distributed in seeds of legumes, some cereal grains and high-oil containing seeds. It has not been observed in bacteria and a true lipoxidase system as such has not been observed in animal tissues although similar reactions are catalyzed by heme-containing proteins.

Crystalline soybean lipoxidase with a molecular weight of 102 000 contains no unusual amino acids in its protein. Moreover, it has no metal porphyrin systems and indeed does not appear to bind or require any metal cations for activity. Substrates which are readily attacked include linoleic acid, α-linolenic acid and arachidonic acid. Inactive substrates include oleic acid, and polyunsaturated fatty acids with *trans* double bonds. Free as well as esterified fatty acids are susceptible to attack. Inhibitors of the enzyme include anti-oxidants such as α-tocopherol and hydroquinone, which act as chain breakers in the free-radical chain reaction. Inhibitors such as azide, cyanide and metal chelators are ineffective; *p*-hydroxymercuribenzoate and iodoacetate do not inhibit the enzyme. It is thus clear that both transition metals and SH groups are not requirements for the reaction catalyzed by soybean lipoxidase.

A mechanism which has been proposed by Tappel[2] and others suggests the sequence of events given in Fig. 1 to explain the formation of the final product.

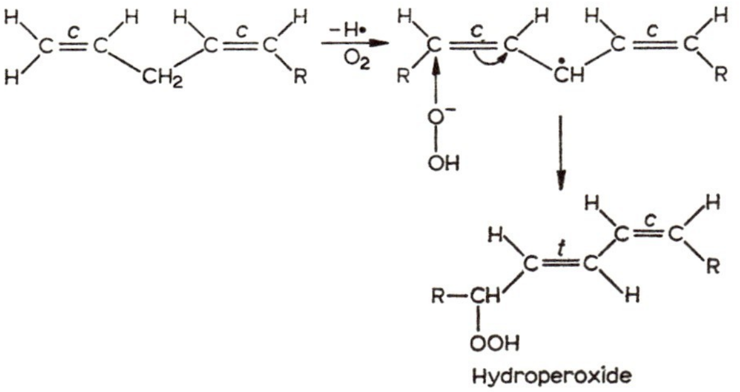

Fig. 1. Formation of the final product of a lipoxidase-catalyzed reaction.

Recent research employing the highly sensitive and new technique of mass spectrometry has allowed a closer examination of the mechanism of this reaction. By using these techniques Hamberg and Sammuelsson[3] determined the nature of the products of lipoxidase-catalyzed reactions employing a number of polyunsaturated fatty acids as substrates. Their results show a highly specific structural requirement by the soybean lipoxidase for a

References p. 291

cis,cis-1,4-pentadiene group in the polyunsaturated fatty acid with the methylene group between the two double-bond systems being located in the $\omega 8$ position. The oxygen function is always introduced at the $\omega 6$ carbon position with the subsequent rearrangement of the non-conjugated double-bond system to a *trans*-double bond adjacent to the hydroperoxyl group and the *cis*-double bond retained in the distal position:

$$CH_3(CH_2)_4-CH\stackrel{c}{=}CH-CH_2-CH\stackrel{c}{=}CHR \rightarrow \qquad (2)$$
$$\uparrow \qquad \uparrow$$
$$\omega 6 \qquad \omega 8$$

$$CH_3(CH_2)_4-\underset{H}{\overset{\overset{\displaystyle H}{\overset{\displaystyle O}{\overset{\displaystyle O}{|}}}}{C}}-CH\stackrel{t}{=}CH-CH\stackrel{c}{=}CHR \qquad (3)$$

Dutton and his colleagues[4,5] using approaches rather similar to those described by Sammuelsson in 1965, have fully confirmed and extended the results of Sammuelsson. When, for example, linoleic acid is auto-oxidized by chemical catalysis, equal proportions of the 9- and 13-hydroperoxyl products accumulate. Depending on the source of soybean lipoxidase, Dutton observed in some cases either the exclusive formation of the 13-hydroperoxyl product or varying but small amounts of the 9-product with major amounts of the 13-isomer. His results suggest the possibility of at least two forms of lipoxidase, one being specific for the 9 and the other for the 13 position. Further results from Dutton's laboratory showed that $^{18}O_2$ is introduced into the molecule and that $H_2^{18}O$ is excluded (Dolev et al.[5]).

These observations are of considerable interest since they indicate a high enzyme specificity for a precise part of a polyunsaturated fatty acid.

The function of lipoxidase in plants with high concentrations of the enzyme is not clearly understood. Since a large number of plants do not contain any lipoxidase activity, it is obvious that the enzyme does not play a critical role in the general metabolism of polyunsaturated fatty acids. However, two acids, 13-hydroxy-*cis*-9,*trans*-11-octadecadienoic acid and 9-hydroxy-*trans*-10,*cis*-12-octadecadienoic acid have been isolated from plants[4]. Reactions can be postulated to suggest a possible participation of lipoxidase in the biosynthesis of these acids.

Of further interest is the recent observation by Zimmerman[6] that flaxseed extracts can catalyze the conversion of linoleic acid to a ketohydroxyunsaturated acid presumably by the following reactions:

$$\text{linoleic acid} \xrightarrow{\text{lipoxidase}} \text{13-hydroperoxyl-9,11-octadecadienoic acid} \xrightarrow{\text{flaxseed extract}} \text{12-keto-13-hydroxy-}cis\text{-9-octadecenoic acid} \quad (4)$$

Added soybean lipoxidase generates a mixture of hydroperoxides from linoleic acid and apparently a hydroperoxide isomerase converts the substrate(s) to the final product according to the reactions:

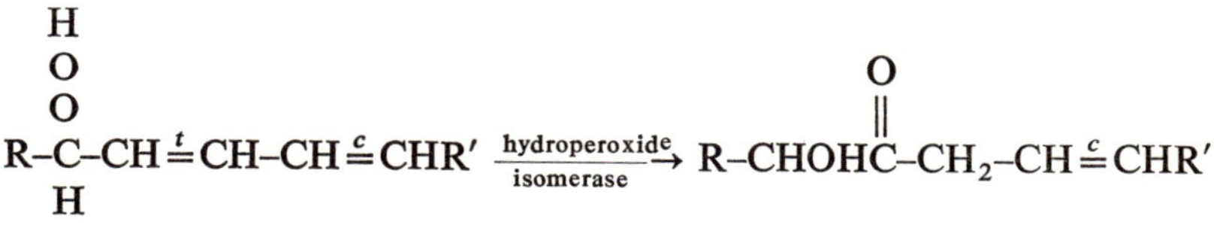

$$\underset{\text{H}}{\overset{\text{HOO}}{\text{R-C-CH}}} \overset{t}{=} \text{CH-CH} \overset{c}{=} \text{CHR'} \xrightarrow{\text{hydroperoxide isomerase}} \text{R-CHOH}\overset{\overset{\text{O}}{\|}}{\text{C}}\text{-CH}_2\text{-CH} \overset{c}{=} \text{CHR'} \quad (5)$$

It is entirely conceivable that a β-oxidation system can degrade this product to lower fatty acids and acetyl-CoA.

It is thus becoming apparent that lipoxidase may very well be part of a specialized mechanism at the disposal of some plant species for the metabolism of polyunsaturated fatty acids by a rather unique pathway. Research in the next few years will undoubtedly reveal the details of these systems.

(b) α-Oxidation

First observed in crude extracts of germinated peanut cotyledons in 1952 by Newcomb and Stumpf[7], the α-oxidation system has been further described by Martin and Stumpf[8] in peanut cotyledon extracts and by Hitchcock and James[9,10] in young leaf tissue.

α-Oxidation is defined as a series of reactions by which a free fatty acid of chain length ranging from C_{18} to C_{13} is oxidatively degraded with the simultaneous release of carbon dioxide from the carboxyl carbon and the formation of a free fatty acid containing one less carbon atom. The only required cofactor is NAD^+.

The peanut system has components which differ from that of the leaf

References p. 291

system. Thus the peanut system requires an H_2O_2 generating system and a NAD^+-linked long-chain fatty aldehyde dehydrogenase. The only intermediate(s) Martin et al.[8] detected were long-chain fatty aldehydes. In turn, when long-chain fatty aldehydes were added to the peanut extract, they were readily oxidized by a NAD^+-requiring dehydrogenase to the corresponding free fatty acid. Addition of imidazole prevented the formation of the aldehydes suggesting that the site of inhibition was at an oxidative decarboxylation site. The cycle depicted by Martin and Stumpf is presented in Fig. 2.

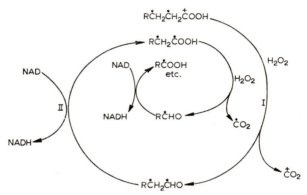

Fig. 2. α-Oxidation of long-chain fatty acids. I, long-chain fatty acid peroxidase. II, aldehyde dehydrogenase.

Hitchcock and James[10] have provided evidence that the leaf system requires molecular oxygen rather than hydrogen peroxide. The first observable intermediate is the formation and utilization of L-α-hydroxy fatty acids to form a fatty acid with one less carbon atom. Since D-α-hydroxy fatty acid is not metabolized, the system is stereospecific. Of interest is their recent observation (Hitchcock et al.[11,12]) that the D- as well as the L-α-hydroxy fatty acids are formed in the α-oxidation of fatty acids by acetone powders of pea leaves. The D-isomer accumulates whereas the L-isomer is further metabolized. Since leaf cerebrosides contain 2-hydroxypalmitic acid, it is quite probable that this acid is synthesized by the α-oxidation pathway.

The pathway proposed by James and his colleagues is presented below:

$$RCH_2COOH \xrightarrow{O_2} \text{L-}RCHOHCOOH \xrightarrow[O_2]{NAD^+} RCOOH + CO_2 \rightarrow \text{etc.} \quad (6)$$

The physiological significance of the α-oxidation pathway remains to be carefully evaluated. Most likely, the far more active and widespread β-oxidation system is the principal mechanism in the conversion of fatty acids to acetyl-CoA. In addition, in the cleavage of the carbon–carbon bond, the bond energy thereby released is far more effectively trapped by the β-oxidation system than by the α-oxidation pathway. It is quite possible though that, if an odd-chain fatty acid must be made available for further synthetic purposes, such as long-chain aldehydes or 2-hydroxy fatty acids to serve as components of complex lipids, the α-oxidation pathway may well serve such a function. Furthermore, the α-oxidation pathway may very effectively be the mechanism for by-passing blocking groups in a hydrocarbon chain of a fatty acid by shifting the β-oxidation reading-frame by a one-carbon unit thereby allowing effective β-oxidation to occur. Thus the involvement of α- and β-oxidation in the metabolism of a fatty acid has been recently invoked by Steinberg and his colleagues [13] in the breakdown of phytanic acid by mammalian tissues.

(c) *β-Oxidation*

While several workers had provided *in vitro* evidence of a β-oxidation system in higher plants, the existence of this system in an *in vitro* plant preparation was first noted by Stumpf and Barber[14] who described a mitochondrial system obtained from germinating peanut cotyledon mitochondria capable of oxidizing [^{14}C]carboxyl-labelled fatty acids to $^{14}CO_2$. The cofactor requirements were $NADP^+$, NAD^+, CoA, ATP, Mg^{2+} and a Krebs cycle acid. A large number of fatty acids were readily oxidized by this system; in addition, internally labelled fatty acids also released $^{14}CO_2$ thereby indicating extensive oxidation of the hydrocarbon chain. In 1964, Rebeiz and Castelfranco[15] re-examined the peanut cotyledon system and showed that a considerable fraction of activity of β-oxidation was localized in the soluble protein fraction. Similar results were obtained by Yamada and Stumpf[16,17] studying the oxidation of ricinoleic acid by germinating castor bean homogenates. About 10% of the total β-oxidative capacity resided in the mitochondrial particles in 5-day-old germinating seeds and the remainder (90%) in the supernatant proteins.

However, recently, evidence has been presented that in both maturing and germinating *Ricinus communis* seeds a very significant proportion of the β-oxidative capacity resides in very fragile organelles, called cytosomes (Hutton and Stumpf[78]) or glyoxysomes (Beevers[79]). These organelles are

References p. 291

readily disrupted by conventional preparative techniques which would explain the earlier results of Rebeiz and Castelfranco and Yamada and Stumpf. When isolated by sucrose-gradient techniques the cytosomes exhibit very high specific activities for crotonase (EC 4.2.1.17), β-hydroxyacyl dehydrogenase (EC 1.1.1.35), and β-ketothiolase (EC 2.3.1.16), Of interest, the cytosomes of maturing *Ricinus* seeds are considerably less fragile than those obtained from the germinating tissue but they have as their maximum activities about 10% of the crotonase activity found in the germinating seed at its maximum activity, 25% of the β-hydroxyacyl dehydrogenase and the β-ketothiolase activities. What role β-oxidation could play in developing tissue which is programmed to synthesize very large amounts of fatty acids as future storage lipid is not understood at present. Perhaps the system serves as a scavenger to remove from the developing tissue fatty acids which otherwise may attain levels of concentrations toxic to the tissue.

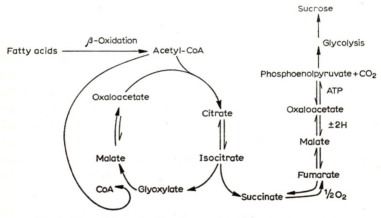

Fig. 3. Glyoxylate cycle, directly coupled to β-oxidation systems.

Directly coupled to β-oxidation systems is the glyoxylate cycle which has been well documented in a number of publications (Beevers[18]). Briefly, the cycle explains the well-known observation that oil-rich seeds during germination rapidly convert fatty acids stored as triglycerides to sucrose (Fig. 3). The two key enzymes responsible in linking the utilization of acetyl-CoA to the glycolytic pathway are malate synthase (EC 4.1.3.2) and isocitrate lyase (EC 4.1.3.1.) These enzymes have in the past been considered either as soluble enzymes or as enzymes associated with mitochondria. However, recently, Breidenbach and Beevers[19] have described a new organelle in oil seeds, namely, the glyoxysome. This organelle contains a high concentration of malate synthase and isocitrate lyase but no fumarase (EC 4.2.1.2) or other Krebs-cycle enzymes. The relationship of this organelle to β-oxidation systems has already been reviewed.

The complete breakdown of ricinoleic acid to acetyl-CoA is an interesting problem since both the cis-9,10-double bond system and the D-12-hydroxy function in ricinoleic acid are barriers for the β-oxidation enzymes. Since the conversion of ricinoleic acid occurs very smoothly in the cotyledonous tissues of germinating castor bean seeds, it is obvious that these barriers are readily circumvented in these tissues. It is quite possible that the mechanism for resolving the cis-9,10-double bond barrier is the same described by Stoffel[20] in mammalian tissues, namely, the cis-3,2-trans enoyl-CoA isomerase. Unsaturated fatty acids such as oleic, linoleic and α-linolenic all with 9,10 and/or 12,13 and 15,16 double-bond systems are readily oxidized by germinating seed tissue. Thus, an enzyme similar to Stoffel's enzyme must be widespread and its identification and characterization will be of interest. The resolution of the D-12-hydroxy function is somewhat more difficult to predict but is now under investigation in the author's laboratory.

(d) Modified β-oxidation

In 1958, Giovannelli and Stumpf[21] described an interesting pathway for the oxidation of propionic acid by a mitochondrial preparation of germinating peanut cotyledons. The pathway readily explains the kinetic experiments obtained with intact tissue incubated with propionic acid labelled in different positions. Experimentally, it has been observed consistently in a number of plant tissues that $^{14}CO_2$ is released in decreasing rates from [1-^{14}C]proprionic, [3-^{14}C]propionic and [2-^{14}C]propionic acids. No Krebs cycle acids can be isolated when [1-^{14}C]propionic acid is fed whereas these acids are readily isolated when [2-^{14}C]propionic and [3-^{14}C]propionic acid are fed (Hatch and Stumpf[22]). Other evidence supports the pathway presented in Fig. 4. This pathway obviates the succinic acid pathway which has been described by Ochoa and his colleagues as the main pathway for propionate metabolism in animal tissues (Kaziro and Ochoa[23]). Since the succinic acid pathway has as one of its key enzymes, methylmalonyl-CoA mutase (EC 5.4.99.2), a cobalamine enzyme, it is of interest that the plant pathway bypasses the necessity for the mutase reaction. It is well known that there

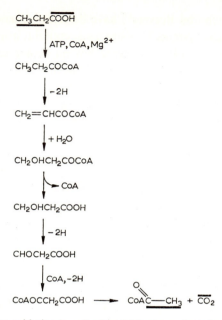

Fig. 4. Propionate oxidation by mitochondrial preparation of peanut cotyledons.

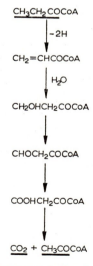

Fig. 5. Propionate metabolism in some bacteria.

is no evidence for the participation of the cobalamine coenzymes in any reactions in higher plants (Evans and Kliewer[25]). The consistent observation that a large number of higher plants always have the β-hydroxylpropionate pathway gives further support to the contention that the cobalamine coenzymes are non-functioning in higher plants.

From the comparative point of view, a similar system has been described by Vagelos and Earl[26] in a number of bacteria, their pathway is given in Fig. 5. Note the different origins of the carbon atoms of acetate in the bacterial and plant systems. The difference hinges on the necessity for hydrolysis of the β-hydroxypropionyl-CoA to free β-hydroxypropionic acid in higher plants whereas in the bacterial system there is a retention of this thioester and an oxidation of the β-hydroxy function to malonyl-CoA.

The function of this pathway in higher plants is difficult to assess. Although propionic acid is very infrequently found in higher plants, it could be generated by allowing a saturated fatty acid to undergo at least one α-oxidation with the remaining oxidation by the β-oxidation pathway as indicated below:

$$CH_3CH_2(CH_2)_{12}CH_2COOH \xrightarrow{\alpha\text{-oxid.}} CH_3CH_2(CH_2)_{12}COOH + CO_2 \xrightarrow{\beta\text{-oxid.}}$$

$$CH_3CH_2COOH + 6\,CH_3COOH \quad (7)$$

Hatch and Stumpf[22] have clearly shown that β-alanine is formed from propionic acid presumably either by a transamination of the semi-aldehyde of malonic acid or the β-amination of acryl-CoA. Thus propionic acid could be generated from the combined efforts of α- and β-oxidation enzymes and then could serve as a source of β-alanine which of course is an essential component of CoA and ACP. Stinson and Spencer[24] have recently isolated a specific transaminase which catalyzes the formation of β-alanine from malonic semi-aldehyde and aspartic acid.

3. Biosynthesis

(a) *General considerations and properties*

Early in 1953 it was established by Newcomb and Stumpf[27] in plants that acetate was the most effective precursor of fatty acids when a number of compounds such as acetate, formate, glucose, and butyrate were incubated with germinating and developing seeds of peanuts. In general, since then, two types of approaches have been used to test the capacity of tissues of higher plants to synthesize fatty acids. The first consists of incubating a number of ^{14}C-labelled compounds for a given length of time under different conditions with intact, chopped or sliced tissues. This approach has been exploited in recent years with considerable success by James et al.[28,29,32]. The advantage of this method is that the tissue undergoes a minimum of damage and permits a rapid scanning of the capacity for tissue to synthesize a given fatty acid. This approach was used successfully to determine that only maturing seed tissue of *Ricinus communis* and not germinating tissue has the capacity to synthesize ricinoleic acid (James[28], Canvin[30], Yamada and Stumpf[31]). A disadvantage is that since the tissue consists of many different types of cells and since each plant cell contains a number of organelles including mitochondria, plastids, and chloroplasts with specialized functions, the information so obtained is a summation of the capacity of many different operational sites. Furthermore, when a fatty acid is incubated with plant tissues the substrate is activated to the CoA ester with subsequent movement into neutral or complex lipids or degradation by β-oxidation. If complex lipids have a low turnover rate, the fatty acid may no longer be available for further metabolic change. In addition, there is growing evidence that long-chain acyl transacylases which would transfer the acyl component of CoA derivatives to acyl-ACP derivatives are missing in higher plant tissues. For example, acetate is readily incorporated into a broad number of fatty acids. However, both free palmitic and stearic acids when incubated with tissue slices are relatively inert for elongation and desaturation but are readily oxidized or incorporated into complex lipids indicating that they have undergone activation to the CoA level but not to the ACP level.

In addition, on a time scale, the incubation of either acetate or a fatty acid with intact tissues tests the capacity of that tissue to synthesize fatty acids at that given period of development. For example, a developing seed will take approximately 40–60 days to develop prior to dormancy. A time sequence is programmed in the developing tissue by which the enzymes appear for the synthesis of a number of fatty acids, the total of which is reflected in the final composition of the lipids characteristic of the seed. When the capacity of such tissue is assayed by incubating acetate for a period of 2–6 h, the results reflect only the capacity of that tissue in a very brief period in the developmental stage of that seed. The data so obtained frequently do not reflect the final fatty acid composition of that seed. This shift in capacity is easily demonstrated in two developing seed systems, namely the

TABLE II

COMPOSITION OF SEED FATTY ACIDS AS A FUNCTION OF MATURATION OF SEED

Days after fertilization	Lipid (mg/100 seeds)	Fatty acid composition (%)				
		16:0	18:1	18:2	18:3	Ricinoleic
(A) Carthamus tinctorius (var. N-10) (Sims et al.[81])[a]						
20	486		36	63		
30	732		25	75		
40	839		20	80		
50	872		20	80		
(B) Ricinus communis L. (Canvain[30])						
6–9	40	43.5	7.5	46	3	0
18–21	549	7.5	11.5	15	3	63
30–33	5 716	2.3	4.8	7.0	0	86.5
42–45	13 223	0.4	3.5	5.5	0	90.6
60	15 384	1.2	3.5	4.6	0	90.7

[a] Saturated fatty acids are minor components. In this section changes in unsaturated fatty acids are emphasized.

TABLE III

[^{14}C]ACETATE INCORPORATION INTO FATTY ACIDS AS A FUNCTION OF THE DEVELOPMENTAL STAGE OF THE CASTOR BEAN

After Glew and Stumpf[82]

Tissue	% Total ^{14}C incorporated as lipids	Fatty acid				
		16:0	18:0	18:1	18:2	Ricinoleate
Maturing ≃ 40 days	33.8	6.93	6.40	29.9	8.56	48.2
Fully mature	23.0	4.88	trace	95.1	0	0
Dry seed (dormant)	10.1	11.1	trace	88.9	0	0
Germinated 1 day	20.8	15.0	6.1	48.0	30.8	0

Trace of 18:3 synthesis in all cases.

safflower and the castor bean. In Table II, it is obvious that depending on the number of days after fertilization of the ovum, drastically different results are obtained. In Table III, when acetate is fed to developing castor bean slices at various stages of development, marked differences in capacity of the seed are observed. Moreover, long periods of incubation—that is, over 6 h of incubation—strongly invite bacterial contamination with the usual accompanying difficulties of interpretation.

References p. 291

Nevertheless, the broad outlines of biosynthesis have been successfully drawn by a number of investigators in recent years using whole-tissue or tissue-slice techniques. Thus [^{14}C]acetate has been repeatedly found to label a large number of fatty acids in a given tissue, be it avocado mesocarp, developing and germinating castor bean seeds, pea seed, barley, castor, spinach, lettuce leaves, etc. Conversion of a given substrate to a derivative with the same carbon skeleton has also been demonstrated. Thus conversion of oleic to linoleic (James et al.[28,32,33]; McMahon and Stumpf[35]), oleic to ricinoleic (Canvain[30]; James et al.[32,34]; Yamada and Stumpf[31]), trans-3-hexadecenoic to palmitic (Bartel et al.[36]), linoleic to linolenic (Harris and James[33,34]) and, oleic to crepenynic acid (Haigh and James[37]) have been readily demonstrated by the in vivo type experiment. That a given tissue can transform one fatty acid to another derivative allows the investigator to select the tissue at the appropriate stage of development in order to obtain cell-free systems.

The second procedure has been to proceed as rapidly as possible to the cell-free level and analyze in detail the components of the fatty acid synthetase(s). Much information can be obtained by this procedure but the difficulties of this approach are also manifold. Thus, the soluble fatty acid synthetase which can be readily obtained from a number of plant extracts, can convert malonyl-CoA in the presence of ACP, NADPH, NADH and acetyl-CoA to palmitic and stearic acids. Frequently, however, the ability to form polyunsaturated fatty acids has been lost. Final products of synthesis never duplicate the endogenous fatty acid content of the cell. Table IV illustrates this shift employing a leaf tissue and a fruit tissue.

In general, the enzymatic synthesis of long-chain fatty acids in cell-free systems of higher plants is identical to that described for bacterial systems. The same components, namely, malonyl-CoA, ACP, NADPH and NADH are required. All systems synthesize de novo palmitic and stearic acids. The soluble systems do not sediment on prolonged ultracentrifugation. Organelles have been isolated which participate in de novo synthesis. Particles from avocado mesocarp and pea seedlings readily incorporate malonyl-CoA into long-chain fatty acids as well as synthesize unsaturated fatty acids, mostly oleic (Yang and Stumpf[39]).

Recently, Macey and Stumpf[38] have examined the capacity for lipid synthesis by low lipid, high carbohydrate containing cotyledons of germinating peas. Fatty acid synthesis from malonyl-CoA occurs in three cellular fractions, namely a $10\,000 \times g$ fraction, presumably mitochondrial, a

TABLE IV

FATTY ACIDS SYNTHESIZED BY VARIOUS PREPARATIONS OF TISSUE WHICH RANGE FROM THE INTACT TO CELL-FREE PREPARATIONS

Type	Percentage					
	14:0	16	18	18:1	18:2	18:3
(A) Avocado mesocarp (Stumpf[83])						
Avocado mesocarp						
Endogenous fatty acids		20	1	60	18	
[^{14}C]Acetate incorporation by:						
Slice		48	10	31	3	
Particles (10 000 $\times g$)		50	30	20	1	
Acetone powders of particles		40	60	0	0	
Soluble synthetase (not derived from particles)		82	18	0	0	
(B) Lettuce leaf tissue (Brooks and Stumpf[52])						
Leaf tissue						
Endogenous fatty acids	1.7	14	0.8	0.5	6.6	71.2
[^{14}C]Acetate incorporation by:						
Leaf tissue	4.8	41.5	3.9	41.8	8	—
Whole chloroplast	1	24	—	73	—	—
Soluble preparation from chloroplast	—	—	98	—	—	—

100 000 $\times g$ fraction, probably microsomal, and the soluble proteins in the final supernatant. The first and third fractions synthesize primarily palmitic and stearic acids whereas the microsomal fraction synthesizes fatty acids ranging from palmitic to octasanoic acids. Acyl-carrier protein is an essential component of all systems. Soluble synthetase systems have also been described in soybean[77], in potato tubers and in spinach and lettuce chloroplasts.

The results obtained from a study of these various preparations strongly suggest that the mechanism of synthesis is very similar to that proposed by Vagelos and Wakil for the soluble *E. coli* system and differs from the mammalian or avian system which functions as a large complex of at least 7 enzymes. The individual enzymes have not been isolated nor characterized and future investigation could profitably pursue the achievement of such knowledge.

However, one enzyme, namely acetyl-CoA carboxylase (EC 6.4.1.2), has been rather extensively examined. Hatch and Stumpf[40,41] in 1961 described the isolation, purification and distribution of this enzyme in higher plants. Employing wheat germ as a source of the enzyme, Hatch was able to purify

References p. 291

the enzyme some 150-fold, demonstrated the complete inhibition of the enzyme by avidin, and the requirement for free SH groups for activity. He has also shown that it was widespread in a number of plant tissues including wheat, pea, safflower, peanut, potato and castor seed (Hatch and Stumpf[41]).

Burton and Stumpf[42] later examined its possible properties as related to activation since Gregolin et al.[43], Vagelos et al.[44], Waite and Wakil[45] and Matsuhoshi et al.[46], studying either the mammalian or avian liver acetyl-CoA carboxylase demonstrated that the enzyme was markedly activated by the presence of citric or isocitric acid as well as a large number of other compounds in an allosteric manner. White and Klein[47] have shown that yeast acetyl-CoA carboxylase is also under allosteric control. Burton and Stumpf[42] showed quite conclusively that when over 30 compounds were tested on the wheat-germ enzyme, no activating effects were observed. Similar results have been observed with the E. coli acetyl-CoA carboxylase system (Alberts and Vagelos[48]). However, in a number of plant extracts Burton and Stumpf[42] detected a heat-stable inhibitor which markedly inhibited the carboxylase activity without interfering with the other enzymes responsible for the total synthesis of the fatty acid molecule. Heinstein and Stumpf[49] have now purified the enzyme some 1000-fold and have shown that the enzyme consists of a 7.35 and 9.45 component. The 7.35 component is probably the enzyme responsible for the transfer reaction and the CO_2–biotinyl protein (9.45) is the actual carboxylating component of the total enzyme system. The reaction can thus be written as follows:

$$CO_2 + ATP + E_{9.45} \xrightleftharpoons{Mg^{2+}} CO_2 \sim E_{9.45} + ADP + PO_4 \qquad (8)$$

$$\text{Acetyl-CoA} + E_{7.35} \rightleftharpoons CoA + \text{acetyl} \sim E_{7.35} \qquad (9)$$

$$\text{Acetyl} \sim E_{7.35} + CO_2 \sim E_{9.45} \rightleftharpoons E_{7.35} + E_{9.45} + \text{malonyl-CoA} \qquad (10)$$

On a molecular basis, the plant acetyl-CoA carboxylase occupies an intermediate level in organization between the animal and bacterial carboxylases. On one extreme, the animal carboxylase is a tight complex which cannot be dissociated, whereas on the other extreme the bacterial system dissociates very readily into two separate components, namely, the biotinyl-carboxylating enzyme and the transfer enzyme which transfers CO_2 from the CO_2–biotinyl enzyme to acetyl-CoA[48].

Since plant acyl-carrier protein is a required component of the plant synthetases, a discussion of its properties is now appropriate. In 1964,

Overath and Stumpf[50], in characterizing the soluble extract of avocado mesocarp, observed that activity was greatly diminished when the extract was fractionated with ammonium sulfate. The activity could be restored on the addition of heated (100°/5 min) crude extracts. The boiled preparations contained a protein which was heat-stable but was destroyed on brief exposure to the proteolytic enzyme, papain (EC 3.4.4.10). Furthermore, the heated extract could be completely replaced by boiled extracts of *E. coli* which was also required for activity with ammonium sulfate fractions of *E. coli* and which was subsequently identified as the acyl-carrier protein by Vagelos and Wakil. Simoni *et al.*[51] purified and characterized the avocado mesocarp heat-stable protein as well as a similar protein first isolated by Brooks and Stumpf[52]. The two plant ACPs were purified by procedures similar to those employed for the preparation of *E. coli* except that final purification was only achieved by starch gel electrophoresis to remove a small contaminating protein which had no ACP properties. Spinach ACP has 88 amino acid residues while the avocado ACP contained a cysteinyl residue and approximately 117 amino acid residues. Both proteins contained one 4′-phosphopantetheine residue connected to the hydroxyl of a serine residue of the polypeptide chain *via* the phosphate bridge. Recent work by Matsumura and Stumpf[53] showed that the 5 amino acids surrounding the active site, namely, the serine moiety, to which the 4′-phosphopantetheine component is attached, are identical in *E. coli*, spinach, and *Arthrobacter* ACP. In addition, Matsumura and Stumpf[53] have evidence that there are two species of ACP in spinach leaves. The *N*-terminus of the spinach ACP is alanine in contrast to the *E. coli* and *Arthrobacter* ACP *N*-terminus, namely, serine.

Of considerable interest are the cross-reaction experiments in which *E. coli* synthetase reacts in the presence of either *E. coli* ACP or spinach ACP and the spinach synthetase in turn reacts with *E. coli* ACP or with spinach ACP. Table V summarizes the results which show clearly that since β,γ-dehydrase does not occur in plants, no monoenoic acids are formed when the spinach synthetase reacts in the presence of either spinach ACP or *E. coli* ACP. When *E. coli* synthetase reacts in the presence of *E. coli* ACP, the normal predictable products of vaccenic, stearic and palmitic acids are formed but when spinach ACP is substituted for the *E. coli* ACP, the β-hydroxy saturated fatty acids from C_{12} to C_{18} accumulate. Since the branch point of synthesis in *E. coli* fatty acid synthesis occurs at the level of the β-hydroxy C_{10} fatty acid, the evidence suggests that the β-hydroxy acyl-

References p. 291

TABLE V

COMPARATIVE FATTY ACID SYNTHESIS EMPLOYING SPINACH AND *E. coli* ACPs AND SPINACH AND *E. coli* SYNTHETASES[a]

After Simoni et al.[51]

Products	% Fatty acid formed	
	ACPs	
	Spinach	E. coli
(A) Spinach synthetase		
16:0	20	0
18:0	80	100
(B) E. coli synthetase		
3 OH 12:0	10	0
3 OH 12:1 (5)	6	0
3 OH 14:0	8	0
3 OH 14:1	5	0
3 OH 16:0	10	0
3 OH 16:1 (9)	2	0
3 OH 18:0	12	0
3 OH 18:1 (11)	6	0
18:0	33	25
18:1 (11)	8	75

[a] Starting substrate was malonyl-CoA + crude synthetases.

plant ACPs are defective substrates for the α,β-hydroxyacyl dehydrase of *E. coli*.

In a number of plant extracts so far tested, including spinach, lettuce, castor bean, pea, barley and potato, ACP is a critical component of the systems thereby suggesting that ACP readily dissociates from the fatty acid synthetase in higher plants in sharp contrast to the undissociable ACP-type system in yeast and animal extracts (Stumpf[54]).

(b) Photobiosynthesis

In recent years attention has been focused on the role of lipids in photosynthesizing organelles. Since lipids make up over 50% of the dry weight of chloroplasts, it is obvious that the lipid-containing components play an important role in the functions of the chloroplast. It is now firmly established that the four important acyl lipids in chloroplasts are monogalactosyldiglyceride, digalactosyldiglyceride, sulfoquinovosyldiglyceride, and phosphatidyl glycerol[62]. A number of workers have now shown that these lipids are always present in a wide variety of photosynthesizing tissues, although the ratio of one lipid to another may vary somewhat[80]. Another curious observation is that the fatty acids of photosynthetic tissues invariably contain high concentrations of α-linolenic acid and smaller amounts of linoleic, oleic, palmitic acids and *trans*-3-hexadecenoic acid. Triglycerides do not occur in chloroplasts, presumably because they serve as storage lipids and, having no polar groups or charges, are poorly designed to serve as membrane lipids. Clarification of the biosynthesis, metabolism and functional properties of the acyl lipids in chloroplasts would greatly contribute to a sharper understanding of the mechanisms of photosynthesis[62].

Although acetate had been implicated as a precursor of fatty acids in intact leaves by Eberhardt and Kates[55], and confirmed by James and Piper[56], it was Smirnov[57] in 1960 who first demonstrated that isolated chloroplasts could incorporate acetate into long-chain fatty acids and that light was somehow a component of the synthesizing system. In 1962, Mudd and McManus[58], and Stumpf and James[59] essentially confirmed and extended the results of Smirnov. Both groups showed that acetate in the presence of ATP, CO_2 and Mg^{2+} was readily incorporated into long-chain fatty acids providing light was an additional component of the system. Mudd and his group presented evidence which suggested that the principal role of light was the production of NADPH since disrupted chloroplasts were dependent on a NADPH-generating system rather than light. Stumpf and James[59] repeatedly noticed a pronounced light effect—some 20-fold stimulation over a similar system in the dark—and presented evidence that both ATP and an external NADPH-generating system could not replace the light effect. Stumpf *et al.*[60] compared the activity of isolated chloroplasts with the capacity of the system to catalyze photophosphorylation. It was clearly shown that non-cyclic photophosphorylation was required for effective photobiosynthesis of fatty acids since the products of this system were ATP, NADPH and oxygen. Nevertheless, when these compounds were present in ample concentrations, a consistent light effect was still observed.

Further analysis of the system was hampered by the inability to prepare reasonably active preparations of disrupted chloroplasts. In 1965, Brooks and Stumpf[61] recognized two limiting components of the system. As shown in Table VI, acetate, acetyl-CoA and malonyl-CoA, in the presence of ATP, NADPH and NADH, were essentially ineffective as substrates until ACP was supplied. These results are of considerable interest since they strongly

References p. 291

indicate that at least two factors are limiting the rate of synthesis of fatty acids in chloroplasts. The first is the apparent absence of acetyl-CoA carboxylase in the disrupted chloroplast. Since the intact chloroplasts readily utilize acetate as a substrate when CO_2, ATP, and Mg^{2+} were present, intact chloroplasts must have a functioning acetyl-CoA carboxylase; in the disrupted chloroplasts its activity has disappeared.

TABLE VI

REQUIREMENT FOR ACP BY CHLOROPLAST FATTY ACID SYNTHETASE

After Brooks and Stumpf[61]

Enzyme source	Substrate	mμmoles ^{14}C incorporated into long-chain fatty acids	
		Enzyme	Enzyme + ACP
Lettuce	[2-^{14}C]Acetate	0.3	0.7
	[1-^{14}C]Acetyl-CoA	1.7	3.4
	[2-^{14}C]Malonyl-CoA	2.2	8.4
Spinach	[2-^{14}C]Acetate	0.7	1.3
	[1-^{14}C]Acetyl-CoA	0.8	1.9
	[2-^{14}C]Malonyl-CoA	0.9	19.6

Burton and Stumpf[42], as already indicated, observed that when wheat-germ acetyl-CoA carboxylase is added to intact or disrupted chloroplasts, complete inhibition occurs. The inhibitor is heat-stable, is not inactivated by either papain or ribonuclease (EC 2.7.7.16, 17) and evidently does not inhibit the 6 or more enzymes involved in the formation of the hydrocarbon chain. The inhibitor has also been found in other plant tissues. In some manner this inhibitor is non-functional in the intact chloroplast but exerts its effects on the carboxylase in the disrupted chloroplast.

The second important conclusion is that whereas, in the intact chloroplast, ACP is present in sufficient concentration to allow efficient synthesis of fatty acids from acetate, on disruption, its concentration is so reduced by dilution that it no longer serves as an effective component of the system. However, all the enzymes necessary for the *de novo* synthesis namely, acetyl transacylase, malonyl transacylase, the condensing enzyme, β-keto acyl-ACP reductase, β-enoyl dehydrase and the enoyl reductase, are present in sufficient concentrations after disruption to catalyze the necessary reactions once

ACP concentration is restored to the required amount. Thus, although a dilution of protein undoubtedly occurred on the disruption of the chloroplasts, the dilution effect did not lower the concentrations of the fatty acid synthetase enzymes sufficiently to prevent synthesis from malonyl-CoA but did alter the concentration of endogenous ACP. Thus, the conclusion can be drawn that ACP is present in concentrations in the intact chloroplast sufficient for effective fatty acid synthesis, but is present at such a level that any dilution would bring it to suboptimal concentrations. These results suggest that ACP must be structurally organized presumably in the stromal proteins to allow maximum function under minimum concentrations. Devor and Mudd[63] present evidence that spinach lamellar proteins may serve as acyl carriers in fatty acid synthesis. However, Brooks and Stumpf[61] have already shown that when acyl-carrier protein is supplied to a soluble chloroplast synthetase system further addition of grana particles do not stimulate synthesis. Furthermore grana particles do not substitute for ACP.

A thorough discussion of other aspects of photobiosynthesis of fatty acids by chloroplast systems is found in a review by Stumpf et al.[60]. Several problems still remain to be fully explained. Thus, the nature of the acetyl-CoA carboxylase inhibitor in chloroplasts, the site of ACP in the intact chloroplast, the role of light other than its involvement in non-cyclic photophosphorylation, the origin of acetate—whether it is synthesized in cytoplasmic systems or directly in the chloroplasts, the function of lipid biosynthesis in the mature chloroplast—these questions remain for future investigations.

(c) Developmental aspects

One of the challenging problems in plant-lipid biosynthesis is a better understanding of the mechanisms by which the regulatory systems in oil-seed program the appearance and disappearance of enzymes responsible for the biosynthesis of a number of fatty acids so that at the end of maturation the full complement of the fatty acids characteristic for that species have been deposited. Table II shows how the seed shifts in capacity for fatty acid synthesis as it matures. Both the developing safflower seed and the castor bean seed have been examined in this respect and the capacity of the castor bean seed to synthesize fatty acids from acetate is shown in Table III.

The enzyme system responsible for the synthesis of ricinoleic acid has been intensively examined and its properties will be reported in Section

References p. 291

4b(p. 288). It has been shown now by several workers that at a given period in the maturation of the seed, a very rapid capacity for synthesis is initiated, continues for some days, and then is discontinued as the seed goes into dormancy. When the seed germinates, β-oxidation as well as synthesis now appear, but the acids synthesized are no longer ricinoleic acid. In other words the capacity for ricinoleic acid synthesis has been completely lost.

Another approach to the problem of defining patterns of biosynthesis has been the exploitation of the phenomenon known as "aging". When small discs (1 mm thick by 8 mm in diameter) are prepared from storage tuber tissue (potato, carrot, Jerusalem artichoke), these fresh discs have relatively little metabolic activity (Laties[65]). It has been known for some time that when fresh discs are permitted to "age" under aerobic conditions for 10–24 h in water, a remarkable increase in metabolic activity occurs. Willemot and Stumpf[66] demonstrated an increased uptake of acetate by aging tissue and an increased capacity for lipid synthesis. Of particular interest is the observation that while the fresh tissue did not synthesize linoleic acid from acetate, aged tissue readily synthesized large amounts of linoleic acid. This capacity for synthesis is lost when the tissue is incubated with cycloheximide, an inhibitor of protein synthesis. This model system may reveal interesting new insights into the regulation of fatty acid synthesis.

4. Unsaturation

(a) Mono- and dienoic acids

The mechanisms of introducing a single double bond as well as additional double bonds to form a non-conjugated system are of considerable importance in the food chain of the animal kingdom. It is the plant kingdom that is predominantly responsible for the introduction of the second double bond in oleic acid to form linoleic acid.

In considering the problem of unsaturation in higher plants, two separate aspects emerge, (1) the introduction of the first double bond in the 9,10-position counting from the carboxyl carbon, and (2) the subsequent introduction of additional double bonds, usually a non-conjugated system, as well as the introduction of other groups such as hydroxy, epoxy, and cyclopropanoid groups. A good survey of the different types of double bonds in seed lipids is presented by Wolff[1].

There is now considerable evidence that there are at least two systems which occur in nature responsible for the introduction of a double bond into a hydrocarbon chain. The first is the anaerobic pathway thoroughly described by Bloch and his colleagues[67]. This system is the principal method of introducing a double bond in bacteria. The second pathway is the aerobic pathway and occurs in animal, yeast and plant tissues. Both systems have been thoroughly described elsewhere in this volume. In order to determine if the anaerobic reaction is of significance in higher plants, Hawke and Stumpf[68] carried out a series of experiments with barley tissue to test this system. They first showed that although carboxyl-labelled octanoic, decanoic, lauric and myristic acids were readily converted to oleic acid when incubated with leaf tissue, the process nevertheless required oxygen for the conversion of these substrates to oleic acid. In another series of experiments, 3-hydroxy decanoic, 3-hydroxy lauric and 3-hydroxy myristic acids, all labelled in the 3-position with tritium, were fed to leaf tissue and incubated in the presence and absence of oxygen. According to the anaerobic pathway, these substrates would not require oxygen and each would respectively yield 11-octadecenoic acid, 9-octadecenoic acid and 7-octadecenoic acid. In all cases, the principal product was oleic acid. Oxygen was still an absolute requirement. The results can only be interpreted to mean that the different 3-hydroxy fatty acids were dehydrated to the 2,3-unsaturated fatty acid, reduced to the saturated fatty acid, elongated to stearic acid or a suitable derivative which then was desaturated by the aerobic pathway.

Direct evidence for the aerobic pathway at present is not complete in higher plants. It has been repeatedly observed that while acetate is readily incorporated by a number of tissue slices into saturated and unsaturated fatty acids, and that the fatty acids up to and including myristic acid are elongated and converted to oleic acid by an oxygen-requiring system, palmitic and stearic acids are not desaturated to corresponding monoenoic fatty acids. Nagai and Bloch[70] clearly showed that in isolated chloroplasts a rather complicated system involving a NADPH reductase, ferredoxin, and a specific unsaturatase is involved. The substrate may be either stearyl-ACP or stearyl-CoA.

As for the introduction of further double bonds, McMahon and Stumpf[35] have reported on the conversion of oleyl-CoA to linoleyl-CoA by a microsomal preparation obtained from developing safflower seeds which requires oxygen and NADH. Free oleic acid is inactive. Longer periods of incubation gave evidence for the synthesis of linolenic acid. Similar results have been obtained by Harris and James[33,34].

There is therefore much indirect evidence that the unsaturation process in

References p. 291

plants is clearly an aerobic desaturase system requiring molecular oxygen and resembles the system described by Bloch from yeast and from *Mycobacterium phlei*, as well as from *Euglena* and isolated chloroplasts.

(b) *Hydroxylation and epoxy acid synthesis*

Indirectly related to the unsaturation process are the reactions by which hydroxy acids and epoxy acids are formed in higher plants.

It has already been mentioned that a typical hydroxy fatty acid in plants is ricinoleic acid. This acid makes up over 90% of the fatty acids in the mature castor bean seed. It is synthesized only in the maturing seed by the hydroxylation of oleyl-CoA to ricinoleyl-CoA. The enzyme is missing in the germinating seed. James[28] was the first to report by *in vivo* studies that oleic acid is probably the precursor of ricinoleic acid. Galliard and Stumpf[69] investigated the enzyme system responsible for this catalysis and found that it was a mixed function microsomal hydroxylase requiring oleyl-CoA as the substrate and molecular oxygen and NADH as the components essential for activity. Galliard was also able to demonstrate that the D-12-[^{3}H]oleyl-CoA was rapidly washed out by the enzyme whereas the L-12-[^{3}H]oleyl-CoA was

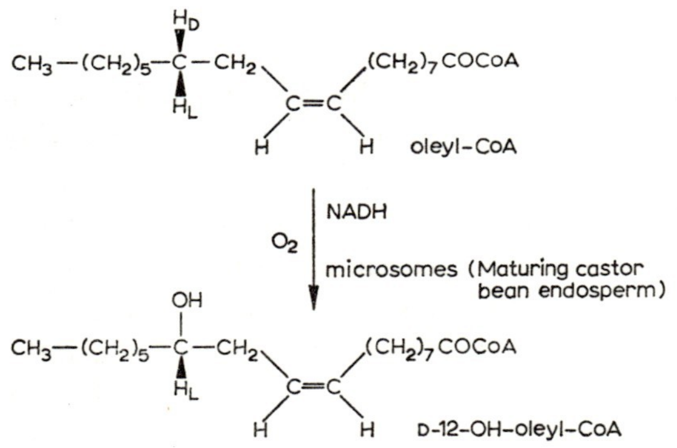

$$CH_3-(CH_2)_5-\overset{H_D}{\underset{H_L}{C}}-CH_2\diagdown_{C=C}\diagup^{(CH_2)_7COCoA}_{H} \quad \text{oleyl-CoA}$$

$$\downarrow O_2 \quad NADH$$

microsomes (Maturing castor bean endosperm)

$$CH_3-(CH_2)_5-\overset{OH}{\underset{H_L}{C}}-CH_2\diagdown_{C=C}\diagup^{(CH_2)_7COCoA}_{H} \quad \text{D-12-OH-oleyl-CoA}$$

Fig. 6.

retained. These results were recently confirmed by Morris[71] working with developing castor bean endosperm slices. Furthermore, the enzyme is highly specific in that no other substrate is hydroxylated by the system. Thus, *cis*-vaccenic, stearic, elaidic, palmitoleic and linoleic acids and their CoA derivatives were inactive. The sequence of events can be depicted as in Fig. 6.

Another acid which is quite novel is cis-9,10-epoxyoctadecanoic acid. Several epoxy acids have now been isolated from seed oils. These include 9,10-epoxyoctadecanoic acid, 9,10-epoxy-12-octadecenoic acid, 12,13-epoxy-9-octadecenoic acid, and 15,16-epoxy-9,12-octadecadienoic acid. The cis-9,10-epoxy-octadecanoic acid has been found in a number of fungal spores in significant concentrations.

Recently, it has been shown by Knoche[72] that red-stem-rust infected wheat plants readily incorporate acetate as well as oleic acid into 9,10-epoxy-octadecanoic acid. Molecular oxygen is required for the formation of the epoxy acid. Little is known concerning the enzymatic mechanism of this interesting reaction.

5. Waxes

Plant waxes are found widespread on the surfaces of plant structures, in particular on the upper surface of leaves. Of great variety and complexity, surface lipids usually contain long-chain hydrocarbons, waxy esters consisting of primary alcohols and long-chain fatty acids, and a large number of minor components, depending on the plant species. Hydrocarbons are usually n-paraffins varying from 20 to 35 carbon atoms; the waxy esters are made up of fatty acids and alcohols ranging from 10 to 30 carbon atoms, usually even-numbered. The hydrocarbons, mono- and diketones, and secondary alcohols commonly have odd-numbered carbon atoms while the fatty acids, primary alcohols, aldehydes, hydroxy and dicarboxylic acids are predominantly even in carbon-numbered chains. Branched hydrocarbons primarily 2- and 3-methyl hydrocarbons, cycloparaffins, terpenes and even flavones are sometimes components of waxes. For further details reviews by Kolattukudy[73] and Eglinton and Hamilton[74] are noteworthy.

Despite the complexity of the waxes, information is now becoming available concerning the biosynthesis of some of the simple hydrocarbons. Kolattukudy[75] has recently provided sound evidence for an elongation–decarboxylation pathway to explain the synthesis of a C_{29} straight-chain hydrocarbon in *Brassica oleracea*. According to this pathway, palmitic acid is synthesized *de novo* from acetate in leaf tissue. This acid is then further elongated by another system which is sensitive to trichloroacetic acid ($10^{-4}M$) until a C_{30} fatty acid is attained which then undergoes a decarboxylation to form a C_{29} hydrocarbon. Evidence in support of this pathway includes the total incorporation of palmitic acid into the hydrocarbon without any degradation, the insensitivity of paraffin biosynthesis to chlorophenyl dimethylurea when palmitate is fed, and great sensitivity to trichloroacetate anion although this inhibitor does not interfere with the *de novo* synthesis of the C_{16} and C_{18} fatty acids. Of further interest is the observation that epidermal layers of cells removed from the leaves of *Senecio odoris* incorporate acetate into both paraffins and into very-long-chain fatty acids whereas internal tissue cells incorporate acetate primarily into C_{16} and C_{18} fatty acids.

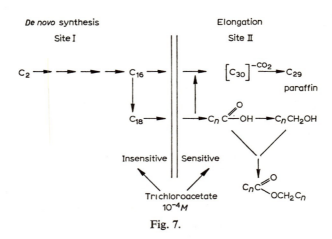

Fig. 7.

Another interesting group of paraffins are the branched-chain hydrocarbons in tobacco leaf wax. The hydrocarbons of this wax contain 17% isoalkanes and 19% anteisoalkanes. Kaneda[76] as well as Kollatukudy[75] has shown that the branched-chain hydrocarbons found in tobacco leaves are derived from the breakdown of valine, leucine and isoleucine to isobutyric, isovaleric, and α-methyl butyric acids respectively which then serve as the primer in the elongation–decarboxylation pathway to yield iso- and anteiso-paraffins.

The further explanation of the mechanisms of biosynthesis of the important components of the plant waxes on an enzyme level will be of considerable interest. Since the very-long-chain fatty acids and their derivatives are of unusually low solubility, the enzymes which catalyze their synthesis probably will be uniquely designed to cope with these physical properties.

REFERENCES

1   IVAN A. WOLFF, *Science*, 154 (1966) 1140.
2   A. L. TAPPEL, in P. D. BOYER, H. LARDY AND K. MYRBÄCK (Eds.), *The Enzymes*, Vol. VIII, Academic Press, New York, 1963, p. 225.
3   M. HAMBURG AND G. SAMMUELSSON, *J. Biol. Chem.*, 242 (1968) 5329.
4   A. DOLEV, W. K. ROHWEDDER AND H. J. DUTTON, *Lipids*, 2 (1967) 28.
5   A. DOLEV, W. K. ROHWEDDER, T. L. MOUNTS AND H. J. DUTTON, *Lipids*, 2 (1967) 33.
6   D. C. ZIMMERMAN, *Biochem. Biophys. Res. Commun.*, 23 (1966) 398.
7   E. H. NEWCOMB AND P. K. STUMPF, in W. D. MCELROY AND B. GLASS (Eds.), *Phosphorus Metabolism*, Vol. II, Johns Hopkins Press, Baltimore, Md., 1952, p. 291.
8   R. O. MARTIN AND P. K. STUMPF, *J. Biol. Chem.*, 234 (1959) 2548.
9   C. H. S. HITCHCOCK AND A. T. JAMES, *J. Lipid Res.*, 5 (1964) 593.
10 C. H. S. HITCHCOCK AND A. T. JAMES, *Biochim. Biophys. Acta*, 116 (1966) 413.
11 C. H. S. HITCHCOCK, L. J. MORRIS AND A. T. JAMES, *Europ. J. Biochem.*, 3 (1968) 419.
12 C. H. S. HITCHCOCK, L. J. MORRIS AND A. T. JAMES, *Europ. J. Biochem.*, 3 (1968) 473.
13 D. STEINBERG, J. H. HERNDON JR., B. W. UHLENDORF, C. E. MIZE, J. AVIGAN AND G. W. A. MILNE, *Science*, 156 (1967) 1740.
14 P. K. STUMPF AND GEORGE A. BARBER, *Plant Physiol.*, 31 (1956) 304.
15 C. A. REBEIZ AND P. CASTELFRANCO, *Plant Physiol.*, 39 (1964) 932.
16 M. YAMADA AND P. K. STUMPF, *Plant Physiol.*, 40 (1965) 653.
17 M. YAMADA AND P. K. STUMPF, *Plant Physiol.*, 40 (1965) 659.
18 H. BEEVERS, *Nature*, 191 (1961) 433.
19 R. W. BREIDENBACH AND H. BEEVERS, *Biochem. Biophys. Res. Commun.*, 27 (1967) 462.
20 W. STOFFEL AND H. CAESAR, *Z. Physiol. Chem.*, 341 (1965) 76.
21 J. GIOVANNELLI AND P. K. STUMPF, *J. Biol. Chem.*, 231 (1958) 411.
22 M. D. HATCH AND P. K. STUMPF, *Arch. Biochem. Biophys.*, 96 (1962) 193.
23 Y. KAIZRO AND S. OCHOA, *Advan. Enzymol.*, 26 (1964) 283.
24 R. A. STINSON AND M. S. SPENCER, *Biochem. Biophys. Res. Commun.*, 34 (1969) 120.
25 H. J. EVANS AND M. KLEIWER, *Ann. N.Y. Acad. Sci.*, 112 (1964) 735.
26 P. R. VAGELOS AND J. M. EARL, *J. Biol. Chem.*, 234 (1959) 2272.
27 E. H. NEWCOMB AND P. K. STUMPF, *J. Biol. Chem.*, 200 (1953) 233.
28 A. T. JAMES, in The Control of Lipid Metabolism, *Proc. Biochem. Soc. Symp.*, No. 24, Oxford, Academic Press, London, 1963, p. 17.
29 A. T. JAMES, H. C. HARDAWAY AND J. P. W. WEBB, *Biochem. J.*, 95 (1965) 448.
30 D. T. CANVAIN, *Can. J. Biochem. Physiol.*, 41 (1963) 1879.
31 M. YAMADA AND P. K. STUMPF, *Biochem. Biophys. Res. Commun.*, 14 (1964) 165.
32 A. T. JAMES, R. V. HARRIS, C. H. S. HITCHCOCK, B. J. B. WOOD AND B. W. NICHOLS, *Fette, Seifen, Anstrichmittel*, 67 (1965) 393.
33 R. V. HARRIS AND A. T. JAMES, *Biochim. Biophys. Acta*, 106 (1965) 456.
34 R. V. HARRIS AND A. T. JAMES, *Biochem. J.*, 94 (1965) 15C.
35 V. MCMAHORN AND P. K. STUMPF, *Biochim. Biophys. Acta*, 84 (1964) 359.
36 C. T. BARTELS, A. T. JAMES AND B. W. NICHOLS, *Europ. J. Biochem.*, 3 (1967) 7.
37 W. G. HAIGH AND A. T. JAMES, *Biochim. Biophys. Acta*, 137 (1967) 391.
38 M. J. K. MACEY AND P. K. STUMPF, *Plant Physiol.*, 43 (1968) 1637.
39 S. F. YANG AND P. K. STUMPF, *Biochim. Biophys. Acta*, 98 (1965) 19.
40 M. D. HATCH AND P. K. STUMPF, *J. Biol. Chem.*, 239 (1961) 2879.
41 M. D. HATCH AND P. K. STUMPF, *Plant Physiol.*, 37 (1962) 121.
42 D. BURTON AND P. K. STUMPF, *Arch. Biochem. Biophys.*, 117 (1966) 604.
43 C. GREGOLIN, E. RYDER, A. K. KLEINSCHMIDT, R. C. WARNER AND M. D. LANE, *Proc. Natl. Acad. Sci. (U.S.)*, 56 (1966) 148.

44 P. R. VAGELOS, A. W. ALBERTS AND D. B. MARTIN, *J. Biol. Chem.*, 238 (1963) 533.
45 M. WAITE AND S. J. WAKIL, *J. Biol. Chem.*, 237 (1962) 2740.
46 M. MATSUHOSHI, S. MATSUHOSHI AND F. LYNEN, *Biochem. Z.*, 340 (1964) 263.
47 D. WHITE AND H. P. KLEIN, *Biochem. Biophys. Res. Commun.*, 20 (1965) 78.
48 A. W. ALBERTS AND P. R. VAGELOS, *Proc. Natl. Acad. Sci. (U.S.)*, 59 (1968) 561.
49 P. HEINSTEIN AND P. K. STUMPF, *Federation Proc.*, 27 (1968) 647.
50 P. OVERATH AND P. K. STUMPF, *J. Biol. Chem.*, 239 (1964) 4103.
51 R. D. SIMONI, R. S. CRIDDLE AND P. K. STUMPF, *J. Biol. Chem.*, 242 (1967) 573.
52 J. L. BROOKS AND P. K. STUMPF, *Arch. Biochem. Biophys.*, 116 (1966) 108.
53 S. MATSUMURA AND P. K. STUMPF, *Arch. Biochem. Biophys.*, 125 (1968) 932.
54 P. K. STUMPF, unpublished observations, 1968.
55 F. M. EBERHARDT AND M. KATES, *Can. J. Botany*, 35 (1957) 907.
56 A. T. JAMES AND E. A. PIPER, *J. Chromatog.*, 5 (1961) 265.
57 B. P. SMIRNOV, *Biokhimiya*, 25 (1960) 419.
58 J. B. MUDD AND T. T. MCMANUS, *J. Biol. Chem.*, 237 (1962) 2057.
59 P. K. STUMPF AND A. T. JAMES, *Biochim. Biophys. Acta*, 70 (1963) 20.
60 P. K. STUMPF, J. M. BOVE AND A. GOFFEAU, *Biochim. Biophys. Acta*, 70 (1963) 260.
61 J. L. BROOKS AND P. K. STUMPF, *Biochim. Biophys. Acta*, 98 (1965) 213.
62 B. W. NICHOLS AND A. T. JAMES, in J. D. PRIDHAM (Ed.), *Plant Cell Organelles*, Academic Press, New York, 1968.
63 K. A. DEVOR AND J. B. MUDD, *Plant Physiol.*, 43 (1968) 853.
64 P. K. STUMPF, J. L. BROOKS, T. GALLIARD, J. C. HAWKE AND R. D. SIMONI, in T. W. GOODWIN (Ed.), *Biochemistry of Chloroplasts*, Vol. II, Academic Press, New York, 1967, p. 213.
65 G. G. LATIES, in B. WRIGHT (Ed.), *Control Mechanisms in Respiration and Fermentation*, Ronald Press, New York, 1963, p. 129.
66 C. WILLEMOT AND P. K. STUMPF, *Can. J. Botany*, 45 (1967) 579.
67 K. BLOCH, in J. K. GRANT (Ed.), *The Control of Lipid Metabolism*, Academic Press, New York, 1963, p. 1.
68 J. C. HAWKE AND P. K. STUMPF, *J. Biol. Chem.*, 240 (1965) 4746.
69 T. GALLIARD AND P. K. STUMPF, *J. Biol. Chem.*, 241 (1966) 5806.
70 J. NAGAI AND K. BLOCH, *J. Biol. Chem.*, 243 (1968) 4626.
71 L. J. MORRIS, *Biochem. Biophys. Res. Commun.*, 29 (1967) 311.
72 H. W. KNOCHE, *Lipids*, 3 (1968) 163.
73 P. E. KOLATTUKUDY, *Science*, 159 (1968) 498.
74 G. EGLINTON AND R. J. HAMILTON, *Science*, 156 (1967) 1322.
75 P. E. KOLATTUKUDY, *Plant Physiol.*, 43 (1968) 375.
76 T. KANEDA, *Biochemistry*, 6 (1967) 2023.
77 R. W. RINNE, *Plant Physiol.*, 44 (1969) 89.
78 D. HUTTON AND P. K. STUMPF, *Plant Physiol.*, 44 (1969) 508.
79 H. BEEVERS, personal communication, 1968.
80 P. G. ROUGHAN AND R. D. BATE, *Phytochem.*, 8 (1969) 363.
81 R. P. A. SIMS, W. G. MCGREGOR, A. G. PLESSERS AND J. C. MES, *J. Am. Oil Chemists' Soc.*, 38 (1961) 276.
82 R. O. GLEW AND P. K. STUMPF, unpublished observations, 1968.
83 P. K. STUMPF, *Nature*, 194 (1962) 1158.

Chapter VII

Lipid Metabolism in Nervous Tissue

A. N. DAVISON

Biochemistry Department, Charing Cross Hospital Medical School, London (Great Britain)

1. Introduction

Morphological examination of the central nervous system reveals an exceedingly complicated fine structure. The neurones with their dendrites and long myelinated axons together with various types of glial cells are closely intertwined in varying architectural patterns. This close packing of different cells, the inert mass of myelin and the non-dividing neurone contribute to some of the special features of the biochemistry of the brain and spinal cord. The organization of the peripheral nervous system is less complex for it is essentially composed of myelinated fibre tracts with attendant Schwann cells.

Knowledge of the lipid composition and metabolism of the component parts of the nervous system is consequently of particular value in interpreting lipid metabolism of the whole organ. The first part of this chapter is concerned with the biochemistry of neural lipids; in the few cases where there are known differences between central and peripheral nervous tissue these will be emphasized. The second part of the chapter will relate this more general neurochemistry to anatomical structures within the brain and, to a lesser extent, peripheral nerve. Finally, the possible physiological and pharmacological significance of lipid metabolism in nervous tissue will be discussed.

2. Neural lipid biochemistry

Since classical methods for the complete extraction of lipids were never satisfactory when applied to fresh nervous tissue, Folch and his colleagues introduced an ingenious new technique for the separation of tissue lipids[1]. Their method facilitated the almost complete extraction of lipid from fresh moist tissue by the inclusion of methanol together with chloroform as solvent (acidified chloroform–methanol mixtures have to be used for the extraction of the polyphosphoinositides)[2]. Fresh tissue homogenized in a mixture of 19 volumes of chloroform–methanol (2:1 v/v) is filtered and

TABLE I

NEURAL LIPIDS

Class	Lipid	Probable biological rôle	Mean molecular weight
Sterol	Cholesterol	Membranes myelin	387
	Cholesterol ester	Sterol precursors	—
	Desmosterol		385
Sphingolipids	Ceramide	Precursor	590
	Cerebroside	Myelin	788
Glycolipids	Sulphatide	Myelin	876
	Gangliosides	Neurone cell body	—
Phospholipid	Sphingomyelin	Myelin membrane	756
Phospholipids	Phosphatidyl choline (lecithin)		769
		All phospholipids present in mitochondrial membrane, myelin, etc.	
	Phosphatidal ethanolamine	Myelin	756
Cephalins	Phosphatidyl ethanolamine		
	Phosphatidyl serine		816
	Mono-, di- and tri-phosphoinositides	Polyphosphoinositides in myelin or axolemma	—
	Phosphatidic acid	Metabolic intermediate	—
	Cardiolipin	Mitochondrial membrane	—

The molecular weights quoted are mean figures for human-brain lipids[4].

TABLE II

COMPARISON OF HUMAN NEURAL AND LIVER LIPID COMPOSITIONS

Constituent	Liver[a]	Foetal brain[8] (5 months)	Adult[4] (grey matter)	Adult[4] (white matter)	Adult[9] (peripheral nerve)
Water	75	90	82.3	75.2	44.3
Total lipid	24	21.7	39.6	64.6	65.8
Cholesterol	2.9	4.3	7.2	15.1	14.5
Cholesterol ester	2.3	0[b]	0	0	0
Triglyceride	11.5	0	0.2^{10}	0.4^{10}	4.4
Cerebroside	—	0.016[c]	2.3	12.5	8.3
Sulphatide	—	0.01	0.8	3.0	1.4
Ceramide	—	0.05[d]	0.5	0.8	—
Ethanolamine phospholipid	7.2	6.2	9.2	9.1	10.3
Phosphatidyl serine			2.9	4.2	5.4
Phosphatidyl choline		8.2	9.0	8.2	8.4
Sphingomyelin		0.45	1.9	5.2	8.9
Gangliosides	—	0.9	1.91^{11}	0.39^{11}	—

All values except water are expressed as a percentage of the dry weight.
[a] Kindly supplied by Dr. I. McDonald, Guy's Hospital.
[b] traces of esters found.
[c] galactocerebrosides, a trace of glucocerebroside present.
[d] ceramide dihexosides and trihexosides.

contaminants removed by addition of water or salt solutions and separation of an upper aqueous phase. This upper layer contains water-soluble materials and gangliosides while the chloroform lower phase retains all other lipids. Some of the lipid present in the chloroform-rich lower layer is bound to protein. This so-called proteolipid behaves physically as lipid but disruption of the lipid–protein bonds occurs on taking the extracts to dryness or on treatment with alkaline solutions of high ionic strength. Proteolipid is found largely but not exclusively in the nervous system where it is concentrated in myelin (almost all the myelin protein is chloroform–methanol soluble[3]). A trypsin-resistant protein residue—neurokeratin—is the predominant myelin protein of peripheral nerve myelin. The various lipid species present (Table I) in the lower washed phase may then be conveniently separated by column or thin-layer chromatography before analysis[5-7].

Comparison of the lipid composition of the brain, with that of the liver (Table II) emphasizes some of the differences between nervous and other tissues. Thus, triglycerides are essentially absent from normal nervous tissue and cholesterol esters are found only during development. Although in most other respects immature brain and adult grey matter has a composition similar to that of other organs, the white matter is unusually rich in lipid (30% of the wet weight) and is characterized by its high content of cholesterol and cerebrosides (Table II). It will be noted that the chemical composition of peripheral nerve more closely resembles that of white than grey matter. Cerebrosides are found in only very small amounts in other tissues and these lipids may therefore be regarded as typical of nervous tissue.

3. Cerebrosides and sulphatides

(a) General properties

Cerebrosides are hexose-containing derivatives of *N*-acylsphingosine or dihydrosphingosine[12,13]. They may therefore be classified either as sphingolipids or as glycolipids for they contain both a sugar *e.g.* D-galactose and the long-chain amino dialcohol sphingosine. Cerebrosides and their sulphate esters (sulphatides) are readily distinguished from sphingomyelin by the absence of phosphorus and from gangliosides by the absence of neuraminic acid. Cerebrosides containing cerebronic acid (a C_{24} fatty acid with an α-hydroxyl group) are often referred to as phrenosine and those with the unsubstituted saturated fatty acid (C_{24} lignoceric acid) are called kerasin.

Analysis of the acids linked as amides in cerebroside and sulphatide[14] shows that these sphingolipids have a high proportion of long-chain fatty acids (14–26 carbon atoms) both in the normal and α-hydroxy series. In cerebroside isolated from mature human brain 24:0 (lignoceric, n-tetracosanoic) and 24:1 (nervonic, 15-tetracosanoic) were the major fatty acids in both series although the actual amounts varied with the age of the subject and the anatomical location in the brain.

The sulphatides are now known to be esters of cerebroside with the sulphate group located at C-3 of the galactose moiety[15,16]. Two broad groups of the class of sulpholipid exist corresponding to the phrenosine and kerasine type of cerebrosides. These can be easily recognised by chromatography after removing sulphate by mild acid hydrolysis at room temperature[17]. In all other respects the sulphatides resemble (*e.g.* β-galactosides, fatty acids, etc.) their parent cerebrosides. At first it was thought the sulphate group or the sulphatide molecule was localized on C-6 of the galactose moiety, however, isolation of 2,4,6-tri-*O*-methylgalactoside by gas–liquid chromatography[18], stability to alkaline hydrolysis and resistance to periodic acid oxidation established the sulphate at C-3 of the galactose ring. Suggestions for the possibility of naturally occurring polysulphatides containing 3,6-diester bonds have been advanced[19,20]. Sulphatides in large amounts have been isolated from the brain in cases of metachromatic leucodystrophy where the sulpholipid replaces unesterified cerebroside in myelin. A full account of the chemistry of the cerebrosides and sulphatides is given by Hanahan and Brockerhoff[13] in Volume 6 of this series.

(b) Estimation and separation from nervous tissue

The cerebrosides and sulphatides may be readily separated on thin-layer silicic acid chromatographic plates using, for example, chloroform–methanol–ammonia as developing solvent[21]. Larger quantities can be conveniently fractionated on chromatographic columns (alumina, silicic acid or florisil) using various chloroform–methanol or chloroform–ethanol–water mixtures. Sulphatides of high purity have been isolated from white matter by Svennerholm and Thorin[22]; in their method a crude cerebroside sulphate fraction is separated on DEAE cellulose columns using suitable eluants.

Various methods may be employed for the analysis of cerebrosides. Following acid hydrolysis of lipid extracts released galactose can be estimated by the orcinol reaction[23]. This is a convenient method but care has to be

References p. 325

taken that other substances (*e.g.* gangliosides) do not interfere with the colour reaction. Alternatively after hydrolysis, sphingosine may be estimated as chloroform-soluble nitrogen or directly by reaction with methyl-orange. The latter method of Lauter and Trams[24] has proved to be particularly sensitive. Similar methods of analysis may be applied to sulphatides if separated in a pure state, for example, by thin-layer chromatography[25]. Alternatively it is possible to estimate released sulphate following prolonged acid hydrolysis of whole washed lipid extracts[26], to utilize absorption at 8.02 μ in the infrared region of the spectrum[27] or simply to estimate metachromatic staining with cresyl violet in order to directly analyse sulphatide[28].

(c) Galactolipid metabolism within the nervous system

(i) Biosynthesis of cerebroside and sulphatide

Using labelled intermediates Brady and his colleagues[29] have been able to establish the main pathway for the biosynthesis of sphingosine by tissue microsomal fractions from serine and palmitic aldehyde with pyridoxal phosphate and manganese ions as requisite co-factors (Fig. 1). The primary

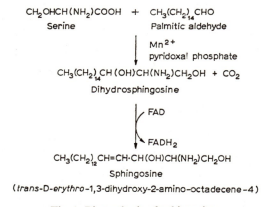

Fig. 1. Biosynthesis of sphingosine.

step in cerebroside synthesis was shown to be the transfer of galactose from uridine diphosphogalactose to sphingosine to give psychosine (Fig. 2). The reaction is catalysed by galactosylsphingosine transferase—an enzyme specific for the natural *erythro* isomer of sphingosine and UDP–galactose. Acylation of psychosine follows by transfer from long-chain fatty acyl coenzyme A derivatives to give the completed galactocerebroside.

The addition of a sulphate group to cerebroside receptor molecule leads to the synthesis of sulphatides through the participation of active sulphate (3'-phosphoadenosine 5'-phosphosulphate or PAPS). This sulphation process is catalysed by a microsomal system from either kidney, liver or brain. The synthetase using active sulphate (PAPS) as donor has been found to be most active in the 18–22 day post-natal rat-brain system when its activity can be correlated with active myelination and with the availability of PAPS in the brain[30,31].

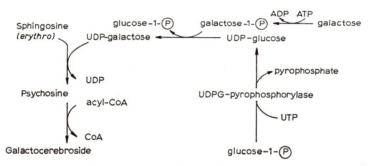

Fig. 2. Synthesis of cerebroside from sphingosine.

(ii) Catabolism of cerebroside and sulphatide

Cerebroside metabolism can be conveniently followed by the incorporation and turnover of [^{14}C]galactose or other labelled precursors taken up into the cerebroside molecule. Such studies show that synthesis like that of the sulphatides is most active in the developing brain at a time coinciding with myelination[32]. Once labelled, most of the brain cerebroside undergoes only slow turnover with a half-life in the order of 100 days[33–35]. Despite the stability *in vivo*, nervous tissue has been found to contain a galactosidase which catalyses the breakdown of cerebrosides. This enzyme has been isolated from rat and pig brain by Hajra *et al.*[36] and shown to promote the hydrolysis of cerebroside:

stearoyl-[^{14}C]psychosine → stearoyl-[^{14}C]sphingosine + galactose
   (cerebroside)   (ceramide)

The same enzyme appears to be active in central nervous tissue *in vivo* for a highly labelled ceramide can be isolated following injection of lignoceroyl-

References p. 325

[^{14}C]psychosine (kerasin) into the brain. The ceramide may then be further degraded by ceramidase to fatty acid and sphingosine as shown by Gatt[37].

[^{35}S]Sulphate is the most suitable precursor for studying sulphatide metabolism. Synthesis of the sulpholipid both *in vitro* and in the intact animal is most active during the period of brain development and again like cerebrosides, the sulphatides are metabolically stable[26,38]. Hydrolysis of the sulphate ester bond on C-3 of the galactose portion of the molecule has been found to be catalysed by an arylsulphatase present in nervous tissue[39]. It is of special interest that reduction in cerebral sulphatase activity[40] is associated with the slow accumulation of cerebroside sulphate leading to the rare lipidosis—metachromatic leucodystrophy. In this disease more sulphatide and less cerebroside is found than is present in normal brain, and the quantity of myelin which can be isolated from the diseased brain by centrifugal techniques is reduced[41]. The myelin composition differs from normal in being abnormally rich in sulphatides and deficient in cerebrosides.

4. Gangliosides

Gangliosides are a class of complex glycolipids containing sialic acid (*N*-acetylneuraminic acid), galactosamine, hexoses and ceramide in varying proportions (see Svennerholm, chapter IV, p. 201, of this volume). The high sugar content of the gangliosides accounts for their greater solubility in water than other types of lipid. Four types of ganglioside structure predominate in nervous tissue although at least ten gangliosides have so far been recognised in human brain[42,43]. One of the commonest gangliosides is the monosialoganglioside isolated and characterised from the brain[44]. This molecule forms the basic structural unit from which addition of one or more *N*-acetylneuraminic acid (NANA) residues gives rise to the other major gangliosides found in human and beef brain. The proposed structures of the less common mono-, di- and trisialogangliosides found in nervous tissue have been described[43]. The fatty acid composition of gangliosides is unusual, for stearate has been found to be the predominant fatty acid; 2-hydroxy fatty acids found in some other brain sphingolipids are not present.

Gangliosides are generally extracted from tissues in combination with other types of lipid. Such solvent systems as chloroform–methanol (1:1, v/v) have proved suitable for extraction of gangliosides from nervous tissue particularly if combined with prior extraction by acetone to increase both yield and purity. The extracted gangliosides may be separated by solvent partition[1], column chromatography[45,46] or thin-layer techniques[47]. Analytical methods for the estimation of neuraminic acid are usually based on an adaptation of Bials test[11].

Metabolism of gangliosides

Burton et al.[48], Suzuki[49] and Rosenberg and Stern[50] have shown that gangliosides accumulate in the rat brain most rapidly from 10–20 days post partum. Little net increase in total ganglioside content occurs 30 days after birth. However, in the human brain the adult pattern shown by the four major gangliosides is not reached until about 30 years (gangliosides type G_3 (G_{D1_a}) and G_4 (G_{M1}) becoming relatively less plentiful during this period).

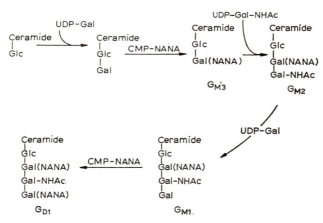

Fig. 3. Postulated route for the biosynthesis of a disialoganglioside. UDP–Gal, uridine diphosphogalactose; Glc, glucose; CMP, cytidine monophosphate; NANA, N-acetylneuraminic acid; Gal, galactose; NHAc, N-acetyl.

Possible routes for the biosynthesis of a disialoganglioside are shown in Fig. 3. Thus, embryonic chick brain has been shown[51] to contain a sialyl transferase which catalyses the addition of cytidine monophosphate neuraminic acid[52] to ceramide dihexoside. Glycolipid galactose transferase catalyses the transfer of galactose from uridine diphosphate galactose to the appropriate acceptors. Glucocerebroside does not appear, however, to act as a ganglioside precursor.

Glucosamine, glucose, galactose and serine have all been used as pre-

References p. 325

cursors for rat-brain gangliosides[48,53–55]. These studies indicate that labelled materials are rapidly incorporated into the gangliosides of different aged rats, maximum radioactivity being found 12–30 h after injection in 15-day-old rats, hence labelled gangliosides undergo quite fast turnover compared to cerebrosides.

5. Cholesterol

Although cholesterol is widely distributed in the various body tissues, it is found in highest concentration within the central and peripheral nervous systems. Only traces of other sterols are present in neural tissue of adult animals.

(a) Separation and analysis of sterols

The complete extraction of sterols from nervous tissue is readily achieved with solvents such as chloroform–methanol (2:1, v/v). Methods for the subsequent isolation of cholesterol from other lipids in the extract and its analysis are also well established. Cholesterol may for example be isolated as its digitonide[56], it may be separated by column chromatography on alumina[57–59], silicic acid[9] or on the micro scale, sterols and their esters can readily be separated by chromatography on thin-layer silicic acid plates. Many colorimetric methods are also available for the estimation of cholesterol, thus the Liebermann–Burchard reaction has been widely adopted for the determination of isolated cholesterol[56,60]. Other procedures allow analysis of cholesterol in total lipid extracts[61,62]. Special methods such as thin-layer or gas-liquid chromatography may be necessary for the separation and analysis of cholesterol in the presence of other sterols (*e.g.* desmosterol).

(b) Biosynthesis of cholesterol

(i) *In vitro*

While adult nervous tissue contains relatively large amounts of cholesterol, much less of the sterol is present during the early stages of development. Accumulation of cholesterol parallels the deposition of myelin and this rather prolonged process is preceded by the accumulation of variable amounts of esterified cholesterol together with small quantities of desmosterol (24-dehydrocholesterol; cholesta-5,24-dien-3β-ol) and zymosterol (5α-cholesta-

8,24-dien-3β-ol) but their appearance is transitory and only traces of these substances are found in mature nervous tissue.

Although the metabolic pathways for the biosynthesis of cholesterol in other tissues are well established not all steps in the biosynthetic pathway have been recognised in neural tissue[63]. Grossi et al.[64] and Garattini et al.[67] have found that slices or fortified homogenates of rat brain will convert [1-^{14}C]acetate or [2-^{14}C]mevalonate to [^{14}C]squalene and [^{14}C]cholesterol [64-66]. Liver slices utilize mevalonate more effectively for cholesterol biosynthesis than they do acetate. On the other hand, brain slices from young rats synthesize relatively less cholesterol from either substrate but, in this tissue, acetate is 4.6 times more effective a precursor than mevalonate. Similar results have been obtained with cell-free preparations, thus eliminating the possibility that mevalonate penetrates less readily into brain cells than into liver cells; and, although in vitro systems were first used in these studies, other experiments on 14-day-old rats in vivo gave substantially similar results[66]. A possible explanation suggested by Paoletti and his colleagues is that there is a deficiency in mevalonic kinase and decarboxylase in the brain preparations of young rats.

Mevalonic lactonase is lacking in the brain, in comparison with the liver in the same animal and as a result potassium mevalonate is far better utilised than mevalono-lactone by brain preparations (Paoletti, personal communication).

Various sterol intermediates have been isolated from the developing brain. Zymosterol ($\Delta^{8,24}$-cholestadienol), $\Delta^{7,24}$-cholestadienol and $\Delta^{5,7,24}$-cholestatrienol have been found. The pattern of sterol synthesis in the developing rat brain indicates the presence of two slow steps. Holstein et al.[67] have demonstrated the slow turnover of ^{14}C activity incorporated into the zymosterol fraction of 5-day rat brain (turnover time of about 27 h) which makes it likely that the rate-controlling step is located along the pathway leading from the 8,9-double bond to the 5,6-double bond compounds (see Fig. 4). The second slow step resembles that observed in chick-embryo brain by Fish et al.[68] in which there is only a slow reduction of the 24,25-double bond precursors of cholesterol. With increasing age slices of brain become less able to synthesize cholesterol from simple precursors such as [2-^{14}C]mevalonate or [U-^{14}C]glucose[69] and it is indeed difficult to demonstrate any cholesterol synthesis at all using adult-brain slices[70,71]. Using suitably fortified minced brain preparations from 1-year-old animals however, Nicholas[72] was able to show utilization of [2-^{14}C]acetate and

References p. 325

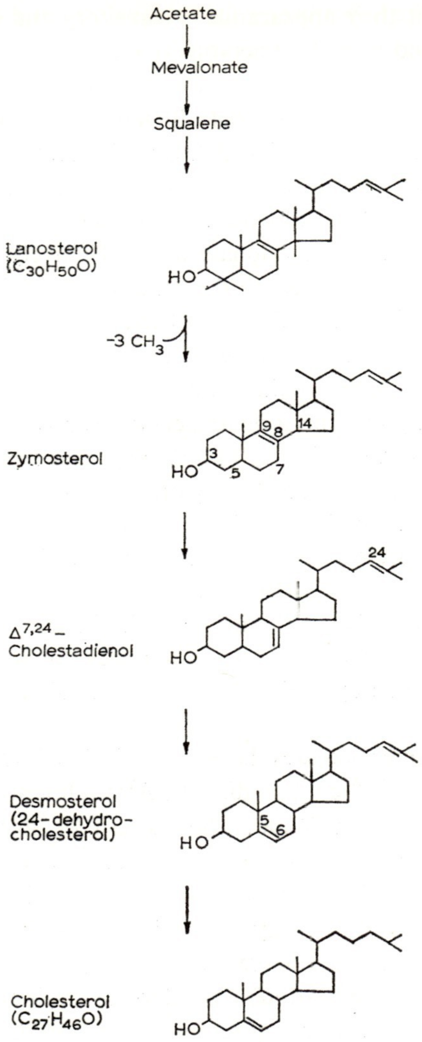

Acetate
↓
Mevalonate
↓
Squalene
↓

Lanosterol
($C_{30}H_{50}O$)

↓ -3 CH_3

Zymosterol

↓

$\Delta^{7,24}$-Cholestadienol

↓

Desmosterol
(24-dehydro-cholesterol)

↓

Cholesterol
($C_{27}H_{46}O$)

Fig. 4. Steps in the pathway of cholesterol synthesis

mevalonate to form labelled nonsaponifiable material. After incubation [^{14}C]squalene and [^{14}C]cholesterol were isolated from the preparations, by precipitation as digitonide and conversion to the dibromide, and thus a definite conversion by adult-rat brain was established. As expected, liver was much more effective in synthesizing cholesterol, but both adult brain and liver slices utilized mevalonate more efficiently than acetate. This therefore indicates an interesting difference in the cerebral metabolism of young and mature rats. It may be concluded that earlier failures to demonstrate cholesterol biosynthesis by adult-brain slices were due to the fact that the very small quantities of labelled cholesterol synthesized are not easily detected.

(ii) In vivo

The early work of Waelsch et al.[73] established that heavy water was readily incorporated into the unsaponifiable lipids of the brain when fed to newborn rats. The extent of such uptake was inversely related to the age of the rat at the time of dosage and directly related to their current rate of myelin deposition. Only traces of deuterium were found in the brain sterols of adult rats fed heavy water for periods of 4–7 days. Other investigators have been unable to detect any labelling of adult-brain cholesterol following administration of precursors *in vivo*[74-76]. However, slight uptake of [^{14}C]-hexose[77], acetate and leucine[78] into cholesterol has been reported in adult mice. Bloch et al.[79] fed cholesterol labelled with deuterium for 3 days to an adult dog; although at the end of 6 days the marked material could be recovered from the other organs examined, none was found in the brain or spinal cord. It was concluded that these observations illustrated the lack or paucity of metabolic interchange between the sterol of the central nervous system and that of the blood. This work on intact animals therefore fully supports the results obtained with brain slices described above.

Tissue from the developing nervous system is capable of synthesizing cholesterol of higher specific activity than of the liver and other tissues and thus it seems likely that cholesterol is largely synthesized within the nervous system. If [4-^{14}C]cholesterol is injected into newborn animals a small amount of the intact lipid is taken up into the brain[80-82]. Dobbing[82] has shown quite convincingly that incorporation of labelled cholesterol parallels rate of accretion of the sterol in the developing brain. Much less [^{14}C]-cholesterol is taken up into the 1- to 5-day-old rat brain when little cholesterol is deposited in the brain, while optimum incorporation in the 15-day-old animal coincides with the period of maximum cholesterol accumulation. It was also noted that a small but definite amount of radioactive cholesterol can be taken up by the adult brain. Although this contradicts the observations of Bloch[79] it is almost certain that methods then available for the detection of deuterated cholesterol of low specific activity would not

References p. 325

be sufficiently sensitive to estimate trace amounts of labelled cholesterol.

Some investigators suggest that failure to incorporate labelled cholesterol and to utilise precursors for the synthesis of the sterol in the brain can be ascribed to a selective barrier between blood and brain or nerve. It has proved possible to isolate actively labelled cholesterol from the brain following intracerebral or intraventricular injections of [^{14}C]acetate or [^{14}C]-mevalonate although this was not possible with equivalent doses given parenterally. It was therefore argued that this procedure by-passed the "blood-brain barrier". Dobbing has, however, pointed out that interpretation of these experiments is exceedingly difficult; for example, it is not possible to equate a dose given directly into the brain to that given by parenteral injection. Thus, for example, precursor of very high specific activity presented locally by injection into the brain will produce labelled cholesterol even if it is poorly utilised[63]. The sum of current evidence supports the view that synthesis of cholesterol in the adult brain is restricted and this in turn suggests that this cholesterol must be metabolically stable, for the neural sterol content remains essentially constant throughout adult life.

(c) *The catabolism of cholesterol in the nervous system*

Isotopic experiments indicate that a small amount of cholesterol in the adult brain undergoes quite rapid exchange equivalent to a complete turnover time of 80 min[83,78]. In addition, some brain cholesterol undergoes rather slower turnover ($t_{0.5}$ approx. 30 days) but a high proportion is metabolically relatively stable. This metabolic heterogeneity of cholesterol metabolism in the brain indicates the complexity of the system and underlines the importance of relating metabolism to anatomical structure (see below).

It will be recalled that cholesterol is readily incorporated into the nervous system during development. It is therefore possible to label neural cholesterol by administration of suitable precursors during this period. Alternatively cholesterol may be specifically labelled by giving [^{14}C]cholesterol. Thus, [4-^{14}C]cholesterol[33] or [^{3}H]cholesterol[84,85] has been injected into the yolk sac of one-day-old chicks and the incorporation and its subsequent persistence in a small amount of labelled cholesterol has been studied. Animals were allowed to survive for various periods up to 420 days thereafter. Radioactive cholesterol taken up into the brain underwent only slow turnover and little loss of ^{14}C occurs over the last 200 days after injection. Essentially

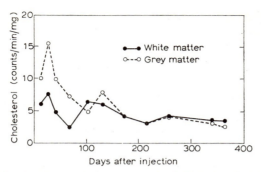

Fig. 5. Radioactivity in white and grey matter of rabbit brain. [4-^{14}C]Cholesterol was injected into 17-day-old rabbits and specific radioactivity (corrected for growth) are shown at intervals after injection. (By courtesy of the *Lancet*.)

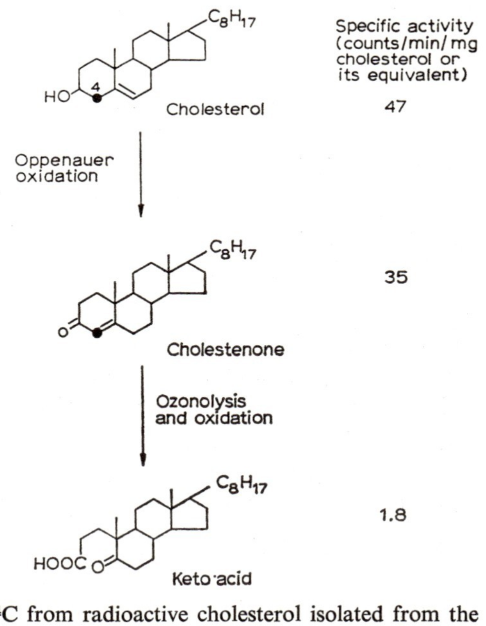

Fig. 6. Loss of ^{14}C from radioactive cholesterol isolated from the brain of a rabbit a year after injection of [4-^{14}C]cholesterol.

similar results (Fig. 5) have been obtained in other species such as rats[86-89] and rabbits[33,80]. It seemed possible that this remarkable persistence of cholesterol in the central nervous system was due to the continual reutilisation of ^{14}C released by degradation. However, radioactive cholesterol

References p. 325

isolated from the brain of animals long after injection of specifically labelled [4-^{14}C]cholesterol was found to retain almost all the label in the C-4 position[90] (Fig. 6). Moreover, it was found that all the radioactivity in the original crude extract of the central nervous system was due to its content of [^{14}C]cholesterol. Nicholls and Rossiter[91] have found incorporation and persistence of [4-^{14}C]cholesterol for periods up to 30 weeks in the sciatic nerve of rats. Four times as much labelled sterol was taken up into nerve after crushing, and again once incorporated there was only very slow catabolism.

These experiments therefore suggest that most of the cholesterol deposited during development and growth of the central and peripheral nervous system undergoes only very slow turnover. The difficulties of labelling brain or nerve cholesterol in adult animals can now be related to this large mass of cholesterol essentially removed from the normal dynamic exchange processes of the body.

6. Neural phospholipids

Phospholipids are found in relatively high concentration in nervous tissue; about 20% of the dry weight of the adult mammalian brain is phospholipid. In order of abundance these phospholipids are: ethanolamine phospholipid, lecithin, the sphingolipid sphingomyelin, serine and inositol phosphatides, together with such phospholipids as cardiolipin and alkyl ether phospholipids[22] which are found only in small amounts. Much of the ethanolamine phospholipid of the nervous system, unlike that of other tissues, is in the form of the aldehydogenic plasmalogen, phosphatidal ethanolamine; only small amounts of serine and choline plasmalogen are to be found. A full account of the chemistry of phospholipids is given by Hanahan and Brockerhoff[13] in volume 6 of this series and also by Van Deenen[92]. The metabolism of phospholipids has been very fully reviewed, for example in the present volume, and by Ansell and Hawthorne[93], Dawson[94] and for the nervous system by Rossiter[95,96]. Sphingomyelin, phosphatidal ethanolamine, phosphatidyl serine and the polyphosphoinositides are 2–3 times more abundant in the white than grey matter. The phospholipid composition of the peripheral nervous system shows interesting differences compared to that of the brain and spinal cord: about twice as much sphingomyelin and less cerebroside and sulphatide are found in the peripheral as opposed to the central nervous system[9]. Brain phospholipids contain a higher proportion of total unsaturated fatty acids than most other tissues and acids so far identified are of even-number chains (C_{16}–C_{26}).

(a) Properties

The special structure of the phospholipids affect both their physico-chemical properties and their metabolic behaviour. Thus, the hydrophobic long-chain fatty acid and sphingosine groups on the one hand and the hydrophilic head or polar group on the other account for the amphipathic character of the phospholipid. Clearly, the net charge existing on each phospholipid-head group is dependent on both the nature and balance of the polar groups present and the extent of their ionization, so that while the choline phospholipids are electrically neutral at physiological pH, others possess a negative charge due to ionization of carboxyl groups (*e.g.* serine) or to incomplete ionization of their amino groups (*e.g.* ethanolamine). These various properties explain the ready formation by phospholipids of both micellar and bimolecular film structures, hence accounting for the important role of these lipids in membrane structures[94].

Differences in the degree of ionization of the various classes of phospholipids are frequently utilised as a means of effecting chromatographic separation. Such fractionation of neural phospholipids into their main classes is readily achieved on columns, silicic acid impregnated paper and thin-layer plates. One widely used procedure is that of Dawson[97,98] who has perfected an ingenious method based on the selective hydrolysis[99] of phospholipids. In this method water-soluble glyceryl phosphoryl esters produced by deacylation[97,98] are separated by two-dimensional paper chromatography or paper chromatography with ionophoresis.

(b) Biosynthesis of phosphatides

In the adult nervous system the continual breakdown of phospholipid is apparently exactly balanced by *de novo* synthesis from phosphate and other simple precursors. The biosynthetic pathways so far identified resemble those described for other tissues. For example, Rossiter and his colleagues[95] have shown that cytidine 5'-triphosphate (CTP) acts as a necessary co-factor in the biosynthesis of the diacylglycerophospholipids of nervous tissue in just the same way as it does for other tissue phospholipids. Other experiments such as these of Ansell and

References p. 325

Spanner[100] suggest that decarboxylation of phosphatidyl serine to phosphatidyl ethanolamine may be effected by brain mitochondria[101]. In developing brain, phosphatidyl serine could also be formed by an exchange reaction between ethanolamine and serine[102]. Relatively less is known about the anabolism of plasmalogens. McMurray[103] has found that CDP choline and CDP ethanolamine labelled with ^{14}C can be incorporated *in vitro* into choline plasmalogen and ethanolamine plasmalogen, a process stimulated by addition of plasmalogenic diglyceride. It therefore seems probable that synthesis is similar to that demonstrated for liver by Kiyasu and Kennedy[104].

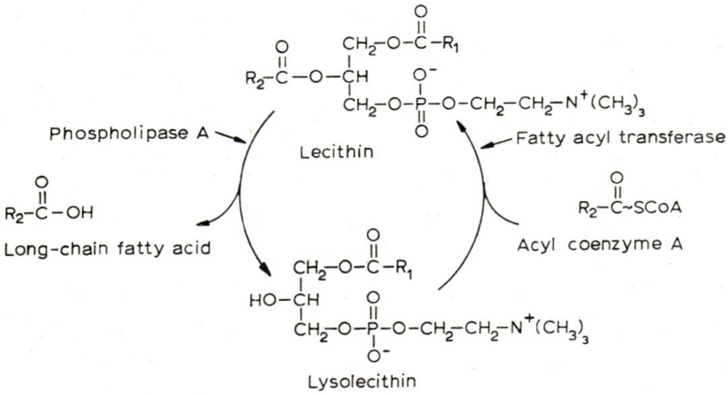

Fig. 7. Deacylation–acylation cycle.

An alternative route for phospholipid metabolism has been described by Webster[105-107] for brain, which like that found by Lands[108] for liver, can account for a high proportion of lecithin turnover (Fig. 7). Lecithin, phosphatidyl ethanolamine and phosphatidyl serine are acted on by brain phospholipase A to give the corresponding lyso derivatives[109]. Ansell and Spanner[110] have identified an enzyme hydrolysing the vinyl ether linkage of ethanolamine plasmalogen to give 2-acyl lysophosphatidyl ethanolamine. It was found by Webster[107] that on incubation of ^{14}C-labelled fatty acids with respiring slices of rat cerebrum that radioactivity could be detected in both 1 and 2 positions of lecithin and ethanolamine phospholipid. These observations support those of others[111-115] and suggest that several different hydrolytic enzymes are present in nervous tissue where they give rise to various isomeric lyso compounds; acylation by long-chain fatty acyl-CoA is then effected by a number of different acyl transferases. There is thus a deacylation–acylation cycle in nervous tissue which can account for much of the fatty acid turnover observed in the brain (234 mμmoles of acid/g of tissue/h are incorporated in an energy-dependent process).

(c) *Sphingomyelin*

Sphingomyelin of the nervous system is thought to be largely synthesised through the acylation of sphingosine by acyl coenzyme A to form ceramide. In the final step choline is transferred from CDP choline to the *N*-acyl-sphingoside to give sphingomyelin. Although Sribney and Kennedy[116] found that *N*-acyl derivatives of *threo*-sphingosine were most effective for the *in vitro* formation of a sphingomyelin (naturally occurring as the *N*-acyl-*erythro*-sphingosyl phosphoryl choline) recent work shows that the intact animal can convert both *erythro*- and *threo*-DL-sphingosine into ceramide and sphingomyelin[117]. It therefore seems probable that there is a system *in vivo* which allows the interconversion and preferential utilisation of *erythro*-sphingosine. However, it has been found by Kanfer and Brady[118] that [1-^{14}C]*N*-stearoyl-sphingosine injected into the brain is not incorporated into sphingomyelin, suggesting the possibility that the major pathway for its biosynthesis in nervous tissue may be through the direct acylation of sphingosine phosphoryl choline and not by addition of phosphoryl choline to a ceramide acceptor[119]. The different enzyme systems leading to the complete catabolism of sphingomyelin have been described for rat brain by Barnholz et al.[120].

(d) *Inositol phospholipids and phosphatidic acids*

Brain phosphatidic acid and phosphoinositol fractions have been found to rapidly incorporate ^{32}P and other precursors both *in vitro* and in experiments *in vivo*[121,122]. This rapid labelling was found by Hokin and Hokin[123,124] and by Larrabee and his colleagues[125,126] to be increased by stimulation. Observations of this type promoted a great deal of interest in the inositol-containing phospholipids so that work intensified on their structure and metabolism. In all, three phosphoinositides, mono-, di- and triphosphoinositides, were identified in the nervous system, the latter two being found in much higher concentration than in other tissues[127]. The pathway for the biosynthesis of the inositol phosphatides differs basically from that of the other acyl glycerylphospholipids. It appears that instead of free 1,2-diglyceride

References p. 325

acting as acceptor for CDP choline, a liponucleotide, cytidine diphosphate diglyceride, reacts directly with inositol[128,129].

Rat-brain homogenates[130] have been found to promote the formation of cytidine diphosphate diglyceride from phosphatidic acid and cytidine 5'-triphosphate (CTP). Inositol is incorporated to form phosphatidyl inositol with the elimination of CMP. Addition of further phosphate groups from ATP is catalysed by phosphatidyl inositol kinase of brain microsomes preparations and by a soluble brain diphosphoinositide kinase. The polyphosphoinositides undergo very rapid catabolism in the nervous system so that nerve tissue has to be frozen to prevent breakdown.

7. Lipid metabolism in relation to anatomical structure

Apart from endothelial cells, the cell population of the central nervous system (Table III) can be conveniently divided into nerve cells or neurones and glial cells (oligodendroglia, astrocytes and microglial). The nerve cell bodies (perikarya) with their arborising dendrites are found in a complex

TABLE III

CELL TYPES IN RAT CEREBRUM

	Number of cells ($\times 10^7$/g wet weight)	
	Grey matter (Cerebral cortex)	White matter (Corpus callosum)
Neurones	2.3	0
Glia[a]	5.8	7.6
Endothelial	1.6	6.2

[a] Oligodendroglia, microglia and astrocytes.

architectural pattern in the grey matter, their slender conducting axons enclosed by the myelin sheath extending into the white matter (Fig. 8). In the cerebral cortex the nerve cell bodies occupy only about one twentieth of the total volume while dendrites and axons account for a much greater volume. The glial cells are, however, more evenly distributed between the two areas of the central nervous system. The work of Hyden[131] and his colleagues in Sweden suggests that the oligodendroglia are particularly closely concerned with the metabolic behaviour of the neurone and it is also now thought they are responsible for forming and sustaining the myelin sheath[132,133]. Since nerve cell bodies with their dendritic fields are absent, the peripheral nervous

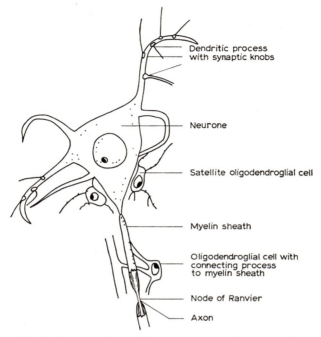

Fig. 8. Representation of the neurone and oligodendroglia.

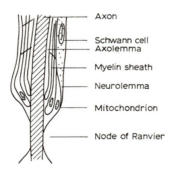

Fig. 9. Representation of structure of peripheral nerve myelin.

system shows resemblances to the white matter of the central nervous system with the exception that the Schwann cell replaces the oligodendroglial cell (Fig. 9).

A high proportion of lipid in the nervous system is found in the white matter where it is concentrated in the myelin sheath enclosing the nerve

References p. 325

axons. Thus treatment of tissue sections with lipophilic reagents such as Sudan black show intense staining of myelinated tracts and nerve bundles in both the central and peripheral nervous systems. The neurones and glia account for much less lipid than the myelin but the detailed composition of these cell types is known with much less certainty[134].

In electron micrographs, myelin appears as a mass of triple-layered unit membranes tightly wound around the nerve fibre. About 25% of the dry weight of myelin is proteolipid protein, the remainder being lipids—cerebrosides, cholesterol and phospholipid (Table IV). Detailed models of the myelin unit membrane structure have been proposed by Finean[135] and by Vandenheuvel[136]. The proposed basic structure is similar to that suggested by Davson–Danielli in which protein layers (30 Å thick) sandwich an inner

TABLE IV

COMPOSITION OF MYELIN FROM VARIOUS MAMMALIAN SPECIES

	Adult human (white matter)			Rat (whole brain)	Ox (white matter)
Reference:	170	4	3	170	170
Protein (mg/100 mg myelin dry wt.):	47	—	21.8	36	29.5
Lipid composition (μ moles/mg dried lipid)					
Cholesterol	0.77	0.65	0.745	0.67	0.622
Total galactolipid	0.295	0.316	0.322	—	0.442
Cerebroside	0.221	0.256	0.263	0.222	0.283
Sulphatide	0.05	0.049	0.043	0.045	0.06
Total phospholipid	0.56	0.481	0.559	0.063	0.580
Phosphatidyl inositol	0.009	—	0.004	0.028	0.018
Serine phospholipid	0.101	0.083	0.056	0.098	0.109
Cardiolipin	0.003	—	—	0.017	—
Sphingomyelin	0.103	0.072	0.105	0.054	0.089
Lecithin	0.132	0.137	0.146	0.165	0.112
Ethanolamine phospholipid	0.251	0.19	0.215	0.283	0.205
Total plasmalogen	0.177	—	0.174	0.310	0.170
Minimal myelin dry wt. (as % of whole tissue)	53			36	36
Minimal myelin lipid (as % of whole tissue)	44			52	40

55 Å lipid bilayer. Polar groups of the component lipids are thought to interact with charged groups of the protein layer and the hydrophobic long chains of the lipid are closely packed in a radial direction. In the detailed molecular model the various lipid myelin components can be shown to interdigitate in a remarkably compact pattern. Stereochemistry of the unsaturated fatty acids can also be shown to contribute to the integrated packing and the stability of the whole complex structure. It now appears possible, however, that the membrane, although appearing histologically continuous, is in fact composed of discrete protein–lipid subunits. Myelin and other membrane systems may be built up of lipoprotein units in which the protein helix forms a continuum in which are inserted the hydrophobic groups of the component lipid (*e.g.*, long-chain fatty acids) with their polar heads on the exterior surface of the subunit[137,138]. It will be appreciated that these various models all emphasise the close and ordered packing of protein and lipid and this concept clearly fits in very well with the observed metabolic behaviour of membrane structure.

(a) *Lipid metabolism and the blood–brain barrier*

Incorporation of labelled glucose and other substrates into the lipids of the nervous system *in vivo* is maximal only during the early stages of development. This reduced uptake of radioactive substrates into the brain lipids of adult animals compared to young has been attributed to the presence of a selective barrier interposed between blood and brain. Furthermore, the concept of the blood–brain or blood–nerve barrier appears to be consistent with the observation that many dyestuffs are unable to penetrate into nervous tissue and some drugs fail to gain access to the central nervous system. However, Dobbing[139] has pointed out that nervous tissue has little extracellular space and consequently protein-bound dyestuff, in the plasma, would fail to penetrate into the brain whereas most other organs with larger extracellular spaces would apparently be dyed. With the low extracellular space, the net of astrocytic end feet around the brain capillaries and the properties of the nerve cell membranes all contribute to the blood–brain barrier effect, but another and important factor must be considered. This is that metabolic requirements of nervous tissue differ in some important respects from those of other tissues; for example, glucose is the main substrate and fatty acids are not easily oxidised by the brain. Moreover, in the adult animal myelin presents an inert mass, inaccessible to the normal dynamic exchange processes of the organ.

References p. 325

It was at first thought possible to relate the restricted lipid synthesis in the adult brain to the failure of substrates to penetrate into the nervous tissue. Various devices were utilized to attempt to correct for this deficiency. Uptake into lipid was measured after intracerebral or intraventricular injection, but as has already been explained it is not possible to calculate the effective dose of labelled substrate in order to establish a precursor–product relationship. Following parenteral administration of labelled substances, correction has been attempted by relating lipid synthesis to the acid-soluble precursors which penetrate into the organ. This allows the calculation of relative specific activities and the determination of rates of incorporation into the various lipid species. The shortcomings of such an approach has been considered by Davison and Dobbing[140]. It was pointed out that metabolism of an individual lipid is more related to anatomical localization and function than to its chemical identity. Hence lecithin in the cytoplasm may well be in a highly dynamic state whereas the same phospholipid may be stable in cell structures. The metabolic heterogeneity of a molecular species therefore makes it difficult to extrapolate from rates of uptake into lipid to meaningful turnover rates, particularly in such a complex organ as the brain. Nevertheless, work such as that of Dawson and Richter[141], and Ansell and Dohmen[121] indicated that some fraction of the brain phospholipids underwent quite rapid metabolism *in vivo* comparable to that shown for other organs. It was shown that [^{32}P]phosphate injected into adult mice was rapidly incorporated into cerebral phosphatidyl inositol and phosphatidic acid, while lecithin and other phospholipids took up ^{32}P rather more slowly. In these experiments the exchange period was over short time intervals (0–7 h) but over longer periods the rapid rate of uptake of ^{32}P was not sustained. Changus *et al.*[142] and later Davison and Dobbing[143,144] studied incorporation of [^{32}P]phosphate into the neural phospholipid of both young and adult rats over periods up to 6 months. Labelling of phospholipids was more extensive in the young 16–20-day-old animals than in the adult, as may be expected from the needs of the developing animal to synthesize and deposit phospholipids into myelin[145]. This provides an explanation for the fact that, particularly in young animals, the fast initial uptake was not mirrored by equally rapid turnover, for although rapid synthesis undoubtedly occurs, much of the labelled lipid is retained in both the central and peripheral nervous system where it undergoes only slow replacement. Similarly phospholipids and other lipids labelled with [1-^{14}C]glycerol[87] undergo only very slow turnover (equivalent to half-life of 100 days) in the brain of both young and adult rats. These experiments can be compared with those of Thompson and Ballou[146,147] who first demonstrated the general stability of brain phospholipids by studying the biological half-lives of the components of a large number of tissues by means of long-term experiments using tritium as an indicator. Later they were able to show that biological half-lives although presenting a wide range of values may be divided grossly into two main groups. Thus, more than half of the total brain constituents of rats exposed to tritium during development have an average half-life as long as 150 days.

It now therefore seems likely[139,148] that the rapid uptake of substrates into lipids of the developing nervous system is related more to the formation of metabolically stable myelin than to a specific blood–brain barrier effect. Other properties of the blood–brain barrier such as the selective staining of the brain of young animals is related more to a larger extracellular space of the neonatal brain than to a barrier phenomenon.

(b) Lipid metabolism at the cellular level

The extent of lipogenesis in nervous tissue depends on many different factors. For example, the rate of synthesis of many lipids is age dependent, being most active during the period of active myelination—a process whose timing is different for each species. The rate of synthesis differs with the substrate supplied and the extent of biosynthesis depends on the region of the nervous system that is studied (Table V). For example, in adult animals

TABLE V

LIPOGENIC ACTIVITY OF NERVOUS, ADIPOSE AND HEPATIC TISSUE AS MEASURED *in vitro*

	Relative incorporation of labelled substrate		
	[^{14}C]acetate (%)	[^{32}P]phosphate (%)	[U-^{14}C]glucose (%)
Grey matter	25	760	221
White matter	10	49	10
Peripheral nerve	100	100	100
Liver	44	—	—
Adipose tissue	12	—	—

The results indicate the degree of incorporation of labelled substrates into the lipids of young adult rats (about 250 g body weight). The specific activity of the sciatic nerve is set at 100 counts/min/100 mg lipid[149].

References p. 325

radioactive phosphate is incorporated more rapidly into cortical grey matter than into peripheral nerve while much less synthesis occurs in the white matter. With acetate as substrate less difference was found by Majno and Karnovsky[147] between the lipid synthetic activity of the cerebral cortex and white matter although in this case preparations of sciatic nerve were found to be most active.

Following injection of isotopic precursors into mice of different ages Torvik and Sidman[150] have used a radioautographic technique to follow lipid metabolism at the cellular level. Nerve cell bodies, neuropil (surface and intracellular membranes belonging to numerous small, densely packed neuronal and glial processes) and white matter were found to have different turnover rates for lipid synthesized from acetate in both young and adult mouse brain. The most rapid incorporation and subsequent replacement of lipid radioactivity occurred in nerve cell bodies, the slowest in white matter. Even after 6 months label was found to persist in the white matter but little radioactivity was localized in the perinuclear cytoplasm of glial cells, suggesting therefore that the relatively stable lipid was localized in the myelin sheath. In the pyramidal cell layer of the hippocampus (nerve cell bodies) and in the stratum radiatum of the same area (neuropil) little radioactivity remained. It was, however, noted by Torvik and Sidman that labelled lipid persisted in areas of neuropil such as the molecular layer of the cerebellum, which are devoid of myelin, suggesting that perhaps some membrane systems other than myelin may be metabolically stable.

(c) *Lipid metabolism of subcellular structures*

About half the lipid of mature brain is present in myelin, the rest being localized in other membrane systems[151] (Table VI). The lipid-rich myelin contaminates both nuclear and mitochondrial fractions when homogenates of nervous material are separated by classical differential centrifugal procedures[152], and it was therefore necessary to devise special techniques in order to study the lipid composition and metabolism of each subcellular fraction. Methods utilizing gradient-density separation have been successfully used[153-155] to remove myelin and isolate mitochondria and other organelles in a reasonable state of purity. Abdel-Latif and Abood[101,156] have studied the incorporation of [^{32}P]phosphate and ^{14}C-labelled serine into lipids and proteins of rat-brain subcellular particles. The total radioactivity in the various fractions at 12 days was higher than at 3 days but

TABLE VI

THE LIPID COMPOSITION OF RAT MYELIN AND BRAIN SUBCELLULAR PARTICLES

Fraction	Cholesterol (mg/g wet wt.)	Molar lipid ratio		
		Cholesterol	Phospholipid	Cerebroside
Myelin	11.4	100	94	42
Nuclei	0.78	100	193[a]	22
Mitochondria	3.26	100	293	6
Microsomes	1.25	100	137	16
Supernatant	0.85	100	154	12
Total homogenate	16.20	100	124	35

Results calculated from various sources, separate analyses of synaptic ending fractions are not included.
[a] primary nuclear fraction probably containing myelin.

TABLE VII

TURNOVER OF RADIOACTIVITY INCORPORATED INTO THE LIPIDS OF ADULT-BRAIN SUBCELLULAR FRACTIONS

	Isotope:	^{35}S	^{35}S
		Fast pool	Slow pool
		Half-lives of labelled lipid ($t_{0.5}$ in days)	
Nuclei		2.6	18.2
Mitochondria		0.35	38.6
Synaptic endings		—	3.0
Microsomes		0.36	10.3
Supernatant		0.27	5.2
Myelin		2.7	138

Fractions were prepared following intraventricular injection of [^{35}S]sulphate into adult rats.

all fractions had similar specific activities 19 h after injection, suggesting that lipid ontogenesis originates from a common source. Total radioactivity was high in the microsomes as may be expected, for this fraction is primarily concerned in both lipid and protein biosynthesis.

It has been suggested[26] that sulphatide forms an integral part of the lipid–protein membrane unit[157] and that metabolism of this sulpholipid may therefore reflect the metabolism of the whole structure. Radioactive sulphate injected into young and adult rats is readily incorporated into sulphatides of nervous tissue so that turnover can then be traced as [^{35}S]lipid content of the various subcellular fractions. It was found by Davison and

References p. 325

Gregson[38] that turnover of sulpholipid of all the subcellular fractions could be broadly divided into fast and slower, more persistent, compartments (Table VII). Similar results have been reported by Khan and Wilson[158] in experiments on rats fed with tritiated water during development. However, further work[89] showed that not all lipids undergo turnover at the same rate (Table VIII) so that for example cholesterol, cardiolipin and sphingomyelin of rat-brain mitochondria appear to be metabolically stable, whereas lecithin and other phospholipids undergo catabolism. Although at present the full significance of these observations is not understood, it is possible that part of the mitochondrial structure acts as a stable framework or that released cholesterol, cardiolipin and sphingomyelin are re-utilized in syntheses of new organelles, whereas other phospholipids break down and newly synthesized lipid is incorporated into the membrane structure.

TABLE VIII

TURNOVER OF RADIOACTIVITY INCORPORATED INTO THE INDIVIDUAL MITOCHONDRIAL LIPIDS OF YOUNG ADULT-RAT BRAIN

Lipids undergoing metabolism ($t_{0.5}$ 11–35 days)	Stable lipids
Phosphatidyl ethanolamine	Cholesterol
Lecithin	Cardiolipin
Phosphatidyl serine	Sphingomyelin
Phosphoinositide	

(d) Lipid metabolism of the myelin sheath

The reduced rate of lipogenesis in adult white matter and the stability of many of the brain lipids led to the conclusion that myelin must be regarded as one of the more metabolically inert components of the body. Thus, following labelling during development, it has been repeatedly shown that there is little turnover and quite remarkable persistence of radioactivity in the lipid and proteolipid protein constituents of myelin from the central and peripheral nervous systems. Fig. 10 shows that [^{35}S]sulphate is rapidly taken up into nervous tissue of 15-day-old rats and that the labelled sulphatide remains with only very slow turnover. Since sulphatides are found in large amounts in myelin this suggested that at least this part of the structure is relatively stable. More direct evidence came when labelled protein and lipid were found to be concentrated in the myelin fraction isolated by differential centrifugation of brain homogenates a year after injection of [^{35}S]methionine into 16-day-old rats. Similar slow turnover of other myelin lipids was demonstrated in other double-labelling experiments with tritium

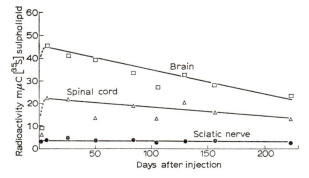

Fig. 10. Incorporation and persistence of [^{35}S]sulphate into the neural lipids of the growing rat brain. 15-day-old rats were given 8 μC of [^{35}S]sulphate by intraperitoneal injection. Data for brain (□), spinal cord (△) and sciatic nerve (●) are shown.

and ^{14}C by Cuzner et al.[89]. The conclusion that myelin, like collagen and certain other body components, is metabolically stable can be correlated with the structure of the sheath, for myelin is built up of many layers of membrane wound around the axon and it is difficult to envisage dynamic exchange processes occurring in such a structure. This view is supported by the lack of enzyme activity found to be associated with myelin[159].

Despite the general validity of the above findings it has been suggested by Eng and Smith[160] that some of the rat-myelin lipids (notably lecithin) undergo turnover (half-lives of 2–4 months) whereas sphingomyelin, ethanolamine phospholipid, cerebrosides, and sulphatides exist as stable complexes with cholesterol. Moreover, it is now established that isotopes can be incorporated into adult myelin and once incorporated undergo slow turnover at the same rate as has been found in myelin from young rats. Rathbone[161] has fed 9–10-week-old rats on sunflower and coconut oil supplemented diets for 34 weeks and showed changes in the relative proportions of linoleic and arachidonic acid in rat-brain myelin, suggesting that some fatty acids may undergo turnover. One possible explanation is that a small number of component molecules can exchange with the largely inert mass of myelin and while some are reincorporated unchanged, others,

References p. 325

such as lecithin, may undergo catabolism and be replaced by newly synthesized phospholipid. This explanation which remains to be verified has also been suggested for mitochondrial membrane (see above).

In addition to the large slowly metabolizing pool of myelin lipid and protein a very small fraction appears to be rapidly metabolized. This compartment may have a specific localization in the surface of the sheath or the axolemma or it may be at the Node of Ranvier where myelin is exposed to glial cytoplasm and where mitochondria are concentrated. Eichberg and Dawson[162] have also found that triphosphoinositide is associated with the myelin fraction and that the phosphate groups of the polyphosphoinositides undergo rapid exchange reactions.

8. Physiological significance of neural lipid metabolism

The dynamic metabolism of some brain lipids and the rapid turnover of a small pool of membrane-bound phospholipid in the brain[121,122] suggests that these particular lipids may have an important metabolic role in nervous tissue. This possibility is supported by the work of Hokin and Hokin[123,124] who showed that labelling of phosphatidic acid and phosphoinositides was increased by addition of acetylcholine to tissue preparations. The effect required the presence of intact microsomal membrane and could be further related to an action on the enzyme diglyceride kinase. The interesting suggestion was made that fast turnover of phosphatidic acid could be correlated with utilization of ATP and the transport of sodium across biological membranes. However, it proved difficult to account for the number of ions actually found to be transported with the sodium ions expected from each turn of the phosphatidic acid cycle. In other experiments Larrabee and his colleagues[125,126] have shown increased labelling of phosphoinositides but not lecithin, phosphatidyl ethanolamines or even phosphatidic acid when ganglia perfused with [^{32}P]phosphate are electrically stimulated. Evidence was obtained that stimulation by pre-synaptic nerve impulses resulted in increased labelling of postsynaptic structure. It was concluded that increased incorporation into phosphatidyl inositol was an effect of the synaptic transmitter on the post-synaptic membrane[126]. Increased labelling of brain phospholipids has been claimed for certain excitants while depressants such as barbiturates and chlorpromazine reduce exchange of [^{32}P]phosphate in brain phospholipids[163].

In addition to the probable role of phospholipids in transmission processes, the gangliosides have also been found to possess interesting properties. McIlwain[164] has shown that gangliosides restore the excitability of brain slices to electrical pulses after inhibition by basic proteins such as protamine. Gangliosides appear to be concentrated within the neurone and synaptic vesicles of the synaptic endings[165]. Since these vesicles appear to be involved in neural transmission mechanisms it is possible that the gangliosides are concerned in these processes.

The participation of lipid in more than semi-permanent structural capacity is also illustrated by the experiments of Geiger and his colleagues[166]. Using cats with an isolated cerebral blood supply it was possible to show that endogenous substrates were utilised when glucose was omitted from the circulating perfusion blood. Further analysis showed that nucleic acids and phospholipids from the brain-microsome fraction appeared to provide the necessary substrate in the absence of exogenous glucose. In the intact animal only a small proportion of [^{14}C]glucose is oxidised to $^{14}CO_2$, much of the label appearing in dicarboxylic amino acids[167,168], proteins and lipids. It seems possible therefore that amino acids and lipids may provide an additional source of energy in the brain to glucose but whether or not this is only of importance in the absence of glucose remains to be established.

These various experiments have entirely altered our concepts on the role of lipids in nervous tissue. It now seems likely that phosphatides and possibly gangliosides are essential participants in ion transport, energy metabolism and conduction mechanisms within the nervous system.

9. Conclusion

The demonstration of the metabolic inertness of myelin has certain important consequences. It has already been pointed out that the mass of stable lipid composing the myelin sheath makes it appear that, taken as a whole brain lipids undergo slow metabolism. However, it is now known that lipid metabolism depends more on the anatomical location and physiological role of the lipid than on its chemical identity. Metabolism of lipid in the nervous system is also related to development, for lipid biosynthesis is most rapid during myelination and restricted in the adult. Myelination may therefore be regarded as a critical period[169]. In man this vulnerable period coincides with just before and just after birth and so proper nutrition of the newborn and particularly the premature infant is of special importance. Certain evidence suggests that intellectual capacity may be affected by faulty myelina-

References p. 325

tion during the critical developmental period. It also follows that demyelination of the adult central nervous system as occurs in multiple sclerosis will have lasting effect, for renewed synthesis of myelin does not generally occur in the central nervous system, although this is not true of peripheral nerve.

In this chapter it has been shown that the biochemical features of lipid metabolism of nervous tissue are similar in many respects to those of other organs. Thus, in general, biosynthetic and catabolic pathways though sometimes slow resemble those identified in other tissues. However, it has been pointed out that the nervous system and particularly brain, present special difficulties to the biochemist, for interpretation of lipid metabolism has to be related to the metabolic stability of myelin, as well as the very complex architecture and physiology of the brain. Moreover, structural and functional development occurs in broad stages during growth and at different times for the various species. Aside from these complications must be added the lack of real biochemical insight into the higher mechanisms of the brain such as memory, thought and nerve conduction. At this level classical biochemistry is not enough, metabolism *must* be related to structure and function.

REFERENCES

1. J. P. Folch, M. Lees and G. H. Sloane-Stanley, *J. Biol. Chem.*, 226 (1957) 497.
2. R. M. C. Dawson and J. Eichberg, *Biochem. J.*, 96 (1965) 634.
3. W. T. Norton and L. A. Autilio, *J. Neurochem.*, 13 (1966) 213.
4. J. S. O'Brien and E. L. Sampson, *J. Lipid Res.*, 6 (1965) 537.
5. B. W. Nichols, L. J. Morris and A. T. James, *Brit. Med. Bull.*, 22 (1966) 137.
6. O. S. Privett, M. L. Blank, D. W. Codding and E. C. Nickell, *J. Am. Oil Chemists' Soc.*, 42 (1965) 381.
7. M. L. Cuzner and A. N. Davison, *J. Chromatog.*, 27 (1967) 388.
8. L. Svennerholm, *J. Neurochem.*, 11 (1964) 839.
9. J. S. O'Brien, E. L. Sampson and M. B. Stern, *J. Neurochem.*, 14 (1967) 357.
10. A. N. Davison and M. Wajda, *Biochem. J.*, 82 (1962) 113.
11. L. Svennerholm, *Acta Soc. Med. Upsalien.*, 62 (1957) 1.
12. I. Zabin, in D. J. Hanahan (Ed.), *Lipide Chemistry*, Wiley, New York, 1966.
13. D. J. Hanahan and H. Brockerhoff, in M. Florkin and E. H. Stotz (Eds.), *Comprehensive Biochemistry*, Vol. 6, Elsevier, Amsterdam, 1965, p. 83.
14. J. S. O'Brien, D. L. Fillerup and J. F. Mead, *J. Lipid Res.*, 5 (1964) 109.
15. P. Stoffyn and A. Stoffyn, *Biochim. Biophys. Acta*, 70 (1963) 218.
16. P. Stoffyn, *J. Am. Oil Chemists' Soc.*, 43 (1966) 69.
17. P. Stoffyn and A. Stoffyn, *Biochim. Biophys. Acta*, 70 (1963) 107.
18. T. Yamakawa, N. Kiso, S. Handa, A. Makita and S. Yokoyama, *J. Biochem. (Tokyo)*, 52 (1962) 226.
19. S. J. Hakomori, T. Ishimoda and K. Nakamura, *J. Biochem. (Tokyo)*, 52 (1962) 468.
20. A. N. Davison, *Biochem. J.*, 91 (1964) 3P.
21. H. G. Muldner, J. R. Wherret and J. N. Cumings, *J. Neurochem.*, 9 (1962) 607.
22. L. Svennerholm and H. Thorin, *J. Lipid Res.*, 3 (1962) 483.
23. L. Svennerholm, *J. Neurochem.*, 1 (1956) 42.
24. C. J. Lauter and E. G. Trams, *J. Lipid Res.*, 3 (1962) 136.
25. G. Rouser, C. Galli, E. Lieber, M. L. Blank and O. S. Privett, *J. Am. Oil Chemists' Soc.*, 41 (1964) 836.
26. A. N. Davison and N. A. Gregson, *Biochem. J.*, 85 (1962) 558.
27. F. J. Witmer and J. H. Austin, *Mikrochim. Acta*, 4 (1960) 502.
28. L. Svennerholm, *Acta Chem. Scand.*, 17 (1963) 1170.
29. R. O. Brady, in R. M. C. Dawson and D. N. Rhodes (Eds.), *Metabolism and Physiological Significance of Lipids*, Wiley, New York, 1964, p. 95.
30. G. M. McKhann, R. Levy and W. Ho, *Biochem. Biophys. Res. Commun.*, 20 (1965) 109.
31. A. S. Balasubramanian and B. K. Bachhawat, *Indian J. Biochem.*, 2 (1965) 212.
32. R. M. Burton, M. A. Sodd and R. O. Brady, *J. Biol. Chem.*, 233 (1958) 1053.
33. A. N. Davison, R. S. Morgan, M. Wajda and G. Payling Wright, *J. Neurochem.*, 4 (1959) 360.
34. M. L. Cuzner, A. N. Davison and N. A. Gregson, *Biochem. J.*, 101 (1966) 618.
35. M. E. Smith and L. F. Eng, *J. Am. Oil Chemists' Soc.*, 42 (1965) 1013.
36. A. K. Hajra, D. M. Bowen, Y. Kishimoto and N. S. Radin, *J. Lipid Res.*, 7 (1966) 379.
37. S. Gatt, *J. Biol. Chem.*, 238 (1963) PC 3131.
38. A. N. Davison and N. A. Gregson, *Biochem. J.*, 98 (1966) 915.
39. J. H. Austin, D. Armstrong and L. Shearer, *Arch. Neurol.*, 13 (1965) 593.
40. E. Mehl and J. Jatzkewitz, *Biochem. Biophys. Res. Commun.*, 19 (1965) 407.

41 W. T. NORTON AND S. PODULSO, in G. B. ANSELL (Ed.), *Chemical Composition of the Nervous System*, Pergamon, Oxford, 1966, p. 82.
42 L. SVENNERHOLM, *J. Lipid Res.*, 5 (1964) 145.
43 R. LEDEEN, *J. Am. Oil Chemists' Soc.*, 43 (1966) 57.
44 R. KUHN AND H. WIEGANDT, *Chem. Ber.*, 96 (1963) 866.
45 L. SVENNERHOLM, *Nature*, 177 (1956) 524.
46 G. ROUSER, G. KRITCHEVSKY, D. HELLER AND E. LIEBER, *J. Am. Oil Chemists' Soc.*, 40 (1963) 425.
47 J. R. WHERRETT AND J. N. CUMINGS, *Biochem. J.*, 86 (1963) 378.
48 R. M. BURTON, L. GARCIA-BUNUEL, M. GOLDEN AND Y. MCBRIDE BALFOUR, *Biochemistry*, 2 (1963) 580.
49 K. SUZUKI, *J. Neurochem.*, 12 (1965) 969.
50 A. ROSENBERG AND N. STERN, *J. Lipid Res.*, 7 (1966) 122.
51 S. BASU, B. KAUFMAN AND R. ROSEMAN, *J. Biol. Chem.*, 240 (1965) PC4115.
52 J. N. KANFER, R. S. BLACKLOW, L. WARREN AND R. O. BRADY, *Biochem. Biophys. Res. Commun.*, 14 (1964) 287.
53 N. S. RADIN, F. B. MARTIN AND J. R. BROWN, *J. Biol. Chem.*, 224 (1957) 499.
54 Y. KISHIMOTO, W. E. DAVIES AND N. S. RADIN, *J. Lipid Res.*, 6 (1965) 525.
55 Y. KISHIMOTO, W. E. DAVIES AND N. S. RADIN, *J. Lipid Res.*, 6 (1965) 532.
56 W. M. SPERRY AND M. WEBB, *J. Biol. Chem.*, 187 (1950) 97.
57 D. N. RHODES AND CH. LEA, *Biochem. J.*, 65 (1957) 526.
58 C. LONG AND D. A. STAPLES, *Biochem. J.*, 78 (1961) 179.
59 A. N. DAVISON AND M. WAJDA, *J. Neurochem.*, 4 (1959) 353.
60 A. N. DAVISON, J. DOBBING, R. S. MORGAN AND G. PAYLING WRIGHT, *J. Neurochem.*, 3 (1958) 89.
61 H. K. HANEL AND H. DAM, *Acta Chem. Scand.*, 9 (1955) 677.
62 A. ZLATKIS, B. ZAK AND G. J. BOYLE, *J. Lab. Clin. Med.*, 41 (1953) 486.
63 A. N. DAVISON, *Advan. Lipid Res.*, 3 (1965) 171.
64 E. GROSSI, P. PAOLETTI AND R. PAOLETTI, *J. Neurochem.*, 6 (1960) 73.
65 S. GARATTINI, P. PAOLETTI AND R. PAOLETTI, *Arch. Biochem. Biophys.*, 80 (1959) 210.
66 S. GARATTINI, P. PAOLETTI AND R. PAOLETTI, *Arch. Biochem. Biophys.*, 84 (1959) 253.
67 T. J. HOLSTEIN, W. A. FISH AND W. M. STOKES, *J. Lipid Res.*, 7 (1966) 634.
68 W. A. FISH, J. E. BOYD AND W. M. STOKES, *J. Biol. Chem.*, 237 (1962) 334.
69 S. KOREY AND A. STEIN, in S. S. KETY AND J. ELKES (Eds.), *Regional Neurochemistry*, Pergamon, Oxford, 1961, p. 175.
70 P. A. SRERE, I. L. CHAIKOFF, S. S. TREITMAN AND L. S. BURSTEIN, *J. Biol. Chem.*, 182 (1950) 629.
71 R. J. ROSSITER, in D. RICHTER (Ed.), *Metabolism of the Nervous System*, Pergamon, Oxford, 1957, p. 355.
72 H. J. NICHOLAS, *J. Kansas Med. Soc.*, 62 (1961) 358.
73 H. WAELSCH, W. M. SPERRY AND V. A. STOYANOFF, *J. Biol. Chem.*, 135 (1940) 297.
74 K. BLOCH, in E. S. GORDON (Ed.), *Symposium on Steroid Hormones*, Univ. Wisconsin Press, Lancaster, Pa., 1950.
75 J. T. VAN BRUGGEN, J. T. HUTCHENS, C. K. CLAYCOMB AND E. S. WEST, *J. Biol. Chem.*, 200 (1953) 31.
76 D. L. AZARNOFF, G. L. CURRAN AND W. P. WILLIAMSON, *J. Natl. Cancer Inst.*, 21 (1958) 1109.
77 H. W. MOSER AND M. L. KARNOVSKY, *J. Biol. Chem.*, 234 (1959) 1990.
78 J. J. KABARA AND G. T. OKITA, *J. Neurochem.*, 7 (1961) 298.
79 K. BLOCH, B. N. BERG AND D. RITTENBERG, *J. Biol. Chem.*, 149 (1943) 511.

80 A. N. DAVISON, J. DOBBING, R. S. MORGAN AND G. PAYLING WRIGHT, *Lancet*, *i* (1959) 658.
81 R. CLARENBURG, I. L. CHAIKOFF AND M. D. MORRIS, *J. Neurochem.*, 10 (1963) 135.
82 J. DOBBING, *J. Neurochem.*, 10 (1963) 739.
83 E. T. PRITCHARD, *J. Neurochem.*, 10 (1963) 495.
84 D. KRITCHEVSKY AND V. DEFENDI, *Nature*, 192 (1961) 71.
85 D. KRITCHEVSKY AND V. DEFENDI, *J. Neurochem.*, 9 (1962) 421.
86 P. J. MCMILLAN, G. W. DOUGLAS AND R. A. MORTENSEN, *Proc. Soc. Exptl. Biol. Med.*, 96 (1957) 738.
87 M. L. CUZNER, A. N. DAVISON AND N. A. GREGSON, *Ann. N.Y. Acad. Sci.*, 122 (1965) 86.
88 M. E. SMITH AND L. F. ENG, *J. Am. Oil Chemists' Soc.*, 42 (1965) 1013.
89 M. L. CUZNER, A. N. DAVISON AND N. A. GREGSON, *Biochem. J.*, 101 (1966) 618.
90 A. N. DAVISON AND M. WAJDA, *Nature*, 183 (1959) 1606.
91 D. NICHOLLS AND R. J. ROSSITER, *J. Neurochem.*, 11 (1964) 813.
92 L. L. M. VAN DEENEN, *Progr. Chem. Fats Lipids*, 8 (1965) 12.
93 G. B. ANSELL AND J. N. HAWTHORNE, *Phospholipids*, BBA Library Vol. 3, Elsevier, Amsterdam, 1964.
94 R. M. C. DAWSON, in P. N. CAMPBELL AND G. D. GREVILLE (Eds.), *Essays in Biochemistry*, Vol. 2, Academic Press, New York, 1966, p. 69.
95 R. J. ROSSITER, in R. M. C. DAWSON AND D. N. RHODES (Eds.), *Metabolism and Physiological Significance of Lipids*, Wiley, New York, 1964, p. 511.
96 R. J. ROSSITER, in K. RODAHL AND B. ISSEKUTZ (Eds.), *Nerve as a Tissue*, Harper and Row, New York, 1966, p. 175.
97 R. M. C. DAWSON, *Biochem. J.*, 75 (1960) 45.
98 R. M. C. DAWSON, N. HEMINGTON AND J. B. DAVENPORT, *Biochem. J.*, 84 (1962) 497.
99 G. SCHMIDT, J. BENOTTI, B. HERSHMAN AND S. J. THAMHAUSER, *J. Biol. Chem.*, 166 (1946) 505.
100 G. B. ANSELL AND S. SPANNER, *Biochem. J.*, 84 (1962) 12P.
101 A. A. ABDEL-LATIF AND L. G. ABOOD, *J. Neurochem.*, 13 (1966) 1189.
102 L. F. BORKENHAGEN, E. P. KENNEDY AND L. FIELDING, *J. Biol. Chem.*, 236 (1961) PC 28.
103 W. C. MCMURRAY, *J. Neurochem.*, 11 (1964) 287.
104 J. Y. KIYASU AND E. P. KENNEDY, *J. Biol. Chem.*, 235 (1960) 2590.
105 G. R. WEBSTER AND R. J. ALPERN, *Biochem. J.*, 90 (1964) 35.
106 G. R. WEBSTER, *Biochem. J.*, 98 (1966) 19P.
107 G. R. WEBSTER, *Biochem. J.*, 102 (1967) 373.
108 W. E. M. LANDS, *J. Biol. Chem.*, 235 (1960) 2233.
109 J. GALLAI-HATCHARD, W. L. MAGEE, R. H. S. THOMPSON AND G. R. WEBSTER, *J. Neurochem.*, 9 (1962) 545.
110 G. B. ANSELL AND S. SPANNER, *Biochem. J.*, 94 (1965) 252.
111 H. VAN DEN BOSCH, N. M. POSTEMA, G. H. DE HAAS AND L. L. M. VAN DEENEN, *Biochim. Biophys. Acta*, 98 (1965) 657.
112 W. E. M. LANDS AND I. MERKL, *J. Biol. Chem.*, 238 (1963) 898.
113 I. MERKL AND W. E. M. LANDS, *J. Biol. Chem.*, 238 (1963) 905.
114 W. E. M. LANDS AND P. HART, *J. Biol. Chem.*, 240 (1965) 1905.
115 S. GATT, *J. Biol. Chem.*, 241 (1966) 3724.
116 M. SRIBNEY AND E. P. KENNEDY, *J. Biol. Chem.*, 233 (1958) 1315.
117 J. N. KANFER AND A. E. GAL, *Biochem. Biophys. Res. Commun.*, 22 (1966) 442.

118 J. N. KANFER AND R. O. BRADY, in B. W. VOLK (Ed.), *Third International Symposium on Cerebral Sphingolipidoses*, Pergamon, Oxford, 1966.
119 R. O. BRADY, R. M. BRADLEY, O. M. YOUNG AND H. KALLER, *J. Biol. Chem.*, 240 (1965) PC 3693.
120 Y. BARNHOLZ, A. ROITMAN AND S. GATT, *J. Biol. Chem.*, 241 (1966) 3731.
121 G. B. ANSELL AND H. DOHMEN, *J. Neurochem.*, 2 (1959) 1.
122 R. M. C. DAWSON, *Biochem. J.*, 57 (1954) 237.
123 L. E. HOKIN AND M. R. HOKIN, *Biochim. Biophys. Acta*, 16 (1955) 229.
124 L. E. HOKIN AND M. R. HOKIN, in R. M. C. DAWSON AND D. N. RHODES (Eds.), *Metabolism and Physiological Significance of Lipids*, Wiley, New York, 1964, p. 423.
125 M. G. LARRABEE, J. D. KLINGMAN AND W. S. LEICHT, *J. Neurochem.*, 10 (1963) 549.
126 M. G. LARRABEE AND W. S. LEICHT, *J. Neurochem.*, 12 (1965) 1.
127 A. SHELTAWAY AND R. M. C. DAWSON, *Biochem. J.*, 100 (1966) 12.
128 B. W. AGRANOFF, R. M. BRADLEY AND R. O. BRADY, *J. Biol. Chem.*, 233 (1958) 1077.
129 H. PAULUS AND K. P. KENNEDY, *J. Biol. Chem.*, 235 (1960) 1303.
130 W. THOMPSON, K. P. STRICKLAND AND R. J. ROSSITER, *Biochem. J.*, 87 (1963) 136.
131 H. HYDEN, in J. BRACHET AND A. E. MIRSKY (Eds.), *The Cell*, Vol. 4, Academic Press, New York, 1960, p. 215.
132 M. B. BUNGE, R. P. BUNGE AND G. D. PAPPAS, *J. Cell Biol.*, 12 (1962) 448.
133 A. N. DAVISON, in F. LINNEWEH (Ed.), *Fortschritte der Pädologie*, Springer, Berlin, 1968, p. 65.
134 P. V. JOHNSTON AND B. I. ROOTS, *Biochem. J.*, 98 (1966) 157.
135 J. B. FINEAN, *Ann. N.Y. Acad. Sci.*, 122 (1965) 51.
136 F. A. VANDENHEUVEL, *J. Am. Oil Chemists' Soc.*, 40 (1963) 455.
137 A. A. BENSON, *J. Am. Oil Chemists' Soc.*, 43 (1966) 265.
138 D. F. H. WALLACH AND P. H. ZAHLER, *Proc. Natl. Acad. Sci. (U.S.)*, 56 (1966) 1552.
139 J. DOBBING, *Physiol. Rev.*, 41 (1961) 130.
140 A. N. DAVISON AND J. DOBBING, *Nature*, 191 (1961) 844.
141 R. M. C. DAWSON AND D. RICHTER, *Proc. Roy. Soc. (London), Ser. B*, 127 (1950) 252.
142 G. W. CHANGUS, I. L. CHAIKOFF AND S. RUBEN, *J. Biol. Chem.*, 126 (1938) 493.
143 A. N. DAVISON AND J. DOBBING, *Biochem. J.*, 73 (1959) 701.
144 A. N. DAVISON AND J. DOBBING, *Biochem. J.*, 75 (1960) 565.
145 B. A. FRIES, G. W. CHANGUS AND I. L. CHAIKOFF, *J. Biol. Chem.*, 132 (1940) 23.
146 R. C. THOMPSON AND J. E. BALLOU, *J. Biol. Chem.*, 208 (1954) 883.
147 R. C. THOMPSON AND J. E. BALLOU, *J. Biol. Chem.*, 223 (1956) 795.
148 J. DOBBING, *Guy's Hosp. Rept.*, 112 (1963) 267.
149 G. MAJNO AND M. L. KARNOVSKY, *J. Exptl. Med.*, 107 (1958) 475.
150 A. TORVIK AND R. L. SIDMAN, *J. Neurochem.*, 12 (1965) 555.
151 L. M. SEMINARIO, N. HREN AND C. J. GOMEZ, *J. Neurochem.*, 11 (1964) 197.
152 T. M. BRODY AND J. A. BAIN, *J. Biol. Chem.*, 195 (1952) 685.
153 S. R. KOREY, in D. RICHTER (Ed.), *Metabolism of the Nervous System*, Pergamon, Oxford, 1957, p. 87.
154 V. P. WHITTAKER, *Progr. Biophys. Mol. Biophys.*, 15 (1965) 39.
155 R. H. LAATSCH, M. W. KIES, S. GORDON AND E. C. ALVORD, *J. Exptl. Med.*, 115 (1962) 777.
156 A. A. ABDEL-LATIF AND L. G. ABOOD, *J. Neurochem.*, 12 (1965) 157.
157 M. L. G. GENT, N. A. GREGSON, D. B. GAMMACK AND J. H. RAPER, *Nature*, 204 (1964) 553.
158 A. A. KHAN AND J. E. WILSON, *J. Neurochem.*, 12 (1965) 81.
159 C. W. M. ADAMS, A. N. DAVISON AND N. A. GREGSON, *J. Neurochem.*, 10 (1963) 383.

160 L. F. ENG AND M. E. SMITH, *Lipids*, 1 (1966) 296.
161 L. RATHBONE, *Biochem. J.*, 97 (1965) 620.
162 J. EICHBERG AND R. M. C. DAWSON, *Biochem. J.*, 96 (1965) 644.
163 G. B. ANSELL, *Advan. Lipid Res.*, 3 (1965) 139.
164 H. MCILWAIN, *Chemical Exploration of the Brain*, Elsevier, Amsterdam, 1963.
165 J. A. LOWDEN AND L. S. WOLFE, *Canad. J. Biochem.*, 42 (1964) 1587.
166 A. GEIGER, *Physiol. Rev.*, 38 (1958) 1.
167 R. VRBA, *Nature*, 195 (1962) 663.
168 R. VRBA, M. K. GAITONDE AND D. RICHTER, *J. Neurochem.*, 9 (1962) 465.
169 A. N. DAVISON AND J. DOBBING, *Brit. Med. Bull.*, 22 (1966) 40.
170 M. L. CUZNER, A. N. DAVISON AND N. A. GREGSON, *J. Neurochem.*, 12 (1965) 469.

Chapter VIII

Fatty Acid Oxidation

RUBIN BRESSLER

Departments of Medicine and Pharmacology, Duke University Medical Center, Durham, N.C. (U.S.A.)

1. Introduction

In the absence of glucose, cellular oxidation of long-chain fatty acids provides the principal alternative energy source of intermediary metabolism. Free fatty acids are generated by the lipolysis of adipose tissue stores of triglycerides[1] and are circulated to other tissues as fatty acid albumin complexes[2-6]. The turnover rate of plasma free fatty acids is approximately 20–40% per min, suggesting avid tissue extraction and/or utilization[6]. Cellular uptake of free fatty acids is determined by the plasma free fatty acid concentration[7,8] and does not appear to be an active metabolic process.

Once inside the cell, free fatty acids are metabolized by a number of diverse pathways, *i.e.*, oxidation, esterification to triglycerides and complex lipids, or conversion to other fatty acids by the processes of carbon-chain elongation, shortening or desaturation. Which cellular pathway predominates at any given time, however, depends on the hormonal, nutritional and thermal environment of the animal. These different metabolic processes may occur at functionally separate sites within the cell, esterification and desaturation being primarily microsomal and β-oxidation, chain elongation, and chain shortening being primarily intramitochondrial[9-11].

Irrespective of the pathway by which they are metabolized, free fatty acids are first activated by esterification to coenzyme A—a prerequisite step for further biochemical alteration of the fatty acyl group. Long-chain fatty acids can be activated both extramitochondrially (70%) and intramito-

References p. 355

chondrially (30%)[12-20] at sites distinct from those of β-oxidation[21]. The preferred site of activation depends not only on the particular tissue, but on the carbon chain length of the fatty acid.

The significance of the spacial separation of the enzymes of fatty acid oxidation has become partially apparent through elucidation of the role of cellular compartmentalization in the regulation of metabolic processes. The rate at which substrates of respiration are utilized depends on a complex interplay of energy requirements, availability of cofactors, and, in addition, selective accessibility barriers. Krebs[22] had introduced the concept of competition for available oxygen to explain the nonadditive stimulation of oxygen uptake by oxidizable substrates. The contributions of the various oxidative processes to the total respiration of the mitochondrion are interdependent and all share the same chain of electron carriers. Therefore, competition is predictable if the capacity of the electron-transport chain is limited[23] by the availability of ADP or P_i. Under other circumstances, however, the mitochondrial oxidation of a substrate may be regulated by the conformation of the enzymes which catalyze its oxidation, by the availability of cofactors such as NAD or CoA, or, more importantly, by the accessibility of the substrate to compartmentalized cofactors or enzymes.

Although it was generally appreciated that the compartmentalization of the enzymes of fatty acid metabolism was probably significant in the regulation of free fatty acid utilization, not until the role of carnitine in mammalian tissues was elucidated was a second major dimension added to the understanding of lipid metabolism[24].

2. Fatty acid oxidation and carnitine

Carnitine (γ-trimethylammonium-β-hydroxybutyrate) (I) is widely distributed in nearly all organisms and all tissues, the highest concentrations residing in muscle[25]. Addition of carnitine to tissues *in vitro* catalytically increases the rate of long-chain fatty acid oxidation, with maximal stimulation occurring in the presence of ATP, coenzyme A, magnesium and high albumin concentrations[26,27]. The oxidation of acetyl, butyryl, hexanoyl, palmityl, or stearyl coenzyme A derivatives however is enhanced by carnitine in the absence of ATP or coenzyme A. Moreover, acylcarnitine derivatives of fatty acids are oxidized by mitochondria as readily as the corresponding acyl-coenzyme A derivatives in the presence of free carnitine[24,25]. Such evidence suggests that a transfer of the fatty acids from coenzyme A to carnitine must occur before the acyl groups can be further metabolized by the intramitochondrial enzymes of β-oxidation.

$$(CH_3)_3N^+CH_2-\underset{\underset{\displaystyle OH}{|}}{CH}-CH_2COOH \qquad \qquad (I)$$

Minor subgroup alterations of the carnitine molecule greatly depress its catalytic effect on fatty acid oxidation. Removal of the hydroxyl group in the β-position abolishes all activity, as does replacement of the carboxyl group with either a cyano, alcohol, or amide grouping or substitution of an amino group for the trimethylammonium moiety of the molecule. Activity is inhibited partially by the substitution of a dimethylammonium group for the trimethylammonium group[28].

In the rat and a number of microorganisms, carnitine is generated by hydroxylation of γ-butyrobetaine[29,30], a reaction localized to the soluble fractions of rat liver and apparently unassociated with other microsomal hydroxylations. This reaction requires molecular oxygen and ferrous ion and is activated by ascorbate in the presence of an NADPH-regenerating system[29]. The biosynthetic origin of γ-butyrobetaine is obscure, although it has been established that the methyl groups are donated by S-adenosylmethionine[31,32]. In mammals, biosynthesis of carnitine is primarily intrahepatic with subsequent distribution to other tissues[32].

(a) Mitochondrial compartmentalization and the role of carnitine in long-chain fatty acid oxidation

The mitochondrial activation of fatty acids to their respective acyl-coenzyme A derivatives occurs at a site outside the "carnitine barrier" as demonstrated by the observation that free fatty acids are not oxidized by isolated mitochondria from the skeletal and cardiac muscle of various animals, whereas the corresponding acylcarnitine derivatives are readily oxidized[33]. Accordingly, most of the mitochondrial acyl-CoA synthetase (EC 6.2.1.3) activity has been localized to the outer membrane fraction[14], although two other activating enzyme systems, one GTP- and the other ATP-dependent, have been identified in intact liver mitochondria. Because long-chain fatty acid oxidation in these preparations is carnitine-independent, these latter enzyme systems can be considered to lie within the mitochondrion beyond the "carnitine barrier"[17-20]. The oxidation of short-chain fatty acids does

References p. 355

not depend on the presence of carnitine as a catalytic substrate[28,34,35], and it would appear that shorter-chain fatty acids are activated by these two latter ATP- and GTP-dependent intramitochondrial enzyme systems[15,36]. *In vitro* depletion of carnitine in the heart, however, decreases the rate of fatty acid oxidation and increases triglyceride synthesis, evidence which indicates that the bulk of fatty acid oxidation *in vivo* is, in fact, carnitine-dependent[32,34].

The inner mitochondrial membrane is permeable to carnitine esters[37,38], although impermeable to acyl-coenzyme A derivatives, free coenzyme A and free carnitine. Esterification of the fatty acid to carnitine, therefore, circumvents the mitochondrial permeability barrier and provides access to the enzymes of β-oxidation.

Compartmentalization of fatty acid oxidation in mitochondria has been demonstrated[21,39] and applies not only to the enzymes of β-oxidation but also to a number of functionally noninterchangeable coenzyme A and carnitine pools. Klingenberg and Pfaff[38] have correlated the inner mitochondrial membrane with a barrier which is impermeable to adenine nucleotides except by an exchange process inhibitable by atractyloside. This membrane is impermeable to free coenzyme A and free carnitine[37] and, therefore, defines two separate coenzyme A and carnitine pools. Although palmityl-coenzyme A may be synthesized at either side of the membrane[18,40], its further oxidation nevertheless requires carnitine. Such observations suggest the existence of at least three different mitochondrial pools of coenzyme A. Two of these pools may be considered to exist within the atractyloside membrane. One of these would be available to the GTP-requiring fatty acyl thiokinase, the α-ketoglutarate dehydrogenase, and the succinyl-CoA synthetase (EC 6.2.1.4, 5), but unavailable to the enzymes of β-oxidation, whereas the second would be associated only with enzymes of β-oxidation. The third pool would lie outside the atractyloside-sensitive membrane, presumably in the space between the inner and outer membranes. Exogenous coenzyme A and carnitine can permeate to less than 70% of the intramitochondrial water, a space which is slightly smaller than the sucrose-permeable space and which defines the area between the mitochondrial membranes[37] (Fig. 1).

In another theory of the compartmentalization of mitochondrial enzymes, two functional barriers are hypothesized[41]. Barrier I, permeable to ATP in the presence of atractyloside and corresponding to Klingenberg's inner mitochondrial membrane, would be permeable to fatty acids, carnitine,

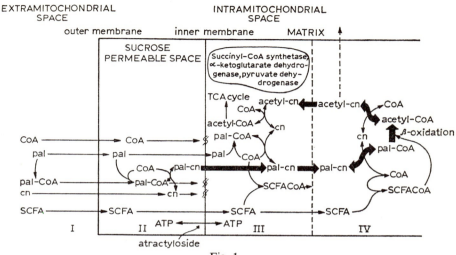

Fig. 1.

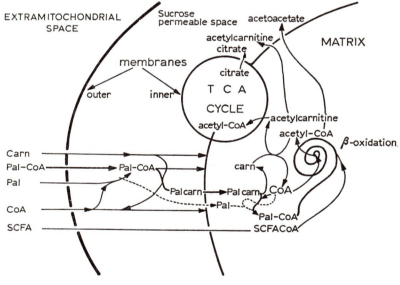

Fig. 2.

acyl-coenzyme A and acylcarnitines, but probably not to coenzyme A. Barrier II would be permeable to acylcarnitines but not coenzyme A or acyl-coenzyme A. As in the former scheme, these barriers would define three pools of coenzyme A. Not only does the formation of acylcarnitine

References p. 355

esters allow the fatty acid substrate to reach the enzymes of β-oxidation, but also provides a mechanism by which acetyl groups formed in the fatty acid oxidation compartment are made available to the tricarboxylic acid cycle for further oxidation[42], a theory consistent with the observation that acetyl-CoA:carnitine O-acetyltransferase (EC 2.3.1.7) is most abundant in those tissues with high rates of fatty acid oxidation[42].

A scheme attempting to integrate the two theories of mitochondrial compartmentalization described above is outlined in Fig. 2. The possible coenzyme A and carnitine pools and their relationship to the enzymes of fatty acid oxidation and activation are shown in Fig. 3. In these schemes, it is

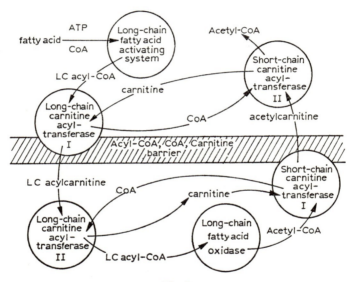

Fig. 3.

assumed that the formation of acetylcarnitine ester not only permits the intramitochondrial shuttling of acetyl groups between the site of formation and the site of oxidation[42,43], but it permits the carnitine which has been carried into the β-oxidation compartment as an acylcarnitine to be returned to other mitochondrial compartments as carnitine.

The concept of compartmentalization together with the proposal that different substrates have different affinities for coenzyme A acquires particular significance in view of observations by Bremer[15,44] and Tubbs et al.[60] that long-chain acylcarnitines markedly suppress the decarboxylation of pyruvate and that succinate, in turn, decreases the oxidation of fatty acids[15]. Using [^{3}H-*Me*]palmitylcarnitine it is possible to demonstrate that the amount of coenzyme A available for acceptance of an acyl group from an acylcarnitine is decreased by succinate. It has been suggested that this phenomenon is the result of either competition for available coenzyme A or inhibition of the enzymes of β-oxidation by succinyl-coenzyme A[15].

(b) Long-chain acyl-CoA–carnitine acyltransferase (LCAT)

The transacylation of the coenzyme A and carnitine esters (Reaction II) is mediated by two separate enzymes, as previously indicated:

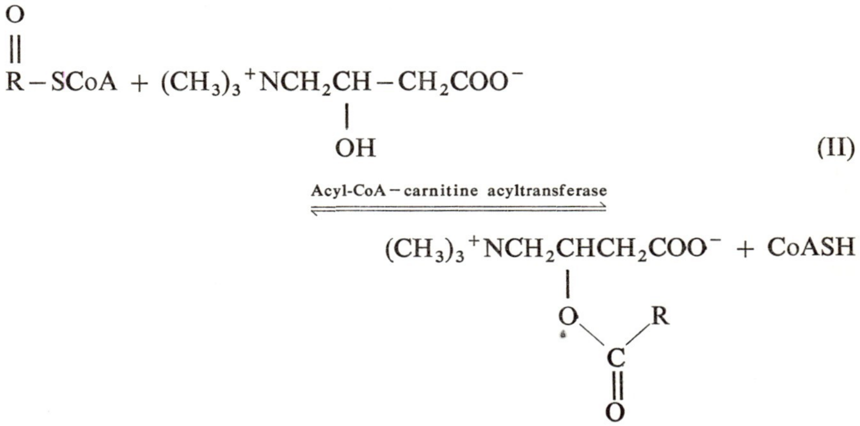

$$\underset{\text{Acyl-CoA – carnitine acyltransferase}}{\overset{\displaystyle R-\overset{O}{\overset{\|}{C}}-SCoA + (CH_3)_3{}^+NCH_2\underset{\underset{OH}{|}}{CH}-CH_2COO^-}{\rightleftharpoons}}$$

$$(CH_3)_3{}^+NCH_2\underset{\underset{\underset{O}{\overset{\|}{C}}\diagdown R}{\overset{O}{|}}}{CH}CH_2COO^- + CoASH \qquad (II)$$

One, acetyl-CoA: carnitine *O*-acetyltransferase (EC 2.3.1.7), is specific for acetyl-coenzyme A and short-chain acyl-coenzyme A's[24,45–47] and will be discussed in the following section. The second, palmityl-CoA:carnitine palmityltransferase (EC 2.3.1.-), is specific for long-chain fatty acyl coenzyme A's[27,48]. Reaction II is freely reversible[48], and there is sufficient mitochondrial enzyme to permit equilibration or near equilibration between the substrate pairs acyl-coenzyme A/coenzyme A and acylcarnitine/carnitine *in vivo*[49]. The tissue levels of acyl-coenzyme A and acylcarnitines (and free coenzyme A and free carnitine), therefore, vary in parallel[33,50–52].

Palmityl-CoA:carnitine palmityltransferase has been partially purified[48] and localized to the inner membrane of the mitochondrion[14,37]. The reaction catalyzed by this enzyme is shown below:

Palmityl-CoA + (−)-carnitine $\rightleftharpoons$ (−)-palmitylcarnitine + CoASH (III)

References p. 355

In the preparations of highly purified carnitine palmityltransferase extracted from calf's liver, the equilibrium constant[48] of Reaction III is 0.45 at pH 7.6, indicating that the reaction is freely reversible and that the ester bond of palmitylcarnitine has an energy content comparable to that of the thioester bond of palmityl-coenzyme A. Carnitine palmityltransferase has a broad pH optimum between 7.0 and 8.2 and is specific for the negative optical isomers of carnitine and palmitylcarnitine. The K_m values for the four substrates involved in the reaction are: palmityl-coenzyme A, $3.1 \cdot 10^{-5} M$; coenzyme A, $4.5 \cdot 10^{-5} M$; (−)-carnitine, $2.1 \cdot 10^{-3} M$; and (−)-palmitylcarnitine, $1.3 \cdot 10^{-4} M$. With increasing length of the acyl carbon chain, the initial velocity of the acyl transfer from acylcarnitine to coenzyme A also increases, indicating the substrate specificity of the enzyme for longer-chain fatty acyl derivatives[48].

The requirement of palmitylcarnitine transferase for carnitine as a substrate is highly specific, and norcarnitine (γ-dimethylammonium-β-hydroxybutyrate) is the only compound chemically analogous to (−)-carnitine capable of serving as an alternative substrate. This strict substrate specificity is further emphasized by the high K_m for norcarnitine; it is approximately three times that found for carnitine. γ-Butyrobetaine (γ-trimethylammoniumbutyrate) competitively inhibits carnitine palmityltransferase, although the K_i is extremely high ($10^{-1} M$)[53]. Of further interest is the competitive inhibition of the enzyme by (+)-palmitylcarnitine which can be demonstrated in intact mitochondria, but not in preparations using the purified enzyme[49,43].

The K_m values of carnitine palmityltransferase for the acylated substrate (palmitylcarnitine, palmityl-CoA) are lower than those for the non-acylated substrates (carnitine, coenzyme A), suggesting that the enzyme possesses specific binding sites for the palmityl group. In partially purified enzyme preparations, the γ-trimethylammonium group, the β-hydroxyl group, and the carboxyl groups are required for binding of the substrate to the enzyme, and substitution of a diethylamino group for the trimethylammonium group also lessens enzyme–substrate affinity. γ-Butyrobetaine, which lacks the β-hydroxy group, has less affinity than carnitine for the enzyme. Moreover, replacement of the carboxyl group of carnitine with a carbamide moiety completely abolishes enzyme–substrate interaction[48].

Palmityl-CoA is both a substrate for the enzyme ($K_m = 3 \cdot 10^{-5} M$) and a potent competitive inhibitor of the second substrate, carnitine ($K_i = 3 \cdot 10^{-6} M$)[54,55]. (+)-Palmitylcarnitine, Tween 80 and other detergents counteract this inhibitory effect of palmityl-CoA and also decrease the

K_m for carnitine, an effect most pronounced in the presence of high concentrations of palmityl-CoA[56]. It is possible that the stimulatory effect of detergents on Reaction III results from the prevention of product inhibition by palmityl-CoA. Although detergents also interfere with the substrate function of palmityl-CoA, the extent of interference is much less[54,55].

Palmitylcarnitine transferase is inhibited by *p*-chloromercuribenzoate (PCMB), *N*-ethylmaleimide (NEM), disulfides, cysteamine and diethylcysteamine. PCMB protects the enzyme against alkylation by NEM[57], as do the disulfides, to a much lesser degree. Inhibition of the enzyme by disulfides is completely reversible by thiols, and inhibition by PCMB is partially reversible. The inhibition by NEM, however, is irreversible and no reactivation by thiols is possible. Protection against inhibition by PCMB or NEM is conferred upon the enzyme by palmityl-CoA, protection against NEM by palmitylcarnitine, and protection against disulfides by free carnitine. Such observations suggest that thiol groups on the enzyme are localized at or near the palmityl-CoA-binding site and that this binding site is near the binding sites for carnitine and palmitylcarnitine[57].

If carnitine acyltransferase is incubated with palmitylcarnitine, coenzyme A and [^{14}C]carnitine, the enzyme will catalyze the incorporation of the labeled carnitine into palmitylcarnitine as shown in Reaction IV.

$$[^{14}C]\text{Palmitylcarnitine} + \text{CoASH} \rightleftharpoons \text{palmityl-CoA} + [^{14}C]\text{carnitine} \quad \text{(IV)}$$

The failure of this reaction to take place in the absence of added coenzyme A militates against the conclusion that an acylthio–enzyme complex participates in the palmityltransferase reaction. The importance of sulfhydryl group integrity on the carnitine acyltransferase enzyme is, nevertheless, well established, even though its precise role has yet to be clarified[57].

(c) *Acetyl-CoA:carnitine acetyltransferase and short-chain fatty acid oxidation*

Carnitine acetyltransferase is located on the mitochondrial inner membrane[58,59] and reversibly catalyzes the transfer of acetyl groupings from coenzyme A to carnitine (Reaction V).

$$[-]\text{-Acetylcarnitine} + \text{CoASH} \rightleftharpoons [-]\text{-carnitine} + \text{acetyl-CoA} \quad \text{(V)}$$

Unlike carnitine palmityltransferase, purification of carnitine acetyltransferase has been extensive and the enzyme has been obtained in crystalline

References p. 355

form[60]. The acyl group of acetyl-CoA, propionyl-CoA and butyryl-CoA are transferred to carnitine at approximately equal rates, whereas the transfer of longer-chain acyl grouping (C_5–C_{16}) occurs only at reduced rates[24,27].

Like the long-chain transferase, acetyltransferase specifically requires carnitine as a substrate, a requirement which may be partially met by norcarnitine (γ-dimethylammonium-β-hydroxybutyrate). Maximal enzyme activity obtains in the pH range of 7.1–8.2 and the equilibrium constant[47] of the reaction at pH 7 is approximately 0.6. The activity of partially purified enzyme preparations from the pig heart is sensitive to inhibition by reagents specific for thiol groups, a condition which can be reversed by the presence of the substrate, acetyl-CoA[59].

The structural requirements for substrates of acetyltransferase are considerably more stringent than those for inhibitors. Whereas the negative optical isomers of carnitine and acetylcarnitine serve as suitable substrates in the reaction, the unnatural positive isomers competitively inhibit the enzyme[59]. The K_m values for (−)-carnitine ($3.1 \cdot 10^{-4} M$), norcarnitine ($3.2 \cdot 10^{-3} M$), and β-hydroxy-γ-aminobutyric ($4 \cdot 10^{-2} M$) would seem to indicate that methylation of the nitrogen facilitates enzyme–substrate interaction. With respect to acetyl-CoA, the K_m of acetyltransferase is considerably lower ($4 \cdot 10^{-5} M$) than the values for carnitine and its analogs.

(+)-Acetylcarnitine, the most potent inhibitor of the transferase, possesses a free carboxyl group, an ester linked to the β-carbon, and a quaternary nitrogen. Hydrolysis of (+)-acetylcarnitines to form the secondary alcohol of (+)-carnitine or hydrolysis of (+)-acetylcarnitine followed by dehydrogenation to form deoxycarnitine diminishes its inhibitory activity 10-fold, while demethylation of the quaternary nitrogen further depresses such activity. Other manipulations, such as deamination of (+)-carnitine to form β-hydroxybutyric acid, esterification of the carboxyl group to form a methoxy ester, or the generation of (+)-norcarnitine amide or (+)-norcarnitol, totally abolish the inhibitory potential of the parent compound[47,59].

Generally, esterification of a fatty acyl derivative to carnitine is a necessary requirement for its admission to the mitochondrial compartment containing the enzymes of β-oxidation. One cannot infer from the previous argument, however, that the formation of acylcarnitine derivatives is the rate-limiting step in fatty acid oxidation, as has been proposed[61]. The maximal rate of formation of palmitylcarnitine from palmityl-CoA and carnitine in mitochondria isolated from rat liver exceeds the maximal capacity for oxidation of palmitylcarnitine by over 6-fold[15,54]. Furthermore, in the presence of palmityl-CoA and added carnitine, mitochondrial uptake rates are equal to those obtained in the presence of palmitylcarnitine[15,54]. Palmitylcarnitine in a concentration of $2 \cdot 10^{-6} M$ results in maximal oxygen consumption by the mitochondria of rat liver[15,54], whereas the *in vivo* concentration of long-chain acylcarnitine derivatives in the liver probably attains[51] the minimal level of $3 \cdot 10^{-5} M$. This evidence would indicate that the formation of palmitylcarnitine is not the rate-limiting step in long-chain fatty acid oxidation, except when low concentrations of extramitochondrial palmityl-CoA obtain[15]. The long-chain acyltransferase is strategically located in the inner mitochondrial membrane and catalyzes the acyl transfer required for this intramitochondrial movement of long-chain fatty acyl derivatives. The primary function of acetyltransferase, on the other hand, is the catalysis of the transfer of acetyl groups and carnitine between the compartment containing the enzymes of β-oxidation and that containing the enzymes of the tricarboxylic acid cycle[15,22,57,58].

(d) The quantitative significance of carnitine in various other disease states

Several lines of evidence indicate that the tissue concentrations of carnitine and the activity of the long-chain carnitine acyltransferase are most critical factors in the regulation of long-chain fatty acid oxidation. The biochemical consequences of carnitine depletion include a depressed rate of long-chain fatty acid oxidation and an unimpaired rate of short-chain fatty acid oxidation[154-158]. The increased cellular accumulation of triglycerides is characteristic of carnitine-depleted tissues, as evidenced by the correction of the long-chain fatty acid oxidation deficit by the addition of exogenous carnitine[154-160].

As one might expect, carnitine deficiency often assumes a critical significance *in vivo*. In the newborn rat heart, an impaired rate of long-chain fatty acid oxidation can be demonstrated during the first few days of life. The newborn animal does not possess sufficient tissue concentrations of carnitine until after the first week of life, at which time fatty acid oxidation normalizes[159]. A second instance of carnitine deficiency occurs in rats treated with diphtheria toxin. The administration of the toxin depresses long-chain fatty acid oxidation, and the tissues become infiltrated with triglycerides. Short-chain fatty acid oxidation, however, which is carnitine-independent, is normal in these animals, and the rate of long-chain fatty acid oxidation by myocardial homogenates of toxin-injected rats is restored

References p. 355

to normal by exogenous carnitine[154-156]. A third instance is the diminished long-chain fatty acid oxidation and decreased tissue carnitine levels in choline deficiency secondary to the creation of a generalized methyl group deficiency[157]. Here again, administration of exogenous carnitine corrects the depressed rates of long-chain fatty acid oxidation. Finally, the administration of hypoglycin (L-α-amino-β-methylenecyclopropanepropionic acid) or 4-pentenoic acid results in the formation of activated coenzyme A derivatives (methylenecyclopropyl acetyl-CoA, 4-pentenoyl-CoA or acryloyl-CoA which can undergo transacylation with carnitine to form carnitine derivatives, but which are not readily oxidized[158,160], thereby leading to a decrease in tissue levels of free carnitine and coenzyme A and subsequent impairment of the numerous processes catalyzed by these cofactors[158-160]. Replacement of coenzyme A and carnitine in these instances also restores the rate of long-chain fatty acid oxidation to normal[161].

3. Fatty acid oxidation and gluconeogenesis

Gluconeogenesis is defined as the production of glucose from compounds other than hexoses. Endogenous synthesis of glucose is a process of great magnitude and assumes particular importance during metabolic periods of relative excess of free fatty acids, *viz.*, fasting, heavy muscular exercise, high dietary lipid intake, uncontrolled diabetes, and hormonally induced adipose tissue lipolysis[6]. Although the kidneys are capable of synthesizing small amounts of glucose, the greatest bulk is synthesized by the liver.

Because of three thermodynamically irreversible reactions[62,63], gluconeogenesis cannot proceed by a simple reversal of glycolysis (Fig. 4). The pyruvate kinase (EC 2.7.1.40) step is irreversible and phosphoenolpyruvate (PEP) must be regenerated by the condensation of pyruvate and CO_2 to form oxaloacetate. A third enzyme, phosphoenolpyruvate carboxykinase, subsequently catalyzes the formation of a high-energy phosphate ester and the loss of CO_2 to yield PEP. Under physiological conditions, the remainder of the reactions of glycolysis are reversible, except for the cleavage of phosphates from fructose 1,6-diphosphate and glucose 6-phosphate—reactions catalyzed by fructose-1,6-diphosphatase (EC 3.1.3.11) and glucose-6-phosphatase (EC 3.1.3.9).

Control over a metabolic pathway can be exerted most precisely by regulation of the irreversible steps in that pathway. As shown in Fig. 5, the significant regulatory steps of glucose metabolism are, in fact, four. Of these

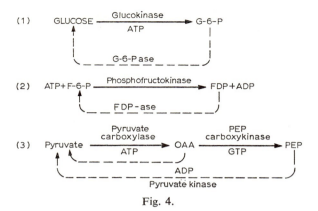

Fig. 4.

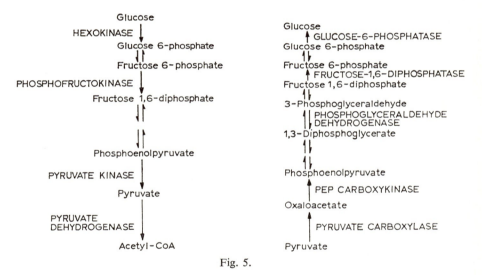

Fig. 5.

4 steps, *in vivo* and *in vitro* measurements of metabolic intermediates have demonstrated the pyruvate carboxylase (EC 6.4.1.1) step to be of singular importance in the changeover from glycolysis to gluconeogenesis, whether gluconeogenesis is stimulated by the acute administration of free fatty acids or glucagon[64-66].

Periods of enhanced gluconeogenesis are characterized by an increased rate of fatty acid oxidation and by an increase in the activity of the enzymes of β-oxidation. In addition, the activities of a number of other hepatic enzymes are increased during these periods of increased gluconeogenesis

References p. 355

and fatty acid oxidation. Administration of small doses of cortisone to rats results in increased hepatic glucose-6-phosphatase, fructose-1,6-diphosphatase, and pyruvate carboxylase activities[67-72], with statistically significant rises assayable within 2–4 h after administration. Actinomycin-D and puromycin prevent these cortisone-induced increases in enzyme activities, suggesting that such increases are secondary to *de novo* protein synthesis[67,69,70]. A similar increase in the activities of these "pacemakers" gluconeogenic enzymes occurs in acute alloxan diabetes and apparently also results from *de novo* enzyme synthesis. Of particular interest, however, is the failure of inhibitors of protein synthesis to block the increase in these enzymes in chronic alloxan diabetes whereas the same inhibitors are capable of reversing the increase in the enzyme activities consequent to chronic steroid administration. All of these observations strongly suggest that *de novo* synthesis of the regulatory enzymes of gluconeogenesis is an important feature of the changeover from glycolysis to gluconeogenesis. That control over the synthesis of these enzymes is indeed of physiological importance is evidenced by the finding that insulin inhibits the increase in gluconeogenic enzyme activities accompanying both steroid therapy and chronic diabetes, suggesting that insulin may act as a physiologic repressor of the hepatic synthesis of the relevant enzymes[67,70].

Because of the considerable latent period required for the effects of steroids on hepatic gluconeogenic enzyme activities[68,71], attention has been directed to factors influencing the immediate acceleration of gluconeogenesis. Fatty acid oxidation and gluconeogenesis are intimately related[73-74], and lipolytic hormones such as epinephrine, glucagon, and growth hormone, which accelerate the release of free fatty acids from adipose tissue, promote increased hepatic free fatty acid uptake and oxidation and a rapid increase in gluconeogenesis in the absence of new enzyme formation[75]. The addition of free fatty acids to kidney slices *in vitro* increases glucose production from lactate[76] and also stimulates gluconeogenesis in several types of liver preparations[77]. Furthermore, (+)-acylcarnitines, which reversibly inhibit long-chain fatty acid oxidation, also inhibit gluconeogenesis[78]. These observations and the demonstration that carnitine enhances gluconeogenesis from rabbit liver and kidney slices *in vitro* emphasize the acute gluconeogenic potential of increased long-chain fatty acid oxidation[66].

Although the liver can metabolize glucose *via* glycolysis at a rate only 2–10% that of muscle[63,79], inhibition of hepatic glycolytic enzymes is nevertheless necessary to achieve a net gain in glucose production. The maximal activity of pyruvic kinase is approximately equal to that of PEP carboxykinase in fasted rat liver[58,63,80], und unless pyruvic kinase were inhibited, the bulk of PEP would be reconverted to pyruvate instead of to glucose (Figs. 4 and 5).

The chemical characteristics of several of the key-regulatory reactions of glucose metabolism illustrate how fatty acids or products of their oxidation might accelerate the process of gluconeogenesis and inhibit glycolysis.

$$\text{Fructose 6-phosphate} + \text{ATP} \rightarrow \text{fructose 1,6-diphosphate} + \text{ADP} \quad (1)$$

This reaction is catalyzed by phosphofructokinase, an enzyme activated by fructose 1,6-diphosphate, fructose-6-phosphate, P_i and AMP and cyclic 3',5'-AMP[81]; it is inhibited by citrate[63] and free fatty acids[67,63,79]. The reverse reaction is catalyzed by a different enzyme, fructose diphosphatase, which is allosterically inhibited by AMP[82].

$$\text{Phosphoenolpyruvate} + \text{ADP} \rightarrow \text{pyruvate} + \text{ATP} \quad (2)$$

Pyruvate kinase catalyzes this reaction and, like glucokinase (EC 2.7.1.2) and phosphofructokinase (EC 2.7.1.11) is inhibited by free fatty acids[67,79]. Pyruvate kinase is also inhibited by NADH, a product of fatty acid oxidation[67], and by ATP and GTP[63,83]. Unlike the PFK reaction, the reversal of this reaction is catalyzed by several enzymes and requires the expenditure of energy.

$$CH_3COCOOH + CoASH + NAD^+ \rightarrow$$
$$CH_3CO\text{-}S\text{-}CoA + CO_2 + NADH + H^+ \quad (3)$$

The oxidative decarboxylation of pyruvate and the subsequent acetylation of coenzyme A is effected by pyruvate dehydrogenase (EC 1.2.4.1), an enzyme complex which is inactivated by phosphorylation in the presence of ATP and reactivated by a magnesium-dependent phosphatase in the absence of elevated mitochondrial levels of ATP[84]. In addition, acetyl-CoA allosterically inhibits the activity of pyruvate dehydrogenase[73,78].

$$\text{Pyruvate} + CO_2 \xrightarrow{\text{ATP}} \text{oxaloacetate} \quad (4)$$

This reaction is catalyzed by pyruvate carboxylase, an enzyme possessing an absolute allosteric requirement of acetyl-CoA for activation[63,85] and is inhibited by malonyl-CoA[63,86,87].

$$\text{Oxaloacetate} \xrightarrow{\text{GTP}} \text{Phosphoenolpyruvate} + CO_2 \quad (5)$$

AMP inhibits the catalysis of this reaction by PEP carboxykinase[82].

References p. 355

1,3-Diphosphoglyceric acid + NADH ⇌
$\quad\quad\quad\quad\quad$ glyceraldehyde 3-phosphate + NAD$^+$ + P$_i$ $\quad$ (6)

Apart from the physiologically irreversible reactions outlined above, the catalysis of this last reaction by glyceraldehyde phosphate dehydrogenase (EC 1.2.1.12) may also be a site for control of glucose metabolism. The predominant direction of this reaction is determined by the cytoplasmic NAD/NADH ratio, and when the ratio is high, the glycolytic pathway will predominate over the gluconeogenic one. A more reduced cytoplasmic environment, therefore, will favor glucose synthesis[63].

The acute acceleration of gluconeogenesis has been ascribed not only to the inhibition of the enzymes of glycolysis by free fatty acids but to the formation of acetyl-CoA, ATP and NADH as products of fatty acid oxidation. Accordingly, a conceptual formulation of the acute stimulation of gluconeogenesis might include the following events: *(1)* adipose-tissue lipolysis, *(2)* elevated plasma free fatty acids, *(3)* enhanced tissue uptake of free fatty acids, *(4)* increased activity of long-chain acyl-CoA:carnitine acyltransferase independent of new protein synthesis[88], *(5)* accelerated long-chain fatty acid oxidation, *(6)* augmentation of the intramitochondrial production of acetyl-CoA, NADH and ATP, *(7)* inhibition of pyruvate dehydrogenase, and *(8)* stimulation of pyruvate carboxylase.

Not only is the presence of acetyl-CoA necessary for activation of pyruvate carboxylase[86] it is apparently also required for the inhibition of the oxidative decarboxylation of pyruvate (pyruvic dehydrogenase)[89]. In addition to its role as an allosteric modifier of these two reactions, acetyl-CoA is the immediate precursor of most energy-yielding reactions in the mitochondria[76]. Sustained gluconeogenesis, therefore, would necessitate steady-state levels of acetyl-CoA sufficient to activate pyruvate carboxylase and meet the oxidative requirements of the cell at the same time. Such levels of acetyl-CoA are provided by the increased oxidation of free fatty acids which also serve to reduce the drain on carbohydrate reserves. Although hepatic fatty acid oxidation transiently increases the levels of intermediates of the tricarboxylic acid cycle, no net glucose production results from long-chain fatty acid oxidation, since the mitochondrial production of citrate requires the condensation of one molecule of oxaloacetate with every molecule of acetyl-CoA. These acutely increased levels of the TCA cycle intermediates generated by long-chain fatty acid oxidation apparently do, however, have a special significance. The majority of evidence indicates that oxidation of fatty acids enhances the formation of glucose from substances entering the gluconeogenesis pathway as triosephosphates[90], an enhancement which cannot be mediated by activation of pyruvate carboxylase. It would appear that this effect may be due, at least acutely, to the transient elevation of intracellular citrate levels, which could inhibit phosphofructokinase[90].

In the liver, an increase in the level of activity of the tricarboxylic acid cycle inhibits the oxidative decarboxylation of pyruvate and stimulates its conversion to oxaloacetate[91-94], an effect not demonstrable in the presence of agents which prevent the intramitochondrial generation of ATP[92,93]. Furthermore, purified preparations of pyruvate dehydrogenase from beef-kidney mitochondria are inactivated in the presence of ATP, whereas an increase in the magnesium concentration of such preparations is able to restore their full activity[84]. It is likely that the oxidative utilization of substrates other than carbohydrate determines the metabolic fate of pyruvate and lactate by regulation of pyruvate dehydrogenase. Increased intramitochondrial production of ATP from such substrates would divert lactate and pyruvate into gluconeogenesis, whereas a paucity of such substrates would allow the mitochondrial oxidation of lactate and pyruvate to proceed[84].

Although much has been hypothesized, little has been definitely established concerning the relative importance of increased acetyl-CoA and NADH concentrations in the acute regulation of gluconeogenesis. In the perfused livers of nonfasted rats, however, the addition of small amounts of caproate, while enhancing gluconeogenesis, does not alter the cytoplasmic NAD/NADH ratio, whereas intracellular levels of acetyl-CoA are elevated. Increasing amounts of caproate do not further increase levels of acetyl-CoA, but do lead to a lower NAD/NADH ratio and a further augmentation of gluconeogenesis[73]. These observations suggest the possibility that the initial result of an increase in fatty acid oxidation is the maximal stimulation of pyruvate carboxylase and the simultaneous inhibition of pyruvate dehydrogenase, mediated *via* the augmented levels of acetyl-CoA[73,78] and ATP[84]. When this maximal stimulation is achieved, the more reduced cytoplasmic environment would then favor a further increase in the production of glucose[73].

The *in vitro* allosteric activation of pyruvate carboxylase by acetyl-CoA has led to attempts to demonstrate this phenomenon in intact cells[95]. Examination of the perfused livers of rats fasted for 24 h, however, has failed to reveal any increase in tissue levels of acetyl-CoA in the face of marked

References p. 355

gluconeogenesis and ketogenesis. Similarly, although gluconeogenesis and ketogenesis in the perfused rat liver can be stimulated not only by the addition of long-chain fatty acids but also by glucagon or cyclic 3′,5′-AMP[73,63–66,74,78,96–98], the stimulation of gluconeogenesis by glucagon or oleic acid is not accompanied by a rise in intracellular acetyl-CoA levels[73,99]. Such observations would indicate that either variations in the activity of pyruvate carboxylase can occur independently of variations in intracellular acetyl-CoA concentrations or that these variations in enzyme activity are regulated by critical changes in the acetyl-CoA content of individual cellular compartments. In diabetic rats, hepatic levels of acetyl-CoA parallel increases in blood acetoacetate concentrations to a point beyond which further elevations of acetoacetate occur without additional increases in hepatic acetyl-CoA[100]. Here again, the data suggest that the acute stimulation of hepatic gluconeogenesis involves the initial activation of pyruvate carboxylase by transient increases in acetyl-CoA levels, whereas the maintenance of accelerated gluconeogenesis requires the participation of other factors, such as a new enzyme synthesis.

Consonant with this formulation of the role of acetyl-CoA in the acute regulation of gluconeogenesis is the finding of highly significant increases of acetyl-CoA in the livers of rats fasted just 6 h, a time when liver citrate is just beginning to decline and plasma-ketone bodies are still normal[101]. This period of increased acetyl-CoA levels might be sufficient to account for the enhanced pyruvate carboxylase activity which is observed after 24 h of fasting. One must remember, however, that hepatic levels of acetyl-CoA are not elevated even after gluconeogenesis and ketogenesis have become well established at 24 h[100]. Despite the body of evidence supporting the central role of fatty acid oxidation in the acute and chronic regulation of gluconeogenesis, the rapid stimulation of glucose production by glucagon in the perfused rat liver cannot be duplicated by perfusion with free fatty acids[102]. It is theorized that increased levels of cyclic 3′,5′-AMP may mediate this effect of glucagon[102].

To recapitulate, the synthesis of glucose does not occur by a simple reversal of glycolysis because of three irreversible reactions which must be circumvented by three different, distinct enzymatic steps. The carboxylation of pyruvic acid to the dicarboxylic acid, oxaloacetate, is the initial step in the gluconeogenic pathway and is catalyzed by an intramitochondrial enzyme, pyruvate carboxylase[103–106]. The subsequent conversion of oxaloacetate to phosphoenolpyruvate is catalyzed by phosphoenolpyruvate carboxykinase, an enzyme which is apparently restricted to the extramitochondrial compartments of mammalian hepatic cells[102-109]. Under a variety of metabolic circumstances, *i.e.*, fasting, diabetes, pancreatectomy, or acute and chronic steroid therapy, the activities of both of these enzymes are greatly enhanced[110,111]. Furthermore, phosphoenolpyruvate carboxykinase has been demonstrated to be sufficiently active to account for the increased rates of gluconeogenesis characteristic of these metabolic disorders[111], as well as under normal physiologic conditions.

These observations make it apparent that all the reactions of gluconeogenesis subsequent to the initial carboxylation of pyruvate take place extramitochondrially. Oxaloacetate, however, cannot diffuse freely across the mitochondrial membrane. Unlike oxaloacetate, aspartate, malate, α-keto-

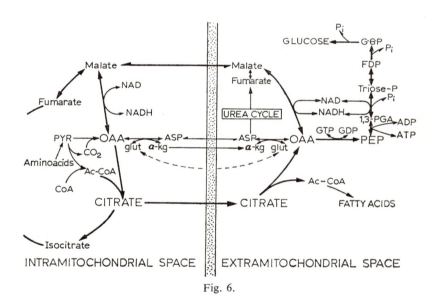

Fig. 6.

glutarate and citrate, to some extent, can permeate the mitochondrial membrane, and it is theorized that the extramitochondrial translocation of oxaloacetate is accomplished by its transamination to form aspartate and/or reduction to malate. In the extramitochondrial compartment of the cell, the oxaloacetate is regenerated by oxidation of the malate and/or transamination of the aspartate, and then converted to phosphoenolpyruvate by the

References p. 355

soluble enzyme, PEP carboxykinase[112]. Sufficient NADH for the intramitochondrial reduction of oxaloacetate to malate is generated by fatty acid oxidation, and the extramitochondrial oxidation of malate is coupled to the reduction of 1,3-diphosphoglyceric acid to 3-phosphoglyceraldehyde. These relationships are shown in Fig. 6.

Because of the inhibition of the key enzymes of glucose oxidation, pyruvate kinase, pyruvate dehydrogenase and phosphofructokinase[67,69,73,113], as well as citrate synthase (EC 4.1.3.7) and several enzymes of the tricarboxylic acid cycle (fumarase, EC 4.2.1.2, and isocitrate dehydrogenase, EC 1.1.1.41) in the presence of increased hepatic uptake and oxidation of long-chain fatty acid, it is probable that free fatty acids enhance gluconeogenesis acutely by virtue of their depressant effect on glucose utilization. It must be emphasized, however, that the inhibition of glycolytic and other enzymes demonstrable in the presence of free fatty acids *in vitro* may be nonspecific and of no physiological significance, since free fatty acids and their acyl-coenzyme A derivatives inactivate numerous enzymes, many of which have no recognizable regulatory role in glycolysis[80,114]. In addition, concentrations of free fatty acids which inactivate glycolytic enzymes in dilute liver homogenates may be ineffective at normal intracellular protein concentrations[115]. Although the available evidence would indicate that free fatty acids *per se* do not directly depress the relevant glycolytic enzymes, it is probable that some product or products of long-chain fatty acid oxidation possess this inhibitory capacity[80]. Such a depression of glycolysis and the tricarboxylic acid cycle would elevate intracellular levels of acetyl-CoA and NADH, which would then stimulate gluconeogenesis and ketosis[64,73,74].

It is valuable to reiterate the biochemical features of the changeover from glycolysis to gluconeogenesis, with special emphasis on the chronological sequence of events. In liver cells utilizing carbohydrate as the primary energy source, the initial consequence of accelerated fatty acid oxidation would be an increase in intramitochondrial citrate, ATP and acetyl-CoA and cellular long-chain acyl-CoA levels. The diffusion of citrate into the cytoplasm and its increasing accumulation there would inhibit phosphofructokinase and augment the free fatty acid-induced depression of glycolysis. The subsequent inhibition of citrate synthase and the other tricarboxylic acid cycle enzymes as a further result of fatty acid oxidation would induce elevations of intramitochondrial acetyl-CoA levels at the expense of citrate levels, which would decline. In addition, the critical allosteric effects of acetyl-CoA on the activities of pyruvate carboxylase and pyruvate kinase would be accentuated. Furthermore, the continued intramitochondrial production of ATP from fatty acid oxidation would inactivate pyruvate dehydrogenase and facilitate the carboxylation of pyruvate to oxalacetate. Finally, increased synthesis of ketone bodies (acetoacetate, β-hydroxy butyrate) would serve to deplete mitochondrial acetyl-CoA levels with the maintenance of the gluconeogenic posture probably dependent on the synthesis or inhibition of the relevant enzymes.

4. Ketogenesis

Ketosis is a physiological phenomenon characterizing metabolic states in which lipids are the predominant energy-yielding substrate. It may also reflect pathological states such as diabetes mellitus. Whenever the availability of carbohydrate is relatively or absolutely depressed in relation to the amount of oxidizable lipid substrate, ketosis will develop. Since the rate of utilization of carbohydrate with respect to lipid is the critical determinant of ketogenesis, those hormones which alter the availabilities of these substrates assume major roles in its regulation[1,116].

Lipid is circulated to the peripheral tissues for oxidative use in two forms, free fatty acids and ketone bodies. The significance of this dual supply is imperfectly understood, but may relate to the lower solubility of the free fatty acid. Ketone bodies serve as normal fuels of respiration of animal tissues and are a quantitatively important source of cellular energy[117]. In the presence of glucose, insulin, or lactate, cardiac muscle preferentially oxidizes acetoacetate[118], as does the renal cortex[117]. In the fasted rat, the amount of substrate provided by blood ketones is equal to that provided by free fatty acids; together, both supply as much oxidizable substrate as does glucose in the fed animal[117].

The extent of ketosis depends on a number of factors, including: (*1*) the rate of mobilization of free fatty acids from adipose-tissue stores, (*2*) the rate of uptake and oxidation of free fatty acids by the liver, (*3*) the rate of carbohydrate utilization in relation to (*1*) and (*2*), and (*4*) the rate of free fatty acid and ketone body (acetoacetate and β-hydroxybutyric acid) utilization by muscle and other extrahepatic tissues.

Of these 4 factors governing the degree of ketosis, the hepatic rate of fatty acid oxidation is apparently the most critical. Diminished tricarboxylic cycle activity secondary to specific blocks of fumarase, isocitrate dehydrogenase and citrate synthase and increased production of acetyl-CoA have

References p. 355

been found in states of increased free fatty acid utilization, such as fasting or alloxan diabetes[64,67]. Furthermore, increased rates of gluconeogenesis in the presence of increased fatty acid oxidation result in a relative and absolute depression of intramitochondrial levels of oxaloacetate[117,119,126]. In the absence of oxaloacetate, liver mitochondria convert fatty acids quantitatively to acetoacetate, whereas in the presence of oxaloacetate fatty acids are converted to tricarboxylic acid cycle intermediates or CO_2. The normal oxidative pathway for acetyl-CoA is an intramitochondrial condensation with oxaloacetate in the presence of citrate synthase to form citric acid. During periods of enhanced gluconeogenesis, however, intramitochondrial oxaloacetate is continually depleted by virtue of its cytoplasmic conversion to glucose, thereby leading to impaired mitochondrial oxidation of acetyl-CoA. The subsequent increase in steady-state levels of acetyl-CoA then results in an accelerated rate of ketogenesis, the only other major metabolic pathway available for the products of β-oxidation. The plausibility of this formulation is supported by determinations of hepatic oxaloacetate levels in severely ketotic diabetic rats[119]. It has been suggested that the observed decrease (0.006 to 0.002 mM) in oxaloacetate concentrations may be metabolically critical since normal hepatic levels of oxaloacetate are below the K_m of the citrate-condensing enzyme (0.015 mM)[121,122]. Such a fall in the intramitochondrial oxaloacetate concentration would, therefore, be expected to diminish the rate of citrate formation from acetyl-CoA. Ketosis, then, may be conceived of as a metabolic posture forced upon the animal because oxaloacetate produced in the liver is utilized to effect the synthesis of glucose. It is important to note that a further drain on the production of energy by the tricarboxylic acid cycle is occasioned by the energy requirements for gluconeogenesis (6 moles of ATP for every mole of glucose formed from lactate)[117].

Hormonal regulation of gluconeogenesis and ketogenesis is manifested both acutely and chronically. Chronic hormonal effects are dependent on *de novo* protein synthesis[64,67], whereas the acute changes are mediated by immediate regulation of the relative availabilities of lipid and carbohydrate[66]. Epinephrine, glucocorticoids, and glucagon all stimulate lipolysis, and may do so by elevating intracellular levels of cyclic 3′,5′-AMP[97,98]. Conversely, insulin is antilipolytic and has been found to depress cellular levels of cyclic 3′,5′-AMP[97,128]. Furthermore, glucagon can activate an hepatic lipase and, thereby, further increase local concentrations of free fatty acid in the liver[124,125].

5. Glucose–fatty acid interactions

In addition to acute effects on hepatic carbohydrate metabolism, elevated plasma free fatty acid levels profoundly influence peripheral glucose utilization. In the isolated perfused rat heart, the presence of free fatty acids and ketone bodies depresses glucose uptake and oxidation[126-129]. Free fatty acids apparently inhibit glucose transport, glycolysis and pyruvate oxidation, thereby enhancing conversion of perfused glucose to cardiac glycogen[126,129]. It is likely that the inhibition of glycolysis by free fatty acids is due to accelerated rates of fatty acid oxidation and the subsequent continued elevations of intracellular citrate[126,127], a potent inhibitor of phosphofructokinase[129]. A second block in peripheral glycolysis due to the diminished phosphorylation of glucose is suggested by the finding of increased intracellular levels of free gluocse during periods of accelerated fatty acid oxidation[127]. This diminished rate of glucose phosphorylation results from an allosteric inhibition of hexokinase (EC 2.7.1.1) by the elevated levels of glucose 6-phosphate which accumulate behind the citrate-induced block of phosphofructokinase[131]. Furthermore, glucose 6-phosphate stimulates glycogen synthetase and facilitates the intracellular deposition of glycogen[132]. It should be emphasized that although the magnitude of the fatty acid-induced inhibition of glucose oxidation in cardiac muscle is well established, the results of studies in other tissue systems are not as conclusive[80,127,128,133,134]. Free fatty acid-induced inhibition of glycolysis has been successfully demonstrated in rat-kidney slices and pigeon-liver homogenates[90,129], but attempts to observe the same phenomenon in the isolated perfused rat liver have been unconvincing[74,78]. Moreover, higher concentrations of free fatty acids actually enhance the *in vitro* oxidation of glucose by adipose tissue. A probable use of the additional glucose from inhibition of glycolysis is the production of α-glycerolphosphate which may then be esterified to fatty acids to form neutral triglycerides[135]. For completeness, it should be mentioned that the hepatic effects of accelerated fatty acid oxidation are similar to those on peripheral tissues, with one major exception. Because peripheral tissues can effect little or no net synthesis of glucose, intramitochondrial levels of oxaloacetate do not diminish. Thus, the major pathway for the metabolism of acetyl-CoA is the condensation with oxaloacetate to form citrate and continued oxidation to CO_2.

A number of physiologic as well as pathologic states, such as diabetes[136],

References p. 355

acromegaly[137], hyperthyroidism[138], pregnancy[139], obesity[140], and starvation[141], are characterized by glucose intolerance as well as elevated plasma free fatty acids. Randle has proposed that the glucose intolerance in these instances may be secondary to the increased utilization of lipid, a metabolic condition which diminishes peripheral oxidation of glucose and impairs the response of muscle to insulin[126,127]. Although such an hypothesis is engaging, *in vivo* studies have thus far yielded conflicting results[80].

Recent studies in dogs, rats and humans have demonstrated depressions of plasma free fatty acids and hypoglycemic effects related to infusion of ketones *in vivo*[142-148]. In dogs, increased levels of pancreatic venous insulin result from such infusion[149], but are of a relatively small magnitude, indicating that the depression of plasma free fatty acids by ketone bodies may not be solely attributable to increased insulin output[148]. Glucose-turnover studies in the dog have revealed that the infusion of ketones decreases hepatic glucose release and increases peripheral glucose utilization[146,150]. In man, such infusions exert a similar effect on plasma glucose and free fatty acid concentrations, but do not elevate peripheral plasma insulin values[142-144]. In the pancreatectomized dog, β-hydroxybutyrate lowers peripheral free fatty acid levels and, in addition, inhibits adipose tissue lipolysis *in vitro*[151,152]. The hypoglycemic effect of increased plasma ketone bodies may result from the release of small quantities of insulin into the portal vein, thereby decreasing hepatic gluconeogenesis, and also from a direct depression of peripheral lipolysis. It is of interest that the lowering of plasma free fatty acid levels *in vivo* has been observed to improve glucose utilization[153].

REFERENCES

1 M. VAUGHAN AND D. STEINBERG, in A. E. RENOLD AND G. F. CAHILL JR. (Eds.), *Handbook of Physiology: Adipose Tissue*, American Physiologic Society, Washington, D.C., 1965, p. 239.
2 V. P. DOLE, *J. Clin. Invest.*, 35 (1956) 150.
3 R. S. GORDON JR., *J. Clin. Invest.*, 36 (1957) 810.
4 R. S. GORDON JR. AND A. CHERKES, *Proc. Soc. Exptl. Biol. Med.*, 97 (1958) 150.
5 D. S. FREDRICKSON AND R. S. GORDON JR., *J. Clin. Invest.*, 37 (1958) 1504.
6 D. S. FREDRICKSON AND R. S. GORDON JR., *Physiol. Rev.*, 38 (1958) 585.
7 M. B. FINE AND R. H. WILLIAMS, *J. Physiol. (London)*, 199 (1960) 403.
8 J. C. SCOTT, L. J. FINKELSTEIN AND J. J. SPITZER, *Am. J. Physiol.*, 203 (1962) 482.
9 S. J. WAKIL, *Ann. Rev. Biochem.*, 31 (1962) 369.
10 B. SHAPIRO, *Ann. Rev. Biochem.*, 36 (1967) 247.
11 J. A. OLSON, *Ann. Rev. Biochem.*, 35 (1966) 559.
12 D. E. GREEN AND S. J. WAKIL, in K. BLOCK (Ed.), *Lipid Metabolism*, Wiley, New York, 1960, p. 1.
13 M. FARSTAD, J. BREMER AND K. R. NORUM, *Biochim. Biophys. Acta*, 132 (1967) 492.
14 K. R. NORUM, M. FARSTAD AND J. BREMER, *Biochem. Biophys. Res. Commun.*, 24 (1966) 797.
15 J. BREMER, in F. C. GRAN (Ed.), *Cellular Compartmentalization and Control of Fatty Acid Metabolism*, Academic Press, New York, 1968, p. 65.
16 C. R. ROSSI AND D. M. GIBSON, *J. Biol. Chem.*, 239 (1964) 1694.
17 S. G. VAN DEN BERGH, in J. M. TAGER, S. PAPA, E. QUAGLIARIELLO AND E. C. SLATER (Eds.), *Regulation of Metabolic Processes in Mitochondria, (BBA Library, Vol. 7)*, Elsevier, Amsterdam, 1966, p. 125.
18 D. W. YATES, D. SHEPHERD AND P. B. GARLAND, *Nature*, 209 (1966) 1213.
19 C. R. ROSSI, L. GALZIGNA, A. ALEXANDRE AND D. GIBSON, *J. Biol. Chem.*, 242 (1967) 2102.
20 C. R. ROSSI, A. ALEXANDRE AND L. SARTORELLI, *Europ. J. Biochem.*, 4 (1968) 31.
21 D. S. BEATTIE, *Biochem. Biophys. Res. Commun.*, 30 (1968) 57.
22 H. A. KREBS, *Biochem. J.*, 29 (1935) 1620.
23 R. J. HASLAM, in J. M. TAGER, S. PAPA, E. QUAGLIARIELLO AND E. C. SLATER (Eds.). *Regulation of Metabolic Processes in Mitochondria, (BBA Library, Vol. 7)*, Elsevier, Amsterdam, 1966, p. 108.
24 I. B. FRITZ, *Advan. Lipid Res.*, 1 (1963) 285.
25 G. FRAENKEL AND S. FRIEDMAN, *Vitamins Hormones*, 15 (1957) 73.
26 I. B. FRITZ, *Am. J. Physiol.*, 197 (1959) 297.
27 I. B. FRITZ AND K. T. N. YVE, *J. Lipid Res.*, 4 (1963) 279.
28 I. B. FRITZ, E. KAPLAN AND K. T. N. YVE, *Am. J. Physiol.*, 202 (1962) 117.
29 G. LINDSTEDT, *Biochemistry*, 6 (1967) 1271.
30 G. LINDSTEDT, S. LINDSTEDT, T. MIDTVEDT AND M. TOFFT, *Biochemistry*, 6 (1967) 1262.
31 D. R. STRENGTH, S. J. YU AND E. J. DAVID, in G. WOLF (Ed.), *Recent Research on Carnitine*, M. I. T. Press, Cambridge, Mass., 1965, p. 45.
32 C. CORREDOR, C. MANSBACH AND R. BRESSLER, *Biochim. Biophys. Acta*, 144 (1967) 366.
33 M. KLINGENBERG AND C. BODE, in G. WOLF (Ed.), *Recent Research on Carnitine*, M.I.T. Press, Cambridge, Mass., 1965, p. 87.
34 R. BRESSLER AND B. WITTELS, *Biochim. Biophys. Acta*, 104 (1965) 39.
35 B. WITTELS AND R. BRESSLER, *J. Clin. Invest.*, 43 (1964) 630.

36 C. R. Rossi and D. M. Gibson, *J. Biol. Chem.*, 239 (1964) 1694.
37 D. W. Yates and P. B. Garland, *Biochem. Biophys. Res. Commun.*, 23 (1966) 460.
38 M. Klingenberg and E. Pfaff, in J. M. Tager, S. Papa, E. Quagliariello and E. C. Slater (Eds.), *Regulation of Metabolic Processes in Mitochondria, (BBA Library, Vol. 7)*, Elsevier, Amsterdam, 1966, p. 180.
39 D. W. Allmann, L. Galzigna, R. E. McCaman and D. E. Green, *Arch. Biochem. Biophys.*, 117 (1966) 413.
40 D. W. Yates and P. B. Garland, *Biochem. J.*, 102 (1967) 40P.
41 G. D. Greville, in J. M. Tager, S. Papa, E. Quagliariello and E. C. Slater (Eds.), *Regulation of Metabolic Processes in Mitochondria, (BBA Library, Vol. 7)*, Elsevier, Amsterdam, 1966, p. 88.
42 I. B. Fritz, *Perspectives Biol. Med.*, 10 (1967) 643.
43 I. B. Fritz, in F. C. Gran (Ed.), *Cellular Compartmentalization and Control of Fatty Acid Metabolism*, Academic Press, New York, 1968, p. 39.
44 J. Bremer, *Biochim. Biophys. Acta*, 116 (1966) 1.
45 K. Norum and J. Bremer, *Biochim. Biophys. Acta*, 78 (1963) 77.
46 J. Bremer, *J. Biol. Chem.*, 237 (1962) 2228.
47 I. B. Fritz, S. K. Schultz and P. A. Skere, *J. Biol. Chem.*, 240 (1963) 2509.
48 K. Norum, *Biochim. Biophys. Acta*, 89 (1964) 95.
49 I. B. Fritz and N. R. Marquis, *Proc. Natl. Acad. Sci. (U.S.)*, 54 (1965) 1226.
50 T. Bøhmer, K. R. Norum and J. Bremer, *Biochim. Biophys. Acta*, 125 (1966) 244.
51 T. Bøhmer, *Biochim. Biophys. Acta*, 144 (1967) 259.
52 D. J. Pearson and P. K. Tubbs, *Biochem. J.*, 105 (1967) 953.
53 K. Norum, *Biochim. Biophys. Acta*, 99 (1965) 511.
54 J. Bremer and K. Norum, *Europ. J. Biochem.*, 1 (1967) 427.
55 J. Bremer and K. Norum, *J. Biol. Chem.*, 242 (1967) 1744.
56 J. Bremer and K. Norum, *J. Biol. Chem.*, 242 (1967) 1749.
57 K. Norum, *Biochim. Biophys. Acta*, 105 (1965) 506.
58 A. M. Beenakkers and M. Klingenberg, *Biochim. Biophys. Acta*, 84 (1964) 205.
59 I. B. Fritz and S. K. Schultz, *J. Biol. Chem.*, 240 (1965) 2188.
60 P. K. Tubbs, D. J. Pearson and J. F. A. Chase, in G. Wolf (Ed.), *Recent Research on Carnitine*, M.I.T. Press, Cambridge, Mass., 1964, p. 117.
61 D. Shepherd, D. W. Yates and P. B. Garland, *Biochem. J.*, (1966) 98.
62 H. A. Krebs, *Proc. Roy. Soc. (London), Ser. B*, 159 (1964) 545.
63 M. C. Scrutton and M. F. Utter, *Ann. Rev. Biochem.*, 37 (1968) 249.
64 J. R. Williamson, E. T. Browning and M. S. Olsen, *Advan. Enzyme Regulation*, 6 (1968) 67.
65 J. R. Williamson, *Advan. Enzyme Regulation*, 5 (1967) 229.
66 E. Struck, J. Ashmore and O. Wieland, *Advan. Enzyme Regulation*, 4 (1966) 219.
67 G. Weber, H. J. Lea, H. J. Hird Covery and N. B. Stamm, *Advan. Enzyme Regulation*, 5 (1967) 257.
68 R. C. Haynes, *Advan. Enzyme Regulation*, 3 (1965) 111.
69 G. Weber, R. L. Singhal, N. B. Stamm, E. A. Fisher and M. A. Mentendick, *Advan. Enzyme Regulation*, 2 (1964) 1.
70 G. Weber, R. L. Singhal and S. K. Srvastava, *Advan. Enzyme Regulation*, 4 (1965) 43.
71 W. Seubert, H. V. Henning, W. Schoner and M. L'Age, *Advan. Enzyme Regulation*, 6 (1968) 153.
72 J. Ashmore, S. R. Wagle and Y. Uete, *Advan. Enzyme Regulation*, 2 (1964) 101.
73 H. D. Soling, B. Willms, D. Fredricks and J. Kleineke, *Europ. J. Biochem.*, 4 (1968) 364.

74 J. R. WILLIAMSON, R. A. KREISBERG AND P. W. FELTS, *Proc. Natl. Acad. Sci. (U.S.)*, 56 (1966) 247.
75 G. WEBER, R. L. SINGHAL, N. B. STAMM, M. A. LEA AND E. A. FISHER, *Advan. Enzyme Regulation*, 4 (1966) 59.
76 H. A. KREBS, R. SPEAKE AND R. HEMS, *Biochem. J.*, 94 (1965) 712.
77 M. BENMILOUD AND N. FREINKEL, *Metabolism*, 16 (1967) 658.
78 R. WILLIAMSON, E. T. BROWNING, R. SCHOLZ, R. A. KREISBERG AND I. B. FRITZ, *Diabetes*, 17 (1968) 194.
79 M. LEA AND G. WEBER, *J. Biol. Chem.*, 243 (1968) 1096.
80 N. B. RUDERMAN, C. J. TOEWS AND E. SHAFRIR, *Arch. Internal Med.*, 123 (1969) 299.
81 O. H. LOWRY AND J. V. PASSONNEAU, *J. Biol. Chem.*, 241 (1966) 2268.
82 W. GEVERS AND H. A. KREBS, *Biochem. J.*, 98 (1966) 720.
83 T. TANAKA, Y. HARANO, T. SUE AND H. MORIMURA, *J. Biochem. (Tokyo)*, 62 (1967) 71.
84 T. C. LINN, F. H. PETTIT AND L. J. REED, *Biochemistry*, 62 (1969) 234.
85 M. F. UTTER AND D. B. KEECH, *J. Biol. Chem.*, 238 (1963) 2603.
86 M. F. UTTER, D. B. KEECH AND M. C. SCRUTTON, *Advan. Enzyme Regulation*, 2 (1964) 49.
87 M. WINTERFELD AND H. DEBUCH, *Z. Physiol. Chem.*, 345 (1966) 11.
88 K. NORUM, *Biochim. Biophys. Acta*, 98 (1965) 652.
89 P. B. GARLAND AND P. J. RANDLE, *Biochem. J.*, 91 (1964) 6C.
90 A. H. UNDERWOOD AND E. A. NEWSHOLME, *Biochem. J.*, 104 (1967) 300.
91 R. J. HASLAM, in J. M. TAGER, S. PAPA, E. QUAGLIARIELLO AND E. C. SLATER (Eds.), *Regulation of Metabolic Processes in Mitochondria, (BBA Library, Vol. 7)*, Elsevier, Amsterdam, 1966, p. 108.
92 D. G. NICHOLLS, D. SHEPARD AND P. B. GARLAND, *Biochem. J.*, 103 (1967) 677.
93 T. KONIG AND G. SZABADOS, *Acta Acad. Sci. Hung.*, 2 (1967) 353.
94 G. B. VON JAGOW, B. WESTERMANN AND O. WIELAND, *Europ. J. Biochem.*, 3 (1968) 512.
95 M. F. UTTER AND N. B. KEECH, *Advan. Enzyme Regulation*, 2 (1964) 49.
96 A. J. GARCIA, J. R. WILLIAMSON AND G. F. CAHILL JR., *Diabetes*, 15 (1966) 188.
97 J. H. EXTON, L. S. JEFFERSON, R. W. BUTCHER AND C. R. PARK, *Am. J. Med.*, 40 (1966) 709.
98 J. H. EXTON AND C. R. PARK, *J. Biol. Chem.*, 243 (1968) 4189.
99 L. A. MENAHAN, B. D. ROSS AND O. WIELAND, *Biochem. Biophys. Res. Commun.*, 30 (1968) 38.
100 D. W. FOSTER, *J. Clin. Invest.*, 46 (1967) 1283.
101 E. HERRERA AND N. FREINKEL, *Biochim. Biophys. Acta*, in the press.
102 J. H. EXTON AND C. R. PARK, *J. Biol. Chem.*, 242 (1967) 2622.
103 M. F. UTTER AND D. B. KEECH, *J. Biol. Chem.*, 235 (1960) P.C.17.
104 M. F. UTTER, *Ann. N.Y. Acad. Sci.*, 72 (1959) 451.
105 A. C. FREEDMAN AND L. KOHN, *Science*, 145 (1964) 58.
106 H. V. HENNING AND W. SEUBERT, *Biochem. Z.*, 340 (1964) 160.
107 M. F. UTTER AND K. KARAHASHI, *J. Biol. Chem.*, 207 (1954) 787.
108 M. F. UTTER AND K. KARAHASHI, *J. Biol. Chem.*, 207 (1954) 821.
109 R. S. NORDLIE AND H. A. LARDY, *J. Biol. Chem.*, 238 (1963) 2259.
110 E. SHRAGO, H. A. LARDY, R. C. NORDLIE AND D. O. FOSTER, *J. Biol. Chem.*, 238 (1963) 3188.
111 H. A. LARDY, D. O. FOSTER, E. SHRAGO AND P. D. RAY, *Advan. Enzyme Regulation*, 2 (1964) 39.

112 H. A. LARDY, V. PATEKAV AND P. WALTER, *Proc. Natl. Acad. Sci. (U.S.)*, 53 (1965) 1410.
113 J. ASHMORE AND G. WEBER, in F. DICKENS, P. J. RANDLE AND W. J. WHELAN (Eds.), *Carbodrate Metabolism and its Disorders*, Academic Press, New York, 1968, p. 336.
114 K. TAKETA AND B. M. POGELE, *J. Biol. Chem.*, 241 (1966) 720.
115 E. SHAFRIR, V. LAURIS AND G. F. CAHILL JR., *Federation Proc.*, 27 (1968) 331.
116 F. L. ENGEL AND T. T. AMATRUDA JR., *Ann. N.Y. Acad. Sci.*, 104 (1963) 753.
117 H. A. KREBS, *Advan. Enzyme Regulation*, 4 (1966) 339.
118 J. R. WILLIAMSON AND H. A. KREBS, *Biochem. J.*, 80 (1961) 540.
119 O. WIELAND, L. WEISS AND I. EGER-NEUFELDT, *Advan. Enzyme Regulation*, 2 (1964) 85.
120 H. D. SOHLING, R. KATTERMANN, H. SCHMIDT AND P. KNEER, *Biochim. Biophys. Acta*, 115 (1965) 1.
121 G. LOFFLER AND O. WIELAND, *Biochem. Z.*, 336 (1963) 447.
122 G. W. KOSICKI AND P. A. SRERE, *J. Biol. Chem.*, 236 (1961) 2560.
123 G. A. ROBINSON, R. W. BUTCHER AND E. W. SUTHERLAND, *Ann. Rev. Biochem.*, 37 (1968) 149.
124 J. R. WILLIAMSON, B. HERCZEA, H. COLES AND R. DANISH, *Biochem. Biophys. Res. Commun.*, 24 (1966) 437.
125 P. D. BEWSHER AND J. ASHMORE, *Biochem. Biophys. Res. Commun.*, 24 (1966) 431.
126 P. J. RANDLE, P. B. GARLAND, C. N. HALES AND E. A. NEWSHOLME, *Lancet*, i (1963) 785.
127 P. J. RANDLE, P. B. GARLAND, C. N. HALES AND E. A. NEWSHOLME, *Recent Progr. Hormone Res.*, 22 (1966) 1.
128 J. C. SHIPP, L. H. OPIE AND D. CHALLONER, *Nature*, 189 (1961) 1018.
129 J. C. SHIPP, *Metabolism*, 13 (1964) 852.
130 E. A. NEWSHOLME AND W. GEVERS, *Vitamins Hormones*, 25 (1967) 1.
131 D. G. WALKER, *Essays Biochem.*, 2 (1966) 33.
132 M. ROSELL-PEREZ, C. VILLAR- PALASI AND J. LARNER, *Biochemistry*, 1 (1962) 763.
133 G. SCHONFELD AND D. M. KIPNIS, *Diabetes*, 17 (1968) 422.
134 G. SCHONFELD AND D. M. KIPNIS, *Am. J. Physiol.*, 215 (1968) 513.
135 B. LEBOEUF AND G. F. CAHILL JR., *J. Biol. Chem.*, 236 (1961) 41.
136 C. W. HALES AND P. F. RANDLE, *Lancet*, i (1963) 790.
137 J. H. KARAM, G. M. GRODSKY, F. C. PAVLATOS AND P. H. FORSHAM, *Lancet*, i (1965) 286.
138 C. N. HALES AND D. E. HYAMS, *Lancet*, ii (1964) 69.
139 R. KALKHOFF, D. S. SCHALCH, J. L. WALKER, P. BECK AND D. M. KIPNIS, *Trans. Assoc. Am. Physicians*, 77 (1964) 270.
140 R. A. KREISBERG, B. R. BOSHELL, J. DIPLACIDO AND R. F. RODDAM, *New Engl. J. Med.*, 276 (1967) 314.
141 G. F. CAHILL JR., M. G. HERRERA, A. F. MORGAN, S. J. SOELDNER, J. STEINKE, P. L. LEVY, G. A. REICHARD AND D. M. KIPNIS, *J. Clin. Invest.*, 45 (1966) 1751.
142 S. S. FAJANS, J. C. FLOYD, R. F. KNUF AND J. CONN, *J. Clin. Invest.*, 43 (1964) 2003.
143 B. SENIOR AND L. LORIDAN, *Nature*, 219 (1968) 83.
144 E. BALASSE AND H. A. OOMS, *Diabetologia*, 4 (1968) 133.
145 E. M. NEPTUNE, *Am. J. Physiol.*, 187 (1956) 451.
146 P. W. FELTS, O. B. CROFFORD AND C. R. PARK, *J. Clin. Invest.*, 43 (1964) 638.
147 H. C. TIDWELL AND E. H. AXELROD, *J. Biol. Chem.*, 172 (1948) 179.
148 E. E. BALASSE, E. COUTUERIR AND J. R. M. FRANCKSON, *Diabetologia*, 3 (1967) 488.
149 L. L. MADISON, D. MEBANE, R. H. UNGER AND A. LUCHNER, *J. Clin. Invest.*, 43 (1964) 408.

150 D. Mebane and L. L. Madison, *J. Lab. Clin. Med.*, 63 (1964) 177.
151 P. Bjorntorp, *J. Lipid. Res.*, 7 (1966) 621.
152 P. Bjorntorp and T. Schersten, *Am. J. Physiol.*, 212 (1967) 683.
153 L. A. Carlson and J. Ostman, *Acta Med. Scand.*, 178 (1965) 71.
154 B. Wittels and R. Bressler, *J. Clin. Invest.*, 43 (1964) 630.
155 R. Bressler, R. Katz and B. Wittels, *Ann. N.Y. Acad. Sci.*, 131 (1965) 207.
156 R. Bressler and B. Wittels, *Biochim. Biophys. Acta*, 104 (1965) 39.
157 C. Corredor, C. Mansbach and R. Bressler, *Biochim. Biophys. Acta*, 144 (1967) 366.
158 M. Entman and R. Bressler, *Mol. Pharmacol.*, 3 (1967) 333.
159 B. Wittels and R. Bressler, *J. Clin. Invest.*, 44 (1966) 1639.
160 C. Corredor, K. Brendel and R. Bressler, *Proc. Natl. Acad. Sci. (U.S.)*, 58 (1967) 2299.
161 K. Brendel, C. Corredor and R. Bressler, *Biochem. Biophys. Res. Commun.*, 34 (1969) 340.

SUBJECT INDEX

Abeta-lipoproteinaemia, failure in LDL formation and in hepatic triglyceride release, 63, 67
—, inhibition of triglyceride release from intestine, 68
Acanthocytosis, and chylomicron formation, 13, 14
Acetic acid kinase, 122
Acetoacetate, infusion, decrease of extraction and oxidation of plasma FFA by heart, 42
—, oxidation in heart muscle, 351
—, preferential oxidation in cardiac muscle, in presence of Glc, insulin or lactate, 351
Acetylcarnitine, formation, 335, 336, 340, 341
(+)-Acetylcarnitine, inhibition of acetyl-transferase, influence of hydrolysis and dehydrogenation and demethylation on inhibitory activity, 340
Acetyl-CoA, activation of pyruvate carboxylase 345–348
—, carboxylase, in chloroplasts, intact, 284, 285
— —, conversion of monomer to trimer by citrate, 144
— —, effect of citrate, 144, 280
— —, inhibitor, in chloroplasts, 284, 285
— —, liver, activation by citric or isocitric acid, 280
— —, plant tissues, 7.35 and 9.45 component, reaction mechanism, 280
— —, —, heat-stable inhibitor, 280
— —, —, occurrence, isolation, 279, 280
— —, stimulation by insulin, 141
—, concentration in livers of fasted rats, 348
—, in gluconeogenesis, 346, 348, 352
—, impaired mitochondrial oxidation by enhanced gluconeogenesis, 352

Acetyl-CoA, (*continued*)
—, inhibition of pyruvate dehydrogenase, 346
—, intracellular concentrations, and gluconeogenesis, 348
—, production, increase in states of increased FFA utilization, 352
Acetyl-CoA:carnitine acetyltransferase, abundance in tissues with high rates of FA oxidation, 336
—, crystallization, 339
—, inhibition by positive isomers of carnitine and acetylcarnitine, 340
—, — by thiol group reagents, reversal by acetyl-CoA, 340
—, intracellular location, 339
—, specific requirement for carnitine, 340
—, transfer of C_2–C_4 acyl groups, 340
—, — of C_5–C_{16} acyl groups, 340
N-Acetylglucosaminyl–N-acetylmuramyl–pentapeptide–lipid complexes, in Gram-positive cell walls, 254, 255
β-N-Acetylhexosaminidase, in brain, particulate and soluble enzymes, 224
—, low activity towards monosialosyl-N-triglycosylceramide, 225
N-Acetylmuramyl–pentapeptide–lipid complexes in Gram-positive cell walls, 254, 255
N-Acetylneuraminic acid, distribution in gangliosides of brain tissue of vertebrates, 205, 206
—, release from gangliosides, 219
Acetyl transacylase, 235
Acetyltransferase, inhibition by (+)-acetylcarnitine, 340
—, norcarnitine as substrate, 340
ACP, see Acyl-carrier protein
Acromegaly, Glc intolerance and elevated plasma FFA, 353

[361]

ACTH, see Adrenocorticotrophic hormone
Actinomycetes, lipid metabolism, 241, 246
Actinomycin D, effect on lipolysis stimulation in adipose tissue, *in vitro*, 33
Acylcarnitine(s), esters, formation, 336
—, and FA oxidation by mitochondria, 332, 333
—, formation, and FA oxidation, 340
—, long-chain, suppression of decarboxylation of pyruvate, 336
—, permeability of inner mitochondrial membrane, 334
(+)-Acylcarnitines, inhibition of long-chain FA oxidation, and of gluconeogenesis, 344
Acyl-carrier protein, (ACP), in activation of FA, in intestinal mucosa, 11
—, concentration in intact chloroplasts, structural organisation in stromal proteins, 285
—, dissociation from FA synthetases in higher plants, 282
—, FA biosynthesis, 235, 236, 281, 282, 284, 285
—, functional group, structure, 236
—, in malonate pathway of FA synthesis, 235, 236
—, mol. wt., 235
—, in phosphatidic acid synthesis, 159
—, plant tissues, amino acids surrounding active site, 281
—, —, 4'-phosphopantetheine component, 281
—, —, requirement in FA synthesizing enzyme systems, 278–281
—, requirement in FA synthetase, chloroplasts, 284, 285
—, spinach leaves and *E. coli*, and FA synthetases thereof, FA synthesis, cross experiments, 281, 282
Acyl-CoA esters, in FA biosynthesis, in bacteria, 236
Acyl-CoA:L-glycerol-3-phosphate *O*-acyl-transferase, acylation of glycerophosphate, 1- and 2- positions, 159
—, — of lysophosphatidic acid, 159

Acyl-CoA–phospholipid acyltransferase(s), and changes of FA composition of phospholipids of erythrocyte membrane by changes in diet, 188, 189
—, erythrocytes, acetylation of lysocholine plasmalogens, 188
—, liver microsomes, preferences in transfer of FA to 1- and to 2- acylglycerylphosphorylcholine, 188
—, microsomes, specificities towards substrates for acylation, 188
—, selectivity for FA, and melting points, 188
Acyl-CoA synthetase(s), general, substrates, 122
—, long-chain, substrates, 122
—, mitochondrial, localization at outer membrane fraction, 334
—, synonyms, 11
—, in triglyceride synthetase complex, 12
Acyl dihydroxyacetone phosphate, formation in liver mitochondria, from DHAP and acyl-CoA, 160
Acyl lipids, in chloroplasts, 282, 283
N-Acylneuraminosylhydrolase, see Neuraminidase
N-Acylsphingosine, see Ceramide
Acyltransacylases, long-chain, absence in higher plant tissues, 276
Adenine nucleotides, permeability of inner mitochondrial membrane, effect of atractyloside, 334
Adenosine cyclic 3',5'-monophosphate, see Cyclic AMP
S-Adenosylhomocysteine, inhibition of cyclopropane FA formation, 243
S-Adenosylmethionine, methylation of phosphatidyl-ethanolamine, 168
—, role in propane ring formation in cyclopropane FA, 242–244
Adenyl cyclase, subcellular preparations of fat cells, activation by lipolytic hormones, 32, 34
Adipose tissue, (brown), and FA storage, 20
—, thermogenic function in new-born animals, 118, 119
—, uncoupling of oxidative phosphorylation in mitochondria, 118, 119

SUBJECT INDEX

Adipose tissue, (white), in albumin-free medium, lipolysis, binding of FFA, 35
—, cell, penetration of FFA, and esterification at the cell membrane, 136, 137
—, entrance of FFA from incubation medium into tissue FFA pool, effect of anaerobiosis and of NaF, 137
—, FA storage as triglycerides, 19, 20
—, FA uptake from blood triglycerides hydrolyzed by clearing-factor lipase, 20
—, fasting rats, *in vitro*, FFA release and lipolysis, effect of added Glc and insulin, 22–24
—, FFA release, hormones stimulating, 141–143
—, histological structure, 20
—, human, lipogenesis, 136
—, increased FA deposition in experimental obesity, 81
—, lipolysis, glycerol release as measure, 21
—, —, hormonally induced, increase of gluconeogenesis, 342
—, location in body, 20
—, rat, homogenate, triglyceride synthesis, requirements, 134, 135
—, triglyceride stores, replenishment from plasma triglycerides, 103
—, — synthesis, 134–137
—, — —, activity of microsomes, mitochondria and cytoplasm, 135
—, — —, different molecular species, 135, 136
—, — —, from FA endogenously made from Glc, 145
—, — —, glycerol phosphate formation in fat cells, 145
—, *in vitro*, pancreatectomized fasted rats, lipolysis and FFA release, effect of Glc and insulin addition, 25
Adrenal glands, effect on FFA mobilization in fasting, 23, 24
—, FFA release, stimulation by ACTH, 43
—, in lipid mobilization, 24, 26–28, 32, 33

Adrenal glands, (*continued*)
—, slices, palmitate incorporation into neutral lipid and phospholipid, effect of ACTH, 43
Adrenal medulla, gangliosides, 206
Adrenalectomized–pancreatectomized rats, effect of administered hormones on blood ketone bodies and lipid, and on liver triglyceride content, 23–29
Adrenaline, effect on clearing-factor lipase and on triglyceride-mobilizing lipase, incubated adipose tissue, 96
—, — on lipolysis, 42, 352
—, — on uptake and utilization of FFA by heart, 42
—, increase of efflux of NEFA from adipose tissue, 141
—, — of FFA in plasma in normal dogs, 37
—, — in gluconeogenesis, in absence of *de novo* enzyme synthesis, 344
—, — of hydrolysis, of triglycerides in adipose tissue, role of cyclic AMP, 141, 352
—, infusion, effect on triglyceride FA utilization of heart *in vivo*, 86
—, lipid-mobilizing effect, effect of adrenalectomy and thyroidectomy, cortisone and thyroxine, 142
Adrenocortical hormones, effects on regulation of plasma triglyceride concentration and tissue clearing-factor lipase, 101
Adrenocorticotrophic hormone, effect on clearing-factor lipase and on triglyceride-mobilizing lipase, incubated adipose tissue, 96
—, — on triglyceride synthesis, 142, 143
—, increase of plasma FFA in normal rabbits, 37
—, in lipid mobilization, 25–27
AEP, *see* 2-Aminoethylphosphonic acid
Age, and activity of UDP–galactose: monosialosyl-*N*-triglycosylceramide galactosyltransferase, 214
Ageing, and changes in sphingosine component of brain gangliosides, 208
—, and FA metabolism, in storage tuber tissues of plants, 286

β-Alanine, formation from propionic acid in plants, 275
Albumin, serum, binding sites for FFA, 35
—, —, requirement in medium for lipolysis in fat cells *in vitro*, 35
—, —, — for release of FFA by adipose tissue and fat cells *in vitro*, 35
Alcohol administration, effect on triglyceride FA utilization of heart *in vivo*, 86
— excess, lipid accumulation in liver, 148
Algae, lipid metabolism, 240, 241
Alimentary lipaemic response, changes by environmental and genetic factors, 71
—, effect of acute dietary load of carbohydrate, 70
—, to fat ingestion, effect of increasing age, 54, 70
—, increase in diabetes, and carbohydrate-induced hyperlipaemic states, 70
—, inhibition by inhibitors of protein synthesis, 13
—, and ischaemic heart disease, 59, 70
—, massive, association with impaired removal of triglyceride FA from blood, 71
Alloxan diabetes, *de novo* synthesis of gluconeogenic enzymes, effect of inhibitors of protein synthesis, 344
Aminoacyl phosphatidylglycerols, *see* Lipoamino acids
2-Aminoethylphosphonic acid, (AEP), discovery, structure, 174, 175
—, formation by rearrangement of lipid-bound PEP, 175
—, in phosphonate lipids, 174, 175
AMP, cyclic, *see* Cyclic AMP
Anterior-pituitary adipokinetic substance, (Fraction H), increase of FFA in plasma, 37
Antibiotics, sensitivity to —, and bacterial surface lipids, 258, 261
Artichoke, FA metabolism, ageing, 286
Arylsulphatase, hydrolysis of sulphatides, 300
ATP:choline phosphotransferase, Mg^{2+} requirement, specificity, 161, 163

Atractyloside membrane, mitochondria, CoA pool, 335, 336
—, —, and permeation of adenine nucleotides, 334
Avocado mesocarp, FA metabolism, 278, 279, 281
Azoproteins, acidic, inhibition of clearing-factor lipase, 90

Bacillus megaterium, lipid metabolism, 240, 241
Bacillus sp., lipid metabolism, 244
Bacterial fatty acids, branched-chain, formation by elongation from short-chain branched FA, ACP esters as intermediates, 244
—, —, from straight-chain FA by dehydrogenation and methylation in mycobacteria, 245
—, —, iso- and anteisomethyl-, occurrence, 244
—, *cis*-oleic series, (table), 239
—, *cis*-vaccenic series, (table), 239
—, cyclopropane, FA unsaturated long-chain, precursors, 241–244
—, hydroxy acids, association with surface and extra-cellular lipids, 246
—, —, synthesis, 246
—, modification of proportion of different straight-chain FA by growth conditions, 235
—, monoenoic, biosynthesis, 238–241
—, peculiar to mycobacteria and corynebacteria, 246–248
—, saturated, formation by biohydrogenation of unsaturated FA, in rumen bacteria, 236, 237
—, —, straight-chain, biosynthesis, 233–237
—, unsaturated, long-chain, formation, 240
—, —, —, synthesis by dehydrogenation of saturated long-chain, 240
—, —, —, — by elongation of short-chain monoenoic FA, 237–241
Bacterial glycolipids, carbohydrates involved, 250
—, occurrence, 250
Bacterial growth, effects of exogenous lipids, 259–261
Bacterial lipases, 259

Bacterial lipids, as carrier or portion of covalently bound intermediates in biosyntheses, 243, 244, 254, 258, 259
—, in cell envelope of Gram-negative bacteria, 255, 257
—, in cell wall and in cytoplasmic membrane of Gram-positive bacteria, 255, 257
—, cellular distribution, 255, 256
—, as energy reserves, 257
—, excretion, 256
—, extracellular, 256
—, functions, 256–259
—, in β-galactoside permease activity, 258, 261
—, increase under stress conditions, 257
—, kinetic function of cytoplasm membrane lipids, 257
—, lipid-bound intermediates, in synthesis of Gram-positive cell walls, 254, 258
—, "lipid" inclusions in EMs, 256
—, in mesosomal structures, 257
—, metabolism, 229–262
—, surface —, and sensitivity to antibiotics, 258
—, and transport of amino acids, 258, 261
—, triglycerides, 257
—, visible, 'lipid' deposits in cytoplasm, composition, 256
Bacterial lipopolysaccharides, in cell envelope of Gram-negative enteric bacteria, immunological activity, 252
—, core polysaccharides, 252, 254
—, and endotoxins, 252
—, functional role of lipids in formation, 254, 258
—, Gram-negative, association with OH-FA, 246
—, and injurious effects of bacteria, (endotoxins), 252
—, lipid A, see Lipid A
—, O-antigen side-chain, 252
Bacterial lipoproteinases, 259
Bacterial phospholipases, 259
Bacterial phospholipids, formation, 248–250
Bacterial toxins, and phospholipid degradation, 176

Barbiturates, effect on exchange of [^{32}P]phosphate in brain phospholipids, 322
Barley, FA metabolism, 282, 287
Beef, brain, ganglioside metabolism, 218, 221, 223
—, spleen, ganglioside metabolism, 206
Bile acids, differential effects on bacteria species, practical applications, 260
Bile salts, absorption in jejunum, intracellular effects on fat absorption, 9
—, catalysis of hydrolysis and of formation of cholesterol esters in intestine, 14, 15
—, emulsification of triglycerides in intestine, 3
—, in intestinal absorption of fats, 2, 3, 14, 15
—, role in formation of intraluminary micellar phase in fat absorption, 3
Bile solubility test, for differentiating pneumococci from streptococci, 260
Blood–brain barrier, and lipid metabolism, 315–317
—, selective, and cholesterol labelling in brain, 305, 306
—, and uptake of radioactive substrates, dyestuffs and drugs into brain lipids, 315–317
Blood plasma, see Plasma
Blood vessels, FFA incorporation into phospholipid and neutral lipid, effect of Glc, 43
Brain–blood barrier, see Blood–brain barrier
Brain, and FFA in blood, 43
—, foetal, lipids composition, 295
—, gangliosides, see Gangliosides, brain
—, growing rat, incorporation and persistence of [^{35}S]sulphate in lipids, (fig.), 321
—, microsomes, lipid composition, 319
—, —, half-life of ^{35}S-labelled lipid, 319
—, mitochondria, half-life of ^{35}S-labelled lipid, 319
—, —, lipid composition, 319
—, —, lipids, turnover of incorporated radioactivity, 320
—, neuraminidase, see Neuraminidase, brain

Brain, (*continued*)
—, phospholipids, labelling, effect of excitants and depressants, 322
—, —, from microsome fraction as endogenous substrate in absence of exogenous Glc, 323
—, rat, cell types in, 312
—, subcellular particles, incorporation of [^{32}P]phosphate and [^{14}C]serine, 318, 319
—, —, lipid composition, (table), 319
—, synaptic endings, half-life of ^{35}S-labelled lipids, 319
Branched-chain fatty acids, bacteria, 244–246
γ-Butyrobetaine, competitive inhibition of palmityltransferase, K_i, 338
—, hydroxylation to carnitine, 333

Caffeine, increase of action of lipolytic fast-acting hormones, 32
Calf, brain, ganglioside metabolism, 214, 218–221, 224
Carbon tetrachloride, administration, effect on plasma triglyceride concentration, and hepatic triglyceride content, 80, 81
—, —, — on plasma VLDL–protein concentration, 66
—, —, lipid accumulation in liver, 148
Cardiolipin, (diphosphatidylglycerol), in bacteria, 249
—, content, in myelin lipid, 314
—, formation from phosphatidylglycerol and CDP diglyceride, 167
—, — from phosphatidylglycerophosphate, prior dephosphorylation, 167
—, incorporation into HeLa cells and rat-liver cells, 194
—, location in mitochondrial membranes, 167
—, rate of metabolism in liver and brain mitochondria *in vivo*, 193, 194
Carnitine, acetyl-, *see* Acetylcarnitine
— acetyltransferase, *see* Acetyl-CoA–carnitine acetyltransferase
— barrier, and long-chain FA oxidation in mitochondria, 334
—, catalysis of long-chain FA oxidation in tissues *in vitro*, 332–339

Carnitine, catalysis (*continued*)
—, — of oxidation of FA-CoA derivatives *in vitro*, 332, 333
— deficiency, and choline deficiency, 342
— —, in newborn rat heart, and impaired rates of long-chain FA oxidation, 341
— —, in rats treated with diphtheria toxin, 341
—, depletion in heart *in vitro*, decrease of FA oxidation, increase of triglyceride synthesis, 334, 341
—, distribution in organisms and tissues, 332
—, ester formation with FA and rate-limiting step in FA oxidation, 340
—, FA esters, *see* Acylcarnitines
—, FA transfer from acyl-CoA, prior to metabolism by intramitochondrial enzymes of β-oxidation, 333
—, formation by hydroxylation of γ-butyrobetaine by soluble fractions of rat liver, requirements, 333
—, increase of gluconeogenesis, in rabbit-liver and kidney slices, 344
—, level in tissues, lowering by hypoglycin and by pentenoic acid, 342
—, palmityl-, *see* Palmitylcarnitine
— palmityltransferase, *see* Palmityl-CoA: carnitine palmityltransferase
—, quantitative significance in disease states, 341, 342
—, role in long-chain FA oxidation, connection with mitochondrial compartmentalization, 333, 334
—, subgroup alterations of molecule, effect on catalytic activity on FA oxidation, 333, 338
Carrot, FA metabolism, ageing, 286
Carthamus tinctorius L. (var. N-10), seed, FA composition, 266, 277
Castor bean, *see Ricinus communis*
Catecholamines, lipolytic effects, 31
—, —, role of β-receptors, 29, 30
CDP-choline, in lecithin biosynthesis, 162
CDP-choline:ceramide cholinephosphotransferase, 173
CDP diglyceride, formation from phosphatidic acid, 249, 312

SUBJECT INDEX

CDP:diglyceride inositol phosphatidate transferase, in phosphatidylinositol biosynthesis, 164
CDP-ethanolamine, in phosphatidylethanolamine synthesis, 163, 164
Ceramidase, 300
Ceramide aminoethylphosphonate, 174
—, biosynthesis from sphingosine by brain homogenate, 207
—, content, in human neural tissue, 295
—, formation from cerebrosides, 207, 208, 299, 300
—, — of sphingomyelin with CDP-choline, 173
—, — from sphingosine, 173, 311
—, release of FA, enzyme from rat brain, 179
Cereal grains, lipoxidases, 267
Cerebral sulphatase activity, and metachromatic leucodystrophy, 297, 300
Cerebronic acid, 296
Cerebroside(s), biosynthesis, 208
—, —, from sphingosine, 298, 299
—, brain, half-life, 299
—, composition, 296, 297
—, content, in human neural tissue, 295
—, —, in myelin lipid, 314
—, FA in, 296, 297
—, formation by acylation of psychosine, 299
— galactosidase, in brain, 299, 300
—, methods for analysis, 297, 298
—, separation from sulphatides, 297
Chaulmoogric acid, content in *P. sativum* seed lipid, 266
Chicken embryonic brain, and ganglioside biosynthesis, 206, 209, 210
Chloroplasts, absence of triglycerides, 283
—, acyl lipids, 282, 283
—, isolated, acetate incorporation in long-chain FA, effect of light, 283
Chlorothiazide, inhibition of organic acid transport mechanism in proximal convoluted tubules of kidney, 40
Chlorpromazine, effect on exchange of [^{32}P]phosphate, in brain phospholipids, 322
Chlorpropamide injection, effect on FFA uptake and triglyceride output by liver, 80

$\Delta^{7,24}$-Cholestadienol, in developing brain, 302–304
$\Delta^{8,24}$-Cholestadienol, in developing brain, 302–304
$\Delta^{5,7,24}$-Cholestatrienol, in developing brain, 302–304
Cholesterol, biosynthesis, in brain, *in vivo*, 305, 306
—, —, in neural tissue *in vitro*, acetate and mevalonate as precursors, 303
—, —, in slices of young and of adult brain, 303, 304
—, in brain, adult, metabolic stability, 305–308
—, —, —, turnover times, 306
—, —, incorporation of ^{2}H from heavy water, 305
—, —, — of labelled — precursors, and blood–brain barrier, 305, 306
—, catabolism, in nervous system, 306–308
—, content, in chylomicrons, 60
—, —, in human neural tissues, 295
—, —, in myelin lipid, 314
— esters, content, in chylomicrons, 60
— —, —, in human neural tissues, 295
— —, hydrolysis in intestine, activation of cholesterol esterase by bile salts, 14, 15
— —, in intestine, 14, 15
—, exchange among erythrocytes and plasma lipoproteins, 191
—, incorporation of labelled —, into developing and adult brain, and blood–brain barrier, 305, 306
—, intestinal absorption, distribution between emulsion and micelles, 14, 15
—, —, esterification in mucosal cell, inclusion in chylomicrons, 15
—, —, inhibition of penetration into mucosal cell, 15
—, —, rate-limiting step, 15
—, isolation from nervous tissue, 302
—, labelled in nervous system during development, persistence, 306–308
—, in lymph chylomicrons, 61
Choline, deficiency, and carnitine deficiency, 342
—, —, fatty liver production, 81

Choline, deficiency, (*continued*)
—, —, reduction in lipoprotein-phospholipid formation, and inhibition of triglyceride release from liver, 67
Cholinephosphate cytidylyl transferase, in lecithin biosynthesis, 162
Cholinephosphotransferase, influence in establishing FA distribution in lecithin, 163
—, in lecithin biosynthesis from CDP-choline and diglyceride, 162
—, reaction in lecithin biosynthesis, reversibility, 162, 163
Chylomicron(s), association of additional lipoprotein components after entry into lymph and plasma, 69
—, chemical composition, 13
—, and complexes derived from —, by triglyceride FA removal from plasma, 61, 62
—, content of cholesterol, cholesterol esters, and phospholipid, 60
—, discharge from mucosal cells, 14
—, EM, 60, 82
—, formation, in organized complex of endoplasmic membrane of mucosal cell, 13, 16
—, —, and protein synthesis, 13, 14
—, lipolysis and reutilization of liberated FA and glycerol during transfer of their FA to extrahepatic tissues, 83
—, lymph, cholesterol and phospholipid, incubated in plasma *in vitro*, 61
—, —, protein of, formation in intestinal cell, 68
—, —, variations in size and availability of phospholipids in intestinal cell, 68
—, penetration into space of Disse through fenestrated endothelial lining of blood capillaries in liver, 84
—, plasma, HDL and LDL components, 60
—, —, primary and secondary particles, 61
—, —, protein content, electrophoretic properties, 60
—, —, secondary particles, origin, 61
—, protein, chemical composition, 13
—, —, effect of inhibitors of protein synthesis on formation, 13

Chylomicron(s), (*continued*)
—, rate of removal of triglyceride FA and cholesterol ester FA from blood by extrahepatic tissues, 82, 83
—, resemblance and differences with VLDL, 52, 69
—, sequestration at capillary endothelial surface and association with clearing-factor lipase, 90
—, size, density, 52, 59
—, stability, role of phospholipid, 60
—, structure changes before removal of triglyceride FA from blood by extrahepatic tissues, 82, 83
—, surface coat, 14
—, thoracic duct lymph, content and nature of protein, 59, 60
—, triglycerides, release from intestine *via* thoracic duct into plasma, 52, 69–71
—, —, trapping in capillary sinusoids, and space of Disse, 84, 87
—, triglyceride FA, in thoracic duct from sources other than diet, 71
—, —, uptake by liver, 54, 80, 84
Citrate, effect on acetyl-CoA carboxylase, 144, 280
—, — on gluconeogenesis, 347
—, inhibition of phosphofructokinase and inhibition of glycolysis during accelerated FA oxidation, 353
—, intracellular, elevation by long-chain FA oxidation, effect on phosphofructokinase and gluconeogenesis, 347
Citrate-cleavage enzyme, level of, rise and fall, in adipose tissue, in feeding and starvation, 143, 144
—, and malic enzyme in NADPH production from NADH, 143, 144
—, stimulation by insulin, 141
Clearing-factor lipase, absence in brain, 86, 87
—, action on artificial emulsions of triglycerides, requirement of incubation with plasma, 89
—, activity, in adipose tissue, control by tissue FFA concentration, 96
—, —, —, during fasting, 53, 91, 97
—, —, —, in fed animal, 54

SUBJECT INDEX

Clearing-factor lipase, activity, in adipose tissue, (*continued*)
—, —, —, hormones affecting, cyclic AMP as mediator, 96
—, —, control in mammary gland, mechanism, 97
—, —, — in muscle, mechanism, 97
—, —, determinant for plasma triglyceride concn., 97–102
—, —, in diabetes, 92
—, —, hormonal control, 55, 96, 97
—, —, in mammary gland during lactation, 55, 91
—, —, in plasma, following heparin injection, as measure of activity of enzyme in tissues, 97, 98
—, —, post-heparin, and ischaemic heart-disease, 100, 101
—, —, — plasma, 89
—, —, — — —, liver diseases, 101
—, —, — — —, pancreatic diseases, 101
—, — of tissues, and determination of plasma triglyceride concentration, 97–102
—, — —, increase of, and synthesis of new enzyme, 95
—, — —, rate of removal of injected triglycerides as a measure, 97, 98
—, — —, regulation, 95–97
—, — —, —, and cyclic AMP, 55, 95–97
—, —, and transitory lipaemia in pregnancy, 91, 99
—, in adipose tissue, 20, 33, 53, 54, 91, 96, 97
—, association with its substrate, 89, 90
—, blood, effect of heparin injection, 88
—, cyclic AMP as regulating factor in adipose tissue, 55, 96, 97
—, directive function in plasma triglyceride transport, 55, 90–92
—, effect of ACTH, 96
—, — of adrenaline, 96
—, — of cycloheximide, 95
—, — of insulin, 96
—, — of noradrenaline, 96
—, — on pattern of uptake of triglyceride FA from blood, 55, 92
—, half-life in tissues, 95
—, hydrolysis of triglyceride emulsions stabilized by phospholipid, 89

Clearing factor lipase, (*continued*)
—, inhibitors, 90
—, in isolated fat cells, 93
—, level in tissues, and hypertriglyceridaemic state, 90
—, localization in tissues, 93, 94
—, preparations, complete hydrolysis of triglycerides to glycerol and FFA, 88, 89
—, —, monoglyceride-lipase activity, 88, 89
—, reciprocal changes in — activity, in muscle and adipose tissue, and in triglyceride FA uptake, 55, 90–92
—, release from tissues, by heparin, mechanism, 88, 90
—, role of HDL in triglyceride hydrolysis, 89
—, site of action, 88
—, stable and unstable states in adipose tissue and heart, 93, 95
—, substrate specificity, 89
—, synonyms, 87
—, tissues containing, 87
—, triglyceride, FA, removal from blood, 87–93, 121
Clostridium perfringens, α-toxin, 259
Clostridium sp., lipid metabolism, 250, 259
CMP–NAN, in ganglioside biosynthesis, 301
CMP–NAN:ganglioside sialosyltransferase, activity, in rat-kidney homogenate, 212
—, chicken-brain homogenate, solubilization, 212
—, chicken embryonic brain, location, 211
CMP–NAN:ganglioside sialosyltransferase A, reaction, substrates, products, 210–212, 217
CMP–NAN:ganglioside sialosyltransferase B, reaction, substrates, products, 211, 215, 217
CMP–NAN:ganglioside sialosyltransferase C, reaction kinetics, substrate, 212, 217
CMP–NGN, incorporation into polymers, 210
CoA ligase (AMP), *see* Acyl-CoA synthetase
Cobalamine coenzymes, non-participation in reactions in higher plants, 275

Compartmentalization, of mitochondrial enzymes, for FA oxidation, barrier I and II, 334, 335
—, mitochondrial, and FA oxidation, CoA and carnitine pools, 333, 334
—, —, and role of carnitine in long-chain FA oxidation, 333, 334
Cornea, eye, uptake and incorporation of oleic acid into triglyceride, 43
Cortisol, in lipid mobilization, 28
Cortisone, induction of *de novo* synthesis of pacemaker gluconeogenic enzymes, effect of actinomycin D and puromycin, 344
—, in lipid mobilization, 26
Corynebacteria, lipid metabolism, 235, 240, 241, 246, 247
Corynomycolenic acid, corynebacteria, structure, formation, 247
Crab, gangliosides from eye and eyestalk, 206
Crepenynic acid, formation from oleic acid by plant tissues, 278
Crotalus atrox, phospholipase A, substrate specificity, 177
Crotonyl-ACP reductase, 234
CTP, inhibition of synthesis of phospholipids in rat-liver homogenates, 129
—, role in lecithin formation, discovery, 161
—, stimulation of tri- and diglyceride synthesis in rat-liver homogenates, 129
CTP:phosphatidic acid cytidylyl transferase, in phosphatidylinositol synthesis, 164
Cyclic AMP, activation of hormone-sensitive triglyceride lipase, 34, 141
—, —, of phosphofructokinase, 344
—, depression of cellular level by insulin, 352
—, effect of prostaglandins on concn. in different cell types, 97
—, inactivation by cyclic nucleotide phosphodiesterase, 32
—, increase in adipose tissue by fast-acting lipolytic hormones, 32, 34
—, role in determination of reserves of triglycerides and of glycogen in the body, 55, 95, 96
—, — in hormonal stimulation of lipolysis, 352

Cyclic AMP, role (*continued*)
—, — in stimulation of triglyceride hydrolysis in adipose tissue, by adrenaline, 142
—, stimulation of gluconeogenesis in perfused rat liver, 348
Cyclic nucleotide phosphodiesterase, inactivation of cyclic AMP, effect of theophylline and caffeine, 32
Cycloheximide, effect on ageing phenomenon in FA synthesis in storage tubers, 286
—, — on clearing-factor lipase activity, 95
—, — on lipolysis stimulation in adipose tissue, *in vitro*, by growth hormone + glucocorticoid, 33
Cyclopropane fatty acids, biotin replacement in microorganism nutrition, 260
—, chain lengths, 241, 242
—, formation, from monoenoic FA in bacteria, 242–244
—, —, stimulation by anionic detergents, 243
—, hydrogens of propane ring, origin, 242
—, in lipid A, and toxicity, 254
—, occurrence, 241, 242
—, in phospholipids, location at β-position, 244
—, role of S-adenosylmethionine in formation, 242–244
—, — of phospholipids in formation, 243, 244
—, structure, 241
Cyclopropane synthetase, specificity with respect to phospholipid substrate, 244
Cytidine diphosphate, *see* CDP
Cytidine monophoshpate, *see* CMP
Cytidine triphosphate, *see* CTP
Cytosomes, *R. communis* seeds, β-oxidation of FA, enzymes, 271–273

Deacylation–acylation cycle, in nervous tissue, and FA turnover in brain, 310, 311
β,γ-Dehydrase, FA, absence in plants, 282, 283
Deoxycorticosterone acetate, in lipid mobilization, 24, 26, 27
Desmosterol, (24-Dehydrocholesterol), in developing nervous tissue, 302–304

Dexamethasone, in lipid mobilization, 28, 31, 32
DHAP, see Dihydroxyacetone phosphate
Diabetes, alloxan, gluconeogenic enzymes, *de novo* synthesis, 344
—, clearing-factor lipase activity, reciprocal changes in muscle and adipose tissue, 92
—, effect on FFA mobilization, 24–29
—, Glc intolerance and elevated plasma FFA, 353
—, increase of ketogenesis, 39
—, — of pyruvate carboxylase and PEP carboxykinase activity, 349
—, ketosis in, 351
—, plasma triglyceride concentration, 100
—, pre-diabetic condition of carbohydrate utilization, and hypertriglyceridaemia, 58
—, pyruvate carboxylase in, 349
—, removal of injected triglyceride FA from plasma, 100
—, triglyceride FA utilization of heart, 86
—, uncontrolled, hepatic triglyceride output, and triglyceride concentration in plasma, 77
—, —, increase of gluconeogenesis, 342
—, —, and VLDL triglyceride concentration in plasma, 59
Diaphragm, rat, lipolysis *in vitro*, effect of hormones, 41
Dibutyryl cyclic AMP, effect on clearing-factor lipase activity of adipose tissue *in vitro*, 95, 96
—, lipolytic stimulation, 32
Dichloroisoproterenol, and FFA mobilization, 29
Dietary lipid intake, high, increase of gluconeogenesis, 342
Digalactosyldiglyceride, in chloroplasts in photosynthesizing tissues, 282, 283
Diglyceride kinase, in membrane-transport processes, 160
Diglycerides, synthesis by rat-liver cell-free systems, effect of ATP and Mg^{2+}, 125–128
Diglyceride transacylase, in triglyceride synthetase complex, 12
1,2-Diglycerides, formed by monoglyceride pathway and by α-glycerophosphate pathway of triglyceride synthesis, lack of equilibration, 12

Dihydrosphingosine, in sphingosine biosynthesis, 171, 172, 298
Dihydroxyacetone phosphate, formation of α-glycero-phosphate from, 159
—, — of phosphatidic acid from, 160
Dimannosyl diglyceride, biosynthesis, in *M. lysodeikticus*, 250, 251
Dipalmitoyl phosphatidyl choline, formation in lung, function, 43
Diphosphatidylglycerol, *see* Cardiolipin
Diphosphoinositide, formation in brain microsomes, 312
— kinase, 165
— —, brain, soluble, formation of triphosphoinositide, 312
—, and regulation of cellular Na and K transport, 165
—, structure, biosynthesis, 165
Diphtheria toxin, carnitine deficiency, in rats treated with, 341
Diplococcus sp., lipid metabolism, 250
Disialosyl-lactosylceramide, (G_{D3}), chemical structure, 202
—, increase in pathological conditions, 215
Disialosyl-*N*-tetraglycosylceramides, (G_{D1a}, G_{D1b}), biosynthesis, 215–217
—, chemical structures, 202, 203
Disialosyl-*N*-triglycosylceramide, (G_{D2}), chemical structure, 202
—, increase in pathological conditions, 215
Disse space, and triglyceride uptake from blood by liver, 84

Endotoxins, and bacterial lipopolysaccharides, 252
Enoyl-ACP hydrase, 235
Epinephrine, *see* Adrenaline
Epoxy acids, in seed oils, formation, 289
cis-9,10-Epoxyoctadecanoic acid, in fungal spores, 289
Epoxyoleic acid, content in *V. anthelmenties* seed lipid, 266
Erythrocyte membrane, changes of FA composition of phospholipids by changes in diet, role of acyltransferase, 188, 189
Esterase, activity, post-heparin plasma, 89

Ethanolaminephosphate cytidylyltransferase, 163
Ethanolaminephosphotransferase, 163
Etherase, glyceryl ethers, liver, function, 184
—, rat-liver microsomes, cleavage of glyceryl ethers, O_2 and cofactor requirements, 183, 184
Ethionine, administration, effect on plasma triglyceride concn. and hepatic triglyceride content, 80, 81
—, inhibition of triglyceride release from intestine, 68
Eubacteria, lipid metabolism, 230, 241, 244

FA, see Fatty acids
Familial hyperchylomicronaemia, (Type I hyperlipoproteinaemia), 99
Fasting, effect on FFA concn. in blood plasma, 22, 37
—, — on FFA and glycerol release from adipose tissue, 22, 23
—, — on Glc concn. in blood plasma, 22
—, — on insulin concn. in blood plasma, 22
—, — on mobilization of FFA, 22, 23
—, Glc intolerance and elevated plasma FFA, 354
—, increase of gluconeogenesis, 342
—, increase of pyruvate carboxylase, and PEP carboxykinase activity, 349
—, and reesterification of FA in adipose tissue, decrease, 23
Fat(s), cells, *in vitro*, fasting rats, FFA release and lipolysis, effect of added Glc and insulin, 22–24
—, —, —, pancreatectomized fasted rats, lipolysis and FFA release, effect of Glc and insulin addition, 25
—, heat and water of combustion, 118
—, intestinal absorption, 1–14
—, —, concn. of enzymes for resynthesis of triglycerides, in upper jejunum, 9
—, —, conflicting theories of the past, 1, 2
—, —, conversion of monoglycerides and FA into triglycerides, 5, 6
—, —, and digestion, biochemical reactions involved, (scheme), 10
—, —, emulsification of triglycerides, 3

Fat(s), intestinal absorption, (*continued*)
—, —, emulsions and micelles in, particle dimensions, 3–5
—, —, intestinal epithelial cell, apical portion, during fat absorption, diagram EM, (fig.), 7
—, —, intracellular metabolism, 9–14
—, —, location in jejunum, 8, 9
—, —, lumen phase, 2–5
—, —, micelle formation, 3, 4
—, —, microscopic and EM investigations, 6–8
—, —, mono-, di- and triglycerides content in micellar phase, 4
—, —, morphological studies, 2, 6, 7
—, —, non-enzymatic, energy-independent penetration of monoglycerides and FA from micelles into mucosal cells, 5, 6
—, —, penetration of intact micelles, or of micelle constituents into mucosa, 6, 8
—, —, — of monoglycerides and FA from micelles into mucosal cells, 5, 6
—, —, — phase, 5–9
—, —, role of bile salts, 2, 3, 14, 15
—, —, — of pancreatic lipase, 3
—, —, — of pinocytosis of unhydrolyzed triglycerides, 8
—, —, thoracic duct lymph, 13
—, —, transfer of FA in monomolecular form into mucosal cells, 6
—, —, triglyceride resynthesis, location in cell, 8
—, storage in animals, 118
—, — in man, 119
—, —, resemblance to dietary fat, 140
Fatty acid(s), activation, in intestinal mucosa, 11
—, —, mitochondria, enzyme systems beyond carnitine barrier for shorter chain FA, 333
—, —, reaction scheme, 122
—, —, thiokinase enzymes, 122
—, adipose tissue, mobilization in stress conditions, effect on triglyceride output from liver, 54, 77
—, asymmetric distribution in glycerides, 124, 125
—, bacteria, see Bacterial fatty acids

Fatty acid(s), (*continued*)
— biosynthesis, chloroplasts, disrupted, ACP requirement, 284, 285
— —, —, isolated, effect of light, role of NADPH production and photophosphorylation, 283
— —, plant tissues, [^{14}C]acetate incorporation into FA as a function of the developmental stage of castor bean, 277
— —, —, acetyl-CoA carboxylase, *see* Acetyl-CoA carboxylase, plant tissues
— —, —, ageing phenomenon in storage tubers, effect of cycloheximide, 286
— —, —, cell-free systems, 278, 279
— —, —, comparative synthesis with spinach and *E. coli* ACP's synthetases, 281, 282
— —, —, composition of seed FA as a function of maturation of seed, 277
— —, —, conversion of FA, to derivative FA with same C skeleton by whole tissue or tissue slices, 278
— —, —, developmental aspects, 285, 286
— —, —, epoxy acid synthesis, 289
— —, —, hydroxylation, 288
— —, —, by intact tissue and by cell-free preparation of Avocado mesocarp and leaf tissue, (table), 278
— —, —, monoenoic acid formation, significance of anaerobic reaction, 286, 287
— —, —, photobiosynthesis, 282–285
— —, —, programming of appearance and disappearance of enzymes by regulatory systems, 285, 286
— —, —, regulatory systems, 285, 286
— —, —, soluble FA synthetase from plant extract, 278, 279
— —, —, studies with cell-free systems, 278–282
— —, —, — with ^{14}C-labelled compounds incubated with intact or sliced tissues, 276–278
— —, —, synthesis of long-chain FA, components required, 278
— —, —, unsaturation, aerobic pathway, 286–289
—, chain elongation and shortening, intracellular location, 331
— : CoA ligase (AMP), *see* Acyl-CoA synthetase

Fatty acid(s), (*continued*)
—, dietary cycling through liver, 80
—, —, direct uptake by liver by portal route, 80
—, distribution between 1 and 2 positions of glycerol in phospholipids and triglycerides, 12, 13, 118, 125, 138–140, 151, 152, 159, 163, 188, 244
—, epoxy acid synthesis, in plant tissues, 289
—, flux, through liver, in fasted state, scheme, 75, 76
—, —, —, in fed state on a high-carbohydrate, low-fat diet, scheme, 73, 78
—, free, *see* Free fatty acids
—, hydroxylation, in plant tissues, 288
—, intracellular metabolic pathways, location in cell, 331
—, long-chain, activation, extramitochondrial, 331
—, —, —, intramitochondrial, sites, 332
—, —, desaturation, intracellular location, 331
—, —, as energy source of intermediary metabolism, 331
—, —, esterification, intracellular location, 331
—, —, increased hepatic uptake and oxidation and inhibition of key enzymes of Glc oxidation, 350
—, —, inhibition of bacterial growth, 259, 260
— metabolism, plant seeds, oil-rich, conversion of FA to sucrose, role of glyoxylate cycle, 272
— —, plant tissues, 265–290
— —, —, modified β-oxidation, 273–275
— —, —, α-oxidation, 269–271
— —, —, β-oxidation, 271–273
— —, —, oxidative systems, 266–275
— —, role of carnitine in FA oxidation, *see* Carnitine
—, mobilization, from adipose tissue during exercise, effect on plasma triglyceride concentration, 54, 77
—, —, — during stress, effect on triglyceride and FFA concentration in plasma, 53, 54, 77
—, monounsaturated, formation in plant tissues, anaerobic pathway, 287
—, non-esterified, (NEFA), *see* Free fatty acids

Fatty acid(s), (*continued*)
—, oxidation, accelerated, effects on Glc metabolism in liver and in peripheral tissues, 353
—, —, —, and increase of intracellular levels of free Glc, 353
—, —, central role in acute and chronic regulation of gluconeogenesis, 348
—, —, decrease by succinate, 337
—, —, ketogenesis, 351
—, —, products, effects on gluconeogenesis and glycolysis, and chemical characteristics of key-regulatory reactions of Glc metabolism, 345
—, —, rate-limiting step, and acylcarnitine formation, 340
—, —, relation with gluconeogenesis, 342–350
—, —, spacial separation of enzymes for 331, 332
—, β-oxidation, intracellular location, 331–335, 340
—, patterns in triglycerides and phospholipids, alteration by reshuffling by transesterification, 125
—, photobiosynthesis, by chloroplast systems, 283–285
—, polyunsaturated, introduction of further double bonds in monoenoic acids in plant tissues, 287, 288
—, positional specificity of saturated and unsaturated FA in phosphatidic acid, 151
—, saturated, straight-chain, biosynthesis, differences with degradation, 233
—, —, —, —, enzymes involved, 235
—, —, —, —, essential steps, (scheme), 233, 234
—, short-chain, activation by enzyme systems beyond carnitine barrier, 333
—, —, — with succinyl-CoA, by thiophorase enzyme, 122
—, sucrose formation from, in plant seeds, 272
—, synthesized in liver, metabolic pathways, 73
— synthetase, chloroplasts, ACP requirement, 284, 285
— —, *E. coli*, activity with plant ACP, 282
— —, higher plants, dissociation of ACP, 282

Fatty acid synthetase, (*continued*)
— —, plant tissues, activity with *E. coli* ACP, 282
— —, soluble, from plant extracts, 278, 279
— thiokinase, *see* Acyl-CoA synthetase
— transport, plasma triglycerides in, *see* Triglycerides, fatty acids, plasma
—, unsaturated, formation in plant tissues, aerobic pathway, 286–288
—, —, long-chain, biotin replacement in microorganism nutrition, 259, 260
—, utilization by bacteria, 259
Fatty aldehydes, incorporation into plasmalogens and glyceryl ether phospholipids, 171
Fatty liver, in advanced liver cirrhosis, 81
—, by alcohol administration, and increased hepatic triglyceride formation, 81
—, in choline deficiency, 81
—, from impairment of protein synthesis by drugs and liver poisons, 148
—, man, by excess alcohol, 148
—, in protein malnutrition, 81
FFA, *see* Free fatty acids
Flax seed extracts, formation of 12-keto-13-hydroxy-*cis*-9-octadienoic acid from linoleic acid, 269
Free fatty acids, addition to perfusing medium, effect on gluconeogenesis and on Glc utilization in liver, 39
—, arterial plasma, uptake by resting muscle, effect of muscular contraction, 41
—, binding sites, in serum albumins, 35
—, blood plasma, concn., effect of fasting, 22
—, concn. in perfusing fluid, and rate of ketogenesis, in perfused livers, 39
—, — in plasma, decrease in carbohydrate feeding to fasting animals, 76
—, — —, — by Glc administration in fasting dogs, 38
—, — —, — by insulin in diabetic dogs, 38
—, — —, increase in diabetic dogs given growth hormone, 37
—, — —, — by fasting and diabetes, 37

Free fatty acids, (*continued*)
—, effect of concn. in medium on uptake and proportion incorporated by liver slices, 39
—, elevated plasma concns., inhibition of peripheral Glc uptake and oxidation, 353
—, enhancement of conversion of perfused Glc to cardiac glycogen in perfused rat heart, 353
—, Glc utilization, in tissues, improvement by lowering of level in plasma, 354
—, incorporation of saturated and unsaturated FA in hepatic neutral lipids and phospholipids, 38
—, increase of *in vitro* oxidation of Glc by adipose tissue, 353
—, inhibition of glucokinase, 345
—, injected into blood stream, esterification in liver, to a non-glyceride ester before triglyceride ester formation, 72
—, —, —, to triglycerides, and appearance in plasma VLDL, 72
—, —, incorporation into phospholipid in liver cell, 72
—, —, — into plasma lipoprotein phospholipids, 72
—, and ketone body concns. in plasma, concurrent increase in fasting and diabetic animals, 39
—, metabolism in muscle, diversion of FFA from oxidation to glyceride formation by Glc, 41
—, mobilization, 20–34
—, —, from adipose tissue, effect of fasting, 21–24
—, —, —, methods for investigation *in vivo* and *in vitro*, 21
—, —, in diabetes, effect of insulin, 24–29
—, —, effect of lipolytic hormones, *see* Hormones, lipolytic
—, —, — of pancreatectomy, 25–29
—, —, fasting animals, role of adrenal gland, 23, 24
—, —, —, — of glucocorticoid, 24
—, —, —, — of growth hormone, 24
—, —, —, — of pituitary gland, 23–29
—, —, —, — of sympathetic nervous system, 24

Free fatty acids, mobilization, (*continued*)
—, —, role of sympathetic nervous system, 29, 30
—, oxidation in isolated diaphragm, effect of Glc and insulin, 41
—, — in liver to CO_2 and ketone bodies, effect of feeding and starvation, 39
—, — — — —, increase in diabetes, 39
—, penetration into adipose tissue cell, and esterification at the cell membrane, 136, 137
—, plasma, depression related to infusion of ketones, *in vivo*, 354
—, —, fractional clearance by liver, 38
—, — pool, disappearance rates of various acids, 37
—, —, uptake by heart, effect of epinephrine, 42
—, —, — by kidney, clearance, 40
—, —, — by proximal convoluted renal tubules, involvement of organic acid, transport mechanism in tubules, 40
—, preferential incorporation of linoleate by liver into plasma cholesterol ester and phospholipid, 37
—, — utilization of palmitic acid by kidney, 37
—, pools, of liver, composition, 38
—, release, from adipose tissue and fat cells, requirement of serum albumin, 35
—, —, and glycerol release, incubated adipose tissue, effect of Glc and insulin, 24
—, —, —, to medium in perfused hearts, increase in diabetes, 42
—, —, and rate of blood flow through adipose tissue, 35
—, transport, 34–37
—, — in blood, distribution between albumin and lipoprotein, dependence on chain length of FA, 35
—, turnover in blood stream, 37
—, uptake from blood, crossing of capillary endothelium and interstitial space, 36
—, — —, by tissues, dependence on concn. of FFA and on molar ratio bound to albumin, 35, 36
—, — by heart, from arterial blood, effect of acetoacetate infusion, 42

Free fatty acids, uptake by heart, (*continued*)
—, — —, route from cell surface to intracellular structures, EM studies, 42
—, — in liver and lipoprotein secretion, effect of puromycin and orotic acid, 38, 39
—, — from medium in incubated adipose tissue from fasted rats, 137
—, — by slices of rat-kidney cortex, oxidation and esterification, 40
—, — and utilization, by rat diaphragm *in vitro*, 41
—, — —, by slices of blood vessels, effect of Glc, 43
—, utilization, 37–43
—, —, by heart muscle, 41, 42
—, —, increased, and diminished tricarboxylic cycle activity, 352
—, —, by kidney, 40
—, —, in liver, of individual —, 38
—, — by lung, 42, 43
—, — in minced brain, 43
—, — by skeletal muscle, 40, 41
Fructose, as carbohydrate in diet, effect on plasma triglyceride concn., 79
Fructose diphosphatase, allosteric inhibition by AMP, 345
—, hepatic, increase of activity by cortisone, 344
Fungal spores, *cis*-9,10-expoxyoctadecanoic acid, 289
Fungi, lipid metabolism, 240, 241

Galactocerebroside, *see* Cerebroside(s)
Galactolipid metabolism, in nervous system, 298–300
β-Galactosidase(s), brain, glycolipid substrates, specificity, 222, 233
—, with specificity toward terminal galactose of Gal(1,4)–Gal(1,4)Glc–Cer in brain, 222, 223
β-Galactoside permease, in bacterial lipids, 258, 261
Galactosylceramide, formation from UDPGal + ceramide by chicken embryonic brain homogenate fraction, 209

Galactosylsphingosine transferase, psychosine synthesis from sphingosine, 298
Ganglia, phosphoinositides, labelling by [^{32}P]phosphate, and synaptic transmitter on post-synaptic membrane, 322
Gangliosidase system, brain, 218
β-Gangliosidase, glucosylceramide hydrolysing in brain and spleen, decrease in Gaucher disease, 223
Gangliosides, adrenal medulla, 206
—, adult pattern in human brain, 301
—, biodegradation, 218–224
—, biosynthesis, 207–217, 301, 302
—, —, elongation of carbohydrate chain, 213, 214
—, — of glucosylceramide and lactosylceramide, 208–210
—, —, and incorporation of labelled precursors *in vivo*, 214
—, —, multiglycosyltransferase systems, 217, 224
—, —, —, intracellular localization, 224
—, brain, changes during early development, 206
—, —, — in sphingosine component with ageing, 208
—, —, content in grey and in white matter, 206
—, —, homogeneity of FA composition, 208
—, —, metabolic pathways, for biosynthesis, (scheme), 215–217
—, — tissue of vertebrates, content, (table), 205
—, —, turnover, half-life, 218
—, —, variations of pattern with age, 206
—, composition, 300
—, content, in human neural tissues, 295
—, definition, 201
—, degradation, intracellular localization of hydrolases, 224
—, excitability restoration of brain slices to electrical impulses, 323
—, extraction, 300, 301
—, FA composition, 300
— glycosidases, 221–224
— —, separation of the different enzymes in brain, 221, 222
—, in invertebrates, 206

Gangliosides, (continued)
—, localization in nervous tissues, 323
—, mammalian brain, generic terms and code systems, 202–204
— metabolism, 201–225
—, occurrence outside nervous system, 201
—, oligosaccharide units, or monosaccharide units, addition to ceramide in biosynthesis, 207
—, peripheral nerve, 206
—, sialic acid bound to terminal galactose, preferential release by neuraminidase, 219
—, spinal cord, 206
—, sugar moiety for characterization, 201
—, topographical distribution in nervous tissues, 206
—, turnover, 302
—, types, 300, 301
—, visceral, composition, 206
—, —, monosialosylgalactosylceramide, 204
Gangliosidosis, generalized, see Monosialosyl-N-tetraglycosylceramide
—, inherited, lack of β-galactosidase specific to lactosylceramide, 223
Gaucher disease, decrease of glucosylceramide hydrolyzing β-glucosidase, 223
Glucagon, activation of hepatic lipase, increase of local concns. of FA in liver, 352
—, effect on liver lipase activity, 73
—, — on triglyceride biosynthesis, in adipose tissue, 142
—, FFA release in isolated fat cells, 31, 32
—, increase in gluconeogenesis, in absence of de novo enzyme synthesis, 344
—, — of hepatic lipase activity, 73
—, stimulation of gluconeogenesis in perfused rat liver, 348
—, — of lipolysis, role of cyclic 3′,5′-AMP, 352
Glucocorticoid, in lipid mobilization, 24, 28, 31–33
—, role in FFA mobilization, in fasting animals, 24
Glucokinase, inhibition, by FFA, 345

Gluconeogenesis, acute acceleration, by formation of acetyl-CoA, ATP and NADH, as products of FA oxidation, 346
—, —, by inhibition of enzymes of glycolysis, by FFA, 346
—, acute increase, and depressant effect of FFA per se or long-chain FA oxidation products on Glc utilization, 350
—, acute regulation, relative importance of increased acetyl-CoA and NADH concn., 347
—, acute stimulation, summary of possible events, involved, 346
—, adrenaline in, 344
—, central role of FA oxidation in regulation, 348
—, changeover from glycolysis, 343, 350
—, definition, 342
—, effect of citrate, 347
—, — of fasting, 342
—, — of FFA addition, 39
—, enhanced, periods of, and increased concomitant rate of FA oxidation and activity of enzymes of β-oxidation, 343
—, enzymes involved, 342
—, factors influencing immediate acceleration, 344
—, in fasting rats, 348
—, hormonal regulation, acute changes, mediation by immediate regulation of relative availabilities of lipid and carbohydrate, 352
—, —, chronic effects, dependence on de novo protein synthesis, 352
—, increased, and increase of activity, of enzymes of β-oxidation, 343
—, —, —, of hepatic fructose-1,6-diphosphatase by administration of cortisone, 344
—, —, —, of hepatic glucose-6-phosphatase by administration of cortisone, 344
—, —, —, of hepatic pyruvate carboxylase by administration of cortisone, 344
—, —, in more reduced cytoplasmic environment, 347
—, —, in rabbit-liver and kidney slices, by carnitine, 344

Gluconeogenesis, (*continued*)
—, inhibition by (+)-acylcarnitines, 344
—, and intracellular acetyl-CoA concn., 348
—, in kidneys, 342
—, from lactate, increase in kidney slices by addition of FFA, 344
—, in liver preparations, increase by addition of FFA, 344
—, metabolic periods in which — is particularly important, 342
—, necessity of inhibition, of hepatic glycolytic enzymes for net gain in Glc production, 344
—, —, of pyruvic kinase for net gain in Glc production in liver, 345
—, pyruvate carboxylase in, 343
—, relation with FA oxidation, 342–350
—, stimulation by glucagon, 348
—, translocation of intramitochondrial formed oxaloacetate across mitochondrial membrane, 349
—, tricarboxylic acid cycle intermediates in, 346
Gluconeogenic enzymes, activities, increase by steroid therapy and chronic diabetes, inhibition by insulin, 344
— —, induction of *de novo* synthesis in alloxan diabetes, 344
Glucose, concn. in blood plasma, effect of fasting, 22
—, free, increase of intracellular levels, during periods of accelerated FA oxidation, 353
—, intolerance, and elevated plasma FFA concns., physiologic and pathologic states with, 353
—, metabolism, key-regulatory reactions, of glycolysis and gluconeogenesis, 345
—, peripheral uptake and oxidation, inhibition by elevated plasma FFA concns., 353
—, production, *see* Gluconeogenesis
—, —, net gain, 345
—, utilization in liver, effect of added FFA, 39
—, — in tissues, improvement by lowering of plasma FFA, 342
Glucose–FA interactions, 353, 354
Glucose-6-phosphatase, hepatic, increase of activity by cortisone, 344

Glucose-6-phosphate dehydrogenase, stimulation by insulin, 141
β-Glucosidase, decrease in Gaucher disease, 223
Glucosylceramide, biosynthesis, 208–210
—, formation from UDPGlc + ceramide by particulate fraction of chicken embryonic brain, 209
—, hydrolysis by β-gangliosidase, 223
Glycerides, *see also* Mono-, Di- *and* Triglycerides
—, mixed ester alkanyl, biosynthesis,138,140
Glycerol, distribution of FA, between 1 and 2 positions in phospholipids and triglycerides, 12, 13, 118, 125, 138–140, 151, 152, 159, 163, 188, 244
—, esterification of 2-position, in milk fat, 140
—, free, formation of α-glycerol phosphate in intestinal mucosal cell, 9
Glycerol kinase, activity, in adipose tissue, in obese individuals, 149, 150
—, —, in intestinal mucosa, 9, 10, 138
—, —, in liver, localization, 132
—, —, in tissues, 132
—, in triglyceride synthesis, 124, 132
Glycerol, metabolism, in liver homogenate, 132, 133
Glycerol phosphate, acylation, distribution of FA, 124, 125, 159
—, formation by ATP-mediated phosphorylation of glycerol, 159
—, — from DHAP, 159
—, — in intestinal mucosal cell, 9
—, — of phospholipids in bacteria, 249
—, pathway, for triglyceride biosynthesis, 9, 11, 12, 124, 159
—, requirement in triglyceride synthesis, in adipose tissue, 135
L-Glycerol 3-phosphate:CMP phosphatidyltransferase, in phosphatidylglycerol formation, 166
Glycerol, phosphorylation, ATP-mediated, 159
Glycerol production, fat pads of rats, effect of fasting, hypophysectomy, and adrenalectomy, (table), 23
—, and FFA release, incubated adipose tissue, effect of insulin, 24
Glycerol release, adipose tissue and lipolysis, 21, 23, 24

SUBJECT INDEX

Glycerophospholipids, intestinal absorption, 16
Glyceryl ether(s), free, degradation in animals, and *in vitro*, 183, 184
—, —, etherase reaction, rat-liver microsomes, O_2 and cofactor requirements, 183, 184
— phospholipids, content in total phospholipid of tissues, 169
— —, definition, 163
— —, desaturation to plasmalogens, 171
— —, ether bond cleavage, 183
— —, formation, 169–171
— —, — of the ether bond, 170
Glycolipids, bacterial, *see* Bacterial glycolipids
Glycolipid galactose transferase, in ganglioside biosynthesis, 301
N-Glycolylneuraminic acid, (NGN), 210
Glycolysis, changeover to gluconeogenesis, 343, 344, 350
—, inhibition by citrate, 353
Glycosidases, ganglioside —, 221–224
Glycosyl diglycerides, in bacteria, 250, 251
Glycosyl glycerides, in *Pneumococci*, biosynthesis, 251
Glycosylsulphate diglyceride, dihydrophytyl diether analogue, biosynthesis, 251, 252
Glycosyltransferases, in ganglioside biosynthesis, 217, 224
Glyoxylate cycle, coupling to β-oxidation systems, role of malate synthase and isocitrate lyase, 272
—, directly coupled to β-oxidation systems of FA in plant seeds, role in conversion of FA to sucrose, 272
Glyoxysomes, *R. communis* seeds, β-oxidation of FA, enzymes, 271–273
Gorlic acid, content in *P. sativum* seed lipid, 266
Gram-negative bacilli, lipid metabolism, 246, 250, 255, 257
Gram-positive bacteria, lipid metabolism, 254, 255, 257, 258
Growth hormone, + glucocorticoid, lipid mobilization, *in vivo*, lag time, 28, 29, 32, 33
—, —, lipolysis stimulation *in vitro*, effect of puromycin, cycloheximide, actinomycin D, and theophylline, 33

Growth hormone, (*continued*)
—, increase of FFA in plasma in diabetic dogs, 37
—, — in gluconeogenesis, in absence of *de novo* enzyme synthesis, 344
—, in lipid mobilization, 24, 26–28, 32, 33
—, role in FFA mobilization, in fasting animals, 24
Guanethidine, and FFA mobilization, 29
Guinea-pig brain, ganglioside metabolism, 214, 218

Halobacterium cutirubrum, lipid metabolism, 251, 252
HDL, *see* High-density lipoproteins
Heart, carnitine depletion, *in vitro*, decrease of FA oxidation, increase of triglyceride synthesis, 334, 341
—, heart–lung preparations, triglyceride FA uptake and utilization, 86
—, ischaemic — disease, *see* Ischaemic heart disease
—, isolated, perfused, triglyceride FA uptake and utilization, 86
—, muscle, acetoacetate oxidation, 351
—, —, reciprocal changes in clearing-factor lipase activity in —, and adipose tissue, 92
—, new-born rat, carnitine deficiency, and impaired rate of long-chain FA oxidation, 341
—, rat, perfused, decrease of Glc uptake and oxidation by addition of FA and ketone bodies, 353
—, —, —, increase of conversion of perfused Glc to glycogen by addition of FFA and ketone bodies, 353
—, substrate oxidation, shifting between carbohydrate and FA, 41, 42
—, triglyceride FA uptake, and clearing-factor lipase activity, correlation, 91, 92
—, — —, and utilization *in vivo*, effect of adrenaline infusion, 86
—, — —, — —, — of alcohol administration, 86
—, uptake and oxidation of plasma FFA, decrease by acetoacetate infusion in dogs, 42
—, utilization of FFA, 41, 42

Heparin-induced lipase, *see*
  Clearing-factor lipase
Heparin, injection, clearing of turbidity of plasma and of added chylomicron suspensions, 87
—, and phospholipase A_1 activity in plasma, 178
Hepatic lipase, *see* Lipase, hepatic
Hexamethonium, and FFA mobilization, 29
Hexokinase, allosteric inhibition by Glc 6-P and block in glycolysis during periods of accelerated FA oxidation, 353
High-density lipoprotein(s), human plasma, physical and chemical characteristics, 56, 57
—, inherited absence of normal formation of — and triglyceride transport in abnormal lipoproteins in plasma, 65
—, role in triglyceride hydrolysis, by clearing-factor lipase, 89
—, — — release from liver, 65, 67, 68, 72
Hormones, and hypertriglyceridaemia, 76, 77
—, imbalance, and changes in plasma triglyceride concns., 101, 102
—, induced adipose tissue lipolysis, increase of gluconeogenesis, 342
—, lipolytic, fast-acting, activation of adenyl cyclase, 32, 34
—, —, —, increase of action by methylxanthines, 32
—, —, —, — of cyclic AMP content of adipose tissue, 32, 34
—, —, —, role of cyclic AMP, 30–32
—, —, —, suppression of effects *in vitro*, by insulin, 31
—, —, promotion of hepatic FFA uptake and oxidation and increase in gluconeogenesis, in absence of *de novo* enzyme synthesis, 344
—, —, slow-acting, 32, 33
Human, brain, ganglioside metabolism, 205, 206, 208, 214, 219
Hydnocarpic acid, content in *H. wightiana* seed lipid, 266
Hydnocarpus wightiana, seed, FA composition, 266
Hydrocarbons, branched-chain, wax of tobacco leaf, formation, 290
—, waxes, formation, 289, 290

β-Hydroxybutyrate, decrease of peripheral FFA in pancreatectomized dogs, 354
—, inhibition of lipolysis of adipose tissue of pancreatectomized dogs *in vitro*, 354
13-Hydroxy-*cis*-9,*trans*-11-octadecadienoic acid, biosynthesis in plants, 268
Hydroxy fatty acids, association with bacterial surface and extracellular lipids, 246
β-Hydroxymyristic acid, in lipid A, 253
9-Hydroxy-*trans*-10,*cis*-12-octadecadienoic acid, biosynthesis in plants, 268
Hyperchylomicronaemia, familial, (Type I hyperlipoproteinaemia), 99
Hyperlipaemia, postprandial, inhibition by inhibitors of protein synthesis, 13
Hyperlipogenesis, obesity, 149, 150
Hyperlipoproteinaemia, type I, 99
Hypertriglyceridaemia, carbohydrate-induced, effect of insulin, 100
—, —, hepatic output of triglycerides, 100
—, caused by hormones, 76, 77
—, and clearing-factor lipase level in tissues, 90
—, in experimental obesity, 81
—, and hormones, 76, 77
—, inherited, tissue clearing-factor lipase activity and plasma triglyceride concns., 99
—, and ischaemic heart disease, 59
—, in man, on high-carbohydrate diet, 79
—, and pre-diabetic condition of carbohydrate utilization, 58
—, and VLDL, 58
Hypoglycaemic effects, related to infusion of ketones *in vivo*, 354
Hypoglycin, (L-α-amino-β-methylenecyclopropanepropionic acid), lowering of tissue levels of carnitine and CoA, 342
Hypophysectomized–pancreatectomized rats, effect of hormones on blood ketone bodies and lipid, and on liver triglyceride content, 23–29

Inositol phospholipids, *see* Phosphoinositides
Insulin, acetyl-CoA carboxylase stimulation, 141
—, action on lipolytic effects of fast-acting hormones *in vitro*, 31, 32

Insulin, (*continued*)
—, administration to diabetic rats, lowering of ketone body production and suppression of FFA mobilization to the liver, 39
—, antilipolytic action, and depression of cellular level of cyclic 3',5'-AMP, 352
—, citrate-cleavage enzyme stimulation, 141
—, concn. in blood plasma, effect of fasting, 22
—, decrease of FFA in plasma, in diabetic dogs, 38
—, deficiency, effect on liver lipase activity, 73
—, effect on carbohydrate-induced hypertriglyceridaemia, 100
—, — on clearing-factor lipase and on triglyceride mobilizing lipase, incubated adipose tissue, 96
—, — on FFA mobilization, in diabetes, 24–29
—, — on glycerol release, in adipose tissue, 24
—, Glc-6-P dehydrogenase stimulation, 141
—, increase of FA and triglyceride synthesis, mechanism, 141, 143
—, — of influx of NEFA into adipose tissue, 141
—, inhibition of increase in gluconeogenic enzyme activities during steroid therapy and chronic diabetes, 344
—, peripheral plasma values, effect of ketone infusion, 354
—, 6-phosphogluconate dehydrogenase stimulation, 141
—, stimulation of Glc-6-P dehydrogenase, 141
—, — of 6-phosphogluconate dehydrogenase, 141
—, suppression of effects of fast-acting lipolytic hormones, *in vitro*, 31
Intermicrovillous spaces, of intestinal mucosal cells, in fat absorption, 4
Intestinal absorption, of fats, *see* Fats, intestinal absorption
Intestinal flora, regulation by the gall bladder, 260
Intestinal mucosa, metabolic reactions with regard to fat absorption, 9–14
Ischaemic heart disease, and high dietary intake of sucrose, 79
—, increase in alimentary lipaemic response, 59, 70
—, and obesity in man, 81
—, plasma triglyceride concn., and post-heparin clearing-factor lipase activity, 100, 101
Isocitrate lyase, in glyoxysomes of oil seeds, 272, 273
Isoproterenol, effect on blood flow in perfused adipose tissue, 30

Kerasin, 296
β-Ketoacyl-ACP reductase, 235
β-Ketoacyl-ACP synthetase, 235
2-Keto-3-deoxyoctonoate, (KDO), attachment of lipid A to core polysaccharide of lipopolysaccharides, 254
Ketogenesis, FA oxidation, 351
—, hormonal regulation, acute changes, 352
—, —, chronic effects, dependence on *de novo* protein synthesis, 352
—, in liver slices, and perfused livers, relation to triglyceride content of liver, increase in diabetes, 39
α-Ketoglutarate, inhibition of organic acid transport mechanism in proximal convoluted tubules of kidney, 40
12-Keto-13-hydroxy-*cis*-9-octadecenoic acid, formation from linoleic acid by flax seed extracts, 269
Ketone body, concns., and FFA concns., in plasma, in fasting and in diabetic animals, 25, 26, 39
—, —, in plasma *in vivo*, relationship with rate of hepatic ketogenesis *in vitro*, 39
—, as normal fuels of respiration in animal tissues, importance as energy source, 351
Ketones, infusion, *in vivo*, decrease of hepatic Glc release and increase of peripheral Glc utilization, 354
—, —, —, depression of plasma FFA and hypoglycaemic effects, 354
—, —, —, effect on peripheral plasma insulin values, 354

Ketones, (*continued*)
—, oxidation, in fasted rats, 351
Ketosis, in diabetes, 351
—, extent of, and hepatic rate of FA oxidation, 351, 352
—, factors determining extent of, 351
—, in metabolic states with lipids as predominant energy-yielding substrate, 351
—, role of hepatic oxaloacetate levels in determining extent of, 352
—, role of hormones altering availabilities of carbohydrate and lipid in metabolism, 351
Kidney, FFA utilization, 40
—, organic acid transport mechanism, in proximal convoluted tubules, inhibition by α-ketoglutarate, probenecid and chlorothiazide, 40
—, palmitic acid uptake, 37, 40
Kinase, fatty acid, *see* Fatty acid kinase
Krebs cycle, *see* Tricarboxylic acid cycle
Kupffer cells, liver, uptake of triglycerides from artificial emulsions, 85
—, sequestration of injected chylomicrons, 85

Lactobacillic acid, carbon of propane ring, origin, 242
Lactonase, mevalonic, in brain, 303
Lactosylceramide, biosynthesis, 208–210
Lanosterol, in cholesterol synthesis, 305
LDL, *see* Low-density lipoproteins
Lecithin, *see* Phosphatidylcholine
Legumes, seeds, lipoxidases, 266
Lettuce leaves, FA metabolism, 278, 282, 284
Leucodystrophy, metachromatic, sulphatide content in brain, 297, 300
Lignoceric acid, 296
Linoleate, incorporation by liver into plasma cholesterol ester and phospholipid, 37
α-Linolenic acid, in photosensitizing tissues, 283
Linum usitatissimum, seed, FA composition, 266
Lipaemia, *see also* Hyperlipaemia *and* Hypertriglyceridaemia

Lipaemia, (*continued*)
—, of pregnancy, and clearing-factor lipase activity, 91
Lipaemic response, alimentary, *see* Alimentary lipaemic response
Lipase(s), bacterial, 259
—, clearing-factor, *see* Clearing-factor lipase
—, heparin-induced, *see* Clearing-factor lipase
—, hepatic, *see* Liver, lipase
—, lipoprotein —, *see* Clearing-factor lipase
—, pancreatic, *see* Pancreatic lipase
—, triglyceride mobilizing, adipose tissue, effect of ACTH, 96
—, —, — of adrenaline, 96
—, —, — of insulin, 96
Lipid(s), accumulation in liver, CCl$_4$ administration, 148
—, — —, in man, by alcohol excess, mechanism, 148
—, bacterial, *see* Bacterial lipids
—, brain, *see* Brain
—, containing C–P bond, formation, 174, 175
— droplets, in renal cortex, formation, 40
—, exogenous, added to cellular systems, methods for dispersion, 151
—, —, effects on bacterial growth, 259–261
—, intestinal absorption, 1–16
— metabolism, bacterial, (*see also* Bacterial), 229–262
— —, and blood–brain barrier, 315–317
— —, in nervous tissue, (*see also* Brain, Nervous tissue *and* Neural lipids), 293–324
— —, rate in brain mitochondria and liver mitochondria of non-growing tissues *in vivo*, 193
— —, — in growing tissues, 194
— —, — *in vivo*, 193–196
— —, role in membrane fabrication, 196
—, mobilization, and adrenal glands, 24, 26–28, 32, 33
—, —, glucocorticoid in, 24, 28, 31–33
—, neural, *see* Neural lipids
—, synthesis, in liver homogenates, effect of ATP, CTP and Mg^{2+} on triglyceride synthesis, and phospholipid synthesis, 145, 146

Lipid(s), synthesis, (*continued*)
—, —, in liver systems, control, 145–147
—, —, rat-liver homogenates, stability of enzymes for di- and triglyceride synthesis and for phospholipid synthesis, 147
—, —, use of labelled FA for quantitation, complications, 151
Lipid A, attachment to core polysaccharide of lipopolysaccharides, 254
—, biosynthesis, 254
—, cyclopropane FA in, and toxicity, 254
—, of *E. coli*, chemical composition, 253
—, —, mol. wt., 253
—, β-hydroxymyristic acid in, 253
—, and toxic effects of lipopolysaccharide, 253
Lipid GPX, synthesis, by rat-liver cell-free systems, 127–129
—, —, — homogenates and slices, effect of puromycin, 147
Lipoamino acids, bacteria, biosynthesis, 250
—, —, function, 250
Lipogenesis, in human adipose tissue, 136
—, liver, in experimental obesity, 81
Lipolysis, adipose tissue, glycerol release as measure, 21
—, —, hormonally induced, increase of gluconeogenesis, 342
—, in chylomicrons, 83
—, effect of adrenaline, 42, 352
—, stimulation by glucagon, 352
Lipolytic hormones, *see* Hormones, lipolytic
Lipopolysaccharides, bacterial, *see* Bacterial lipopolysaccharides
Lipoprotein(s), lipase, *see* Clearing-factor lipase
—, phospholipids, plasma, incorporation of FFA injected into blood stream, 72
—, plasma, *see also* High-density, Low-density, *and* Very-low density lipoproteins
—, —, apoprotein, effect on hepatic triglyceride release, 66, 67
—, —, classes in man, 52, 56
—, —, turnover times of protein and triglyceride constituents *in vivo*, 65, 66

Lipoprotein(s), (*continued*)
—, secretion, effect of puromycin and orotic acid, 38, 39
— triglyceride hydrolase, *see* Clearing-factor lipase
α-Lipoproteins, plasma, *see* High-density lipoproteins
β-Lipoproteins, plasma, *see* Low-density lipoproteins
Lipoproteinases, bacterial, 259
Lipoxidase(s), function in plants, 268, 269
—, plants, occurrence, 266, 267
—, soybean, mechanism, requirement of a *cis,cis*-1,4-pentadiene group in polyunsaturated FA, 266–268
—, —, mol. wt., substrates, inhibitors, 266, 267
Liver, cirrhosis, advanced, fatty liver in, 81
—, diseases, post-heparin plasma clearing-factor lipase activity, 101
—, fatty, *see* Fatty liver
— ketogenesis, *in vitro*, relationship with ketone body concns. in plasma *in vivo*, 39
—, lipase activity, effect of glucagon, and of insulin deficiency, 73
— lipogenesis, in experimental obesity, 81
— —, variations in extent in different development stages, 81
—, triglyceride output into blood, 71–81
—, —, function, 73
—, triglyceride pool, effect of hepatotoxic agents, 73
—, utilization of FFA, 38–40
Lobster, gangliosides in nerves and ganglia, 206
Long-chain acyl-CoA–carnitine acyltransferase, *see* Palmityl-CoA: carnitine palmityltransferase
Long-chain acyltransacylases, absence in higher plant tissues, 276
Low-density lipoprotein(s), *see also* β-Lipoproteins
—, formation in liver, linkage to triglyceride release, 63, 65, 67, 68, 72, 121
—, human plasma, physical and chemical characteristics, 56, 57
—, plasma, decrease by inhibitors of protein synthesis, 13

Lung, formation of dipalmitoyl phosphatidyl choline from plasma palmitate, 43
—, surfactant, function, 43
—, utilization of FFA, 42, 43
Lysocholine plasmalogens, acylation with acyltransferase from erythrocytes, 188
Lysolecithin acyl-hydrolase, temperature sensitivity, 178, 179
Lysolecithin, formation, of lecithin and glycerylphosphorylcholine by rat-liver and yeast microsomes supernatant, 190
—, —, in plasma, 189
Lysophosphatides, acylating enzymes, 188–191
—, exchange from their position in membranes, 192
Lysophosphatidic acid, synthesis, 160
Lysophosphatidyl moiety of phospholipids, intestinal absorption, 15, 16
Lysophospholipase, in microsomes and mitochondria, 178, 179
—, temperature sensitivity, 178, 179

Malate synthase, in glyoxysomes of oil seeds, 272, 273
Malic enzyme, level of, rise and fall, in adipose tissue, in feeding and starvation, 143, 144
Malonyl-CoA, inhibition of pyruvate carboxylase, 345
Malonyl pathway, FA synthesis, bacteria, ACP in, 235, 236
—, —, —, chain length of FA formed, 235
—, —, —, essential steps, 235
—, —, —, regulation of termination, 235
Malonyl transacylase, 235
Mammary gland, triglyceride FA uptake from blood during lactation, 87
Melanocyte-stimulating hormones, lipolytic effects, 31
Membrane components, phospholipids, cellular, assemblence, 196
Membrane fabrication, role of lipid metabolism, 196
Membrane-transport phenomena, role of phosphatidic acid cycle, 160, 181

Metabolic periods, with particular importance of gluconeogenesis, 342
Metachromatic leucodystrophy, sulphatide content in brain, 297, 300
Methionine, in branched-chain FA formation in bacteria, 245
—, in propane ring formation, in cyclopropane FA, 242–244
cis-9,10-Methylene hexadecanoic acid, in bacteria, 242
cis-11,12-Methylene octadecanoic acid, in bacteria, 242
Methyl fatty acids, branched, biosynthesis, methyl transfer from methionine, 245, 246
Methylmalonyl-CoA mutase in succinic acid pathway metabolism of propionic acid, 273
Methylxanthines, increase of activity of fast-acting lipolytic hormones, 32
Mevalonic lactonase, in brain, 303
Micrococcus lysodeikticus, lipid metabolism, 240, 241, 250, 251
— *phlei*, lipid metabolism, 245
— sp., lipid metabolism, 244, 250
Microvilli, of mucosal cells, in intestinal fat absorption, 4, 7
Milk fat, esterification of 2-position of glycerol with saturated FA, 140
—, glycerol positions esterified by various FA, 140
Milk triglycerides, formation from plasma triglycerides, 103
Mitochondrial compartment containing enzymes of β-oxidation, requirement of carnitine ester formation of FA for admission, 340
Mitochondrial membrane(s), inner, permeability to adenine nucleotides, 334
—, —, site of carnitine acetyltransferase, 339
—, —, — of palmityl-CoA:carnitine palmityltransferase, 337
Mitochondria, regulation of substrate oxidation, accessibility of substrate to compartmentalized cofactors or enzymes, 332
—, total respiration, interdependence of various oxidation processes, regulation factors, 332

SUBJECT INDEX

Monogalactosyldiglyceride, in chloroplasts, and photosynthesizing tissues, 282, 283
Monoganglioside, from brain, 300
Monoglyceride(s), acylation, intestinal mucosa, species differences, 121, 140
—, —, by monoglyceride transacylase, 2-monoglyceride as preferred substrate, 12
— pathway, for triglyceride synthesis, in intestinal mucosa, experimental proof, 11, 12
— —, —, in intestine, 137–140
— —, —, location of enzymes, 12
— transacylase, in triglyceride synthetase complex, 12
Monosialosylgalactosylceramide, chemical structure, 203
Monosialosyl-lactosylceramide, (G_{M3}), chemical structure, 202
—, formation, 211, 217
—, and sialosyltransferase C, 217
—, in visceral organs, 204
Monosialosyl-N-tetraglycosylceramide, (G_{M1}), biosynthesis, 215, 217
—, chemical structure, 202
—, enzymes required for complete hydrolysis, 221
—, formation from major brain gangliosides by brain neuraminidase, 221
—, lack of β-galactosidase hydrolyzing terminal Gal in generalized gangliosidosis, accumulation of —, 223
—, resistance to neuraminidase, 213
—, with sialic acid on terminal galactose, absence in normal brain, 217
Monosialosyl-N-triglycosylceramide, (G_{M2}), chemical structure, 202
—, increase in brain in inherited and exogenous diseases of the nervous system, 225
—, resistance to neuraminidase, 214
Multiglycosyltransferase systems, in ganglioside biosynthesis, 217, 224
Multiple sclerosis, demyelination, 324
Muscles, direct uptake and oxidation of chylomicron triglyceride FA for energy needs during exercise in fed state, 103

Muscles, (*continued*)
—, importance of plasma FFA for energy needs in fasting state, 102, 103
—, skeletal, FFA utilization, 40, 41
—, storage of esterified FA for energy needs, 102, 103
Muscular exercise, heavy, increase of gluconeogenesis, 342
—, increase of FA mobilization from adipose tissue, 54
Mycobacteria, lipid metabolism, 235, 240, 241, 245–247
Mycocerosic acids, tubercle bacilli, formation, 245–247
Mycolic acids, structure, occurrence, 247, 248
Myelin, composition in mammalian species, (table), 314
—, content in rat whole brain, 314
—, — in white matter of adult human and ox brain, 314
—, histological details, 312, 313
—, lipid composition, 314, 319
— lipid, phosphatidyl choline in, 314
— —, — ethanolamine in, 314
— —, — inositol in, 314
— —, — serine in, 314
— —, plasmalogen content in, 314
— —, rapidly metabolized fraction, location, 322
—, metabolic stability, 318, 320, 321
—, polyphosphoinositides, phosphate group exchange, 312, 321
—, protein content, 314
—, protein–lipid subunits, 315
— sheath, lipid metabolism, 320–322
—, triphosphoinositide in, 321
—, unit membrane structure, 314, 315
Myelination, critical period in man, 323
—, in developing brain, and sulphatide synthesis, 299, 300
—, distribution of triphosphoinositide, in nervous tissues, 166

NADH, conversion to NADPH, by citrate-cleaving enzyme and malic enzyme, 144
—, increase, and increase of acetyl-CoA concn., relative importance in acute regulation of gluconeogenesis, 347

385

NADPH, production, in control of FA and triglyceride synthesis, 143, 144
Naja naja, phospholipase A, substrate specificity, 177
NAN(A), see *N*-Acetylneuraminic acid
NEFA, (non-esterified fatty acids), see Free fatty acids
Nephrosis, plasma triglyceride concentration, 81
—, triglyceride output from liver, 81
Nervonic acid, 297
Nervous tissue, isolation of subcellular particles, 318
—, lipid metabolism, 293–324
—, —, at the cellular level, 317, 318
—, —, in relation to anatomical structure, 312–322
—, —, of subcellular structures, 318–320
—, lipogenic activity, relative incorporation of labelled substrate, (table), 317
—, role of phospholipids, in transmission processes, 322
Neural lipids, extraction from fresh tissue, 293–297
—, human, composition, comparison with liver lipids, (table), 295
—, mean mol. wts., (table), 294
Neuraminidase, activity, brain tissue, 218, 219
—, brain, degradation of major brain gangliosides, formation of monosialosyl-*N*-tetraglycosylceramide, 221
—, —, ganglioside specificity, 219–221
—, —, preferential action on sialic acid bound to terminal galactose, 219
—, —, purification, detergent requirement, 218, 219
—, —, similarity to *V. cholera* neuraminidase, 220
—, —, soluble and particle-bound, substrate specificities, 220, 221
—, resistance of monosialosyl-*N*-tetraglycosylceramide to, 213
—, — of monosialosyl-*N*-triglycosylceramide to, 214
—, soluble, from pig brain, preparation, 219, 220
—, —, —, rates of hydrolysis of different gangliosides, (table), 219
—, *V. cholera*, 220
Neurokeratin, 296

Neurone, histological details, 312, 313
Neurospora, lipid metabolism, 250
—, mutant strains, methylation of phosphatidylethanolamine to phosphatidylcholine, 168
NGN, see *N*-Glycolylneuraminic acid
Nicotinic acid, injection, effect on FFA uptake, and triglyceride output by liver, 80
Noradrenaline, effect on clearing-factor lipase, and on triglyceride-mobilizing lipase, incubated adipose tissue, 96
—, FFA mobilization, physiological significance, 32
—, lipolysis stimulation in adipose tissue, 29, 30
Norcarnitine, substrate for acetyltransferase, 340
—, — for palmityl-CoA:carnitine palmityltransferase, K_m, 338
Norepinephrine, see Noradrenaline

Obese-hyperglycaemic mice, metabolic alterations correlated with hyper-lipogenesis, 150
— syndrome, in rats, (metabolic obesity), 149, 150
Obese syndrome, aetiological factors, 149
Obesity, comparison of metabolic and regulatory obesity in rats, (table), 149
—, experimental, and increased rate of hepatic lipogenesis, 81
—, —, triglyceride in plasma and tissues, 81
—, Glc intolerance, and elevated plasma FFA, 353
—, hypothalamic, gold–thioglucose induced, (regulatory obesity), 149, 150
— in man, hyperinsulin, hyperthyroid state, 149
— —, and ischaemic heart disease, 81
— —, relation with high serum triglyceride concns., 81
—, metabolic, increase in liver phosphorylase, 149
—, —, in rats and mice, 149, 150
—, and satiety centre, in hypothalamus, 149, 150
Octopus, gangliosides in nerves and ganglia, 206

Oestrogens, effects on regulation of plasma triglyceride concns., and tissue clearing-factor lipase, 101
Oleic acid, and precursors, in bacteria, formation, 237, 238
—, uptake and incorporation into triglyceride, in cornea, 43
cis-Oleic series of FA, in bacteria, (table), 239
Oligodendroglia, histological details, 312, 313
Orotic acid, administration, effect on plasma triglyceride concn. and hepatic triglyceride content, 80, 81
—, effect on FFA uptake and esterification in liver, and on lipoprotein secretion, 39
—, — on triglyceride release and LDL formation in intestine, 68, 69
—, inhibition of hepatic triglyceride release and of LDL formation in liver, 63
Oxaloacetate, conversion to PEP by PEP carboxykinase, 348
—, hepatic levels in severely ketotic diabetic rats, 352
—, intramitochondrial concn., decrease by increase of gluconeogenesis rate, 352
—, —, and extent of ketosis, 352
—, translocation across mitochondrial membrane in gluconeogenesis, 349
α-Oxidation, FA, definition, 269
—, —, long-chain, (scheme), 270
—, —, pea leaves, O_2 requirement, formation of L-α-OH FA, 270
—, —, peanut, peroxide requirement, 269, 270
—, —, in plants, physiological significance, 270, 271
β-Oxidation, FA, circumvention of cis-9,10 double bond and D-12-OH function barriers in ricinoleic acid breakdown, 273
—, —, long-chain, intracellular location, 331–335, 340
—, —, modified, in plant tissues, 273–275
—, —, in plant preparations in vitro, 271, 272

β-Oxidation, FA, (continued)
—, —, plant tissues, direct coupling to glyoxylate cycle, 272
Oxidative systems, FA, plant tissues, 266–275
3-Oxodihydrosphingosine, catabolism to palmitic acid and ethanolamine, 185
—, in sphingosine biosynthesis, 172
Palmitate, free, uptake by kidney, 37, 40
—, supplementation of medium and gluconeogenesis in kidney slices, 40
Palmitylcarnitine, energy content, 338
(+)-Palmitylcarnitine, competitive inhibition of palmityltransferase in intact mitochondria, 338
Palmityl-CoA:carnitine palmityltransferase, (—)-carnitine analogues as alternative substrates, 338
—, effect of detergents, 338
—, incorporation of labelled carnitine into palmitylcarnitine, requirement for added CoA, 339
—, inhibition by γ-butyrobetaine, 338
—, — by thiol group reagents, 339
—, location, K_m values for substrates, 338
—, specific binding sites for palmityl group of palmitylcarnitine and palmityl-CoA, 338
—, specificity for longer-chain FA derivatives, 338
—, — for negative optical isomers of carnitine and palmitylcarnitine, 338
Palmityltransferase, see Palmityl-CoA: carnitine palmityltransferase
Pancreatectomized rats, adrenalectomized, effect of hormones on blood ketone bodies and lipid, and on liver triglyceride content, 25–29
—, fasting, effect on FFA mobilization, and lipolysis in fat cells in vitro, 25
—, hypophysectomized, effect of hormones on blood ketone bodies and lipid, and on liver triglyceride content, 25–29
Pancreatic diseases, post-heparin plasma clearing-factor lipase activity, 101
Pancreatic lipase, in intestinal absorption of fats, 3

Pancreatic lipase, (*continued*)
—, specificity for FA positions in triglycerides, 3
Pathologic states, characterized by Glc intolerance and elevated plasma FFA concns., 353
Pea leaves, α-oxidation of FA, 270
Peanut, FA metabolism, 269–271, 273, 275, 280
Pea seed, FA metabolism, 278–280, 282
Pentadecanoic acid, catabolism of phytosphingosine to, 185
4-Pentenoic acid, lowering of tissue levels of carnitine and CoA, 342
PEP, *see* Phosphoenolpyruvate
Peripheral nerve myelin, histological details, 312, 313
Petroselenic acid, content in *P. sativum* seed lipid, 266
Petroselinum sativum, seed, FA composition, 266
Pharmacological agents, affecting sympathetic nervous system, effects on FA mobilization, 29
Phosphatidate phosphohydrolase, *see* Phosphatidic acid phosphatase
Phosphatides, biosynthesis in nervous tissues, 309, 310
Phosphatidic acid, bacteria, formation of CDP diglyceride, 248, 249
—, —, — of phospholipids, 249, 250
—, biosynthesis, ACP in, 159
—, —, from diglyceride and ATP, 160
—, —, esterification of L-α-glycerophosphate by acyl-CoA, 159
—, —, *via* lysophosphatidic acid synthesis from monoglyceride and ATP, 160
—, —, *in vivo*, positional FA specificity of di- and triglycerides, and of lecithins, 151
—, brain, turnover, relation with Na transport, 322
—, cerebral, rapid ^{32}P incorporation *in vivo* in adult mice, 316
—, cycle, in membrane-transport phenomena, 160, 181
—, formation from DHAP in liver mitochondria, 160
—, natural, FA composition, 124
— phosphatase, in high-speed supernatant from liver and intestinal mucosa, 181

Phosphatidic acid phosphatase, (*continued*)
— —, in lecithin biosynthesis, 161
— —, localization in liver cells, 132
— —, in membrane-transport processes, 160, 181
— —, microsomal and supernatant, substrate specificity, 11
— —, in particulate cell fractions, 181
—, as precursor in phospholipid biosynthesis, 161–168
—, rat-liver, positional specificity of saturated and unsaturated FA, 151, 152
— —, in triglyceride synthesis, 124
Phosphatidyl choline, biosynthesis, 161–163, 169
— —, —, CDP-choline in, 162
— —, —, cholinephosphate cytidylyltransferase in, 162
— —, —, cholinephosphotransferase in, 162, 163
— —, —, methylation pathway in liver of higher animals, 169
— —, —, phosphatidic acid phosphatase in, 161
— —, —, phosphatidyl glycerol phosphate in, 129
— —, —, by rat-liver cell-free systems, effect of ATP and Mg^{2+}, 127, 128
— —, —, rat-liver slices, effect of puromycin, 147
— —, —, role of phosphorylcholine formation in, 161, 162
— — cholinephosphohydrolase, *see* Phospholipase C
— —, content, in human neural tissue, 295
— —, in myelin lipid, 314
— —, FA distribution between 1 and 2 position, and FA distribution in dietary fat, 13
— —, natural, structure, FA distribution, 161, 163
— —, occurrence in bacteria, 248
— — phosphatidohydrolase, *see* Phospholipase D
— —, turnover in brain, deacylation-acylation cycle, 310
— —, — of incorporated radioactivity, 320
— N,N-dimethylethanolamine, in bacteria, formation, 248, 249

Phosphatidyl (*continued*)
— ethanolamine, content, in human neural tissue, 295
— —, —, in myelin lipid, 314
— —, *de novo* biosynthesis, 163, 164
— —, FA distribution, 163
— —, formation by decarboxylation of phosphatidylserine, 164, 168, 310
— —, methylation of ethanolamine, to choline by S-adenosylmethionine, enzymes involved, 168
— —, mitochondrial, preferential hydrolysis by endogenous phospholipase, 177
— —, turnover of incorporated radioactivity, 320
— glycerol, in bacteria, formation, 249
— —, biosynthesis, 166
— —, in chloroplasts in photosynthesizing tissues, 282, 283
— — phosphatase, in phosphatidylglycerol formation, 166
— — phosphate, in lecithin and triglyceride synthesis, in rat-liver systems, 128
— inositol, *see* Phosphoinositide(s)
— N-monoethylethanolamine, in bacteria formation, 248, 249
— serine, in bacteria, 248, 249
— —, content, in human neural tissue, 295
— —, —, in myelin lipid, 314
— —, decarboxylation to phosphatidylethanolamine, 164, 168, 310
— —, formation by exchange, between L-serine and phosphatidylethanolamine, in animal tissues, 166, 169, 310
— —, — from L-serine and CDP-diglyceride in cell-free *E. coli* system, 166
— —, turnover of incorporated radioactivity, 320
Phosphodiesterase, cyclic nucleotide, degradation of cyclic AMP, 32
Phosphoenolpyruvate carboxykinase, activity, increase, 349
— —, conversion of oxaloacetate to PEP, 348
— —, in diabetes, 349

Phosphoenolpyruvate carboxykinase, (*continued*)
— —, extramitochondrial compartments of mammalian hepatic cells, 349
— —, in gluconeogenesis, 342
— —, inhibition by AMP, 345
Phosphofructokinase, activation, 344
—, inhibition by citrate and FFA, 345, 347, 353
6-Phosphogluconate dehydrogenase, stimulation by insulin, 141
Phosphoglyceraldehyde dehydrogenase, predominance of gluconeogenic pathway of reaction over glycolytic pathway with low cytoplasmic NAD/NADH ratio, 346
Phosphoglycerides, catabolism, summary outline, (fig.), 185
Phosphoinositide(s), *see also* Di-, Poly-, *and* Triphosphoinositides
—, in bacteria, 248, 249
—, biosynthesis, 164, 165
—, —, in brain, 165
—, —, from inositol and CDP diglyceride, 311, 312
—, brain, labelling by ^{32}P, increase by addition of acetylcholine in tissue preparations, 322
—, —, rapid incorporation of labelled precursors, increase by stimulation, 311
—, catabolism, summary, (fig.), 186
—, cerebral, rapid ^{32}P incorporation *in vivo* in adult mice, 316
—, cleavage of phosphate-glycerol ester linkage, 180, 181
—, content, in myelin lipid, 314
—, formation from triphosphoinositide, 182
—, ganglia, labelling by [^{32}P]phosphate, increase by electrical stimulation, 322
— inositolphosphohydrolase, 181
— kinase, of brain microsomes, formation of diphosphoinositide, 312
— —, location, 165
—, [^{32}P]phosphate incorporation, and synaptic transmitter on post-synaptic membrane, 322
— phosphomonoesterases, occurrence, 182
—, structure, biosynthesis, 164–166
—, turnover of incorporated radioactivity, 320

Phospholipase(s), activity, post-heparin plasma, 89
—, bacterial, 259
—, endogenous, preferential hydrolysis of mitochondrial phosphatidyl ethanolamine, 177
—, removing fatty acyl groups, 176–180
—, selectivity to fatty acyl groups, *in vivo* and *in vitro*, 189
—, — to nitrogenous base *in vivo* and *in vitro*, 189
—, substrate specificity, 176, 177
Phospholipases A, in animal tissues, heat stability, 177
—, definition, 177, 178
—, mol. wts., 177
—, pancreatic, substrate specificity, 177
—, snake venoms, substrate specificity, 177
—, synonyms, 176
— zymogen, porcine pancreas, trypsin activation, 177
Phospholipase A_1, location, 177, 178
—, in plasma, effect of heparin injection, 178
Phospholipase A_2, definition, 178
Phospholipase B, *see* Lysophospholipase
Phospholipase C, from bacterial toxins, substrates, 180
—, from *C. perfringens*, 259
— type enzyme, in animal tissues, 180, 181
Phospholipase D, (phosphatidyl-cholinephosphatidohydrolase), formation of phosphatidic acid from phosphatides, 182
Phospholipids, *see also* Phosphatidyl compounds
—, amphipathic character, and formation of micellar and bimolecular film structures, 309
—, assemblance into functional membranes, 196
—, bacterial, formation, 248–250
—, biological exchange, and adjusting of membrane composition, 158, 195
—, —, and renewing of structural elements, 158, 194, 195
—, biosynthesis, 158–175
—, —, summary outline, (fig.), 175
—, brain, changes in unsaturation at different environmental temperatures, 195

Phospholipids, brain, (*continued*)
—, —, metabolic stability, 317
—, —, microsome fraction, endogenous substrate for metabolism, in absence of exogenous Glc, 323
—, —, unsaturated FA, 308, 309
—, catabolism, 176–187
—, changes in FA composition, by acyltransferases, 188, 189
—, — —, with changes in amount of polyunsaturated FA in diet, 196
—, content, in chylomicrons, 60
—, —, in neural tissue, 295, 308
—, cyclopropane FA in, 244
—, differences in metabolism rate, in HeLa cells, 194
—, endoplasmic reticulum membrane of rat-liver cells, half-life time, recycling of FA, 194
—, ether, *see* Glyceryl ether phospholipids
—, exchange among erythrocytes and plasma lipoproteins, 191
—, — between mitochondria and microsomes, 192
—, — of FA in β-position, 192
—, — from their position in membranes, 192
—, — of intact molecules between intracellular organelles, 191, 192
—, — processes, 187–193
—, formation from phosphatidic acid, 249, 250
—, interconversions, 168, 169
—, — by direct base exchange, cation dependence, 169
—, intestinal absorption of intact 1-lysophosphatidyl moiety, 15, 16
—, —, complete hydrolysis by digestion, 16
—, liver, rate of metabolism, *in vivo*, 194
—, in lymph chylomicrons, 61, 68
—, in membranes, advantage of continuous replacement for organism, 195
—, —, differences in metabolism rate, 194, 195
—, metabolism, 157–196
—, native, degradation under physiological conditions, 186, 187
—, nervous tissues, fractionation, 309
—, neural, 308–312

Phospholipids, neural, (*continued*)
—, —, incorporation of [^{32}P]phosphate in young and adult rats, 316
—, —, labelled, turnover in brain, 316, 317
—, nitrogen bases exchange *in vivo*, 191, 192
—, patterns of metabolism rates in normal and in regenerating rat liver, 194
—, peroxidation of polyunsaturated FA and metabolic turnover, 195
—, rate of turnover of fatty acyl group in 2-position of phospholipid glycerol, 195
—, role in cyclopropane FA synthesis, 243, 244
—, specific mixture in each cell type, 157
—, turnover, and adaptation of poikilotherms to change in environmental temperature, 195
—, —, and altering of physical properties of membranes, 195
—, —, physiological significance, 194–196
—, unsaturation of FA and fluidity of membranes, 195
Phosphonates, in complex lipids, 174, 175
— lipid of *Tetrahymena pyriformis*, structure, formation, 174, 175
4′-Phosphopantetheine component, in plant ACP, 281
Phosphorus, white, administration, fatty liver production, 81
Phosphorylase, liver, increase in metabolic obesity, 149
Phosphorylcholine, formation, role in lecithin synthesis, 161, 162
Phosphorylethanolamine, formation, 164
Photobiosynthesis, FA, by chloroplast systems, 283–285
—, —, plant tissues, 282–285
Photosynthesizing tissues, absence of triglycerides in chloroplasts, 283
— —, acyl lipids, 282, 283
— —, α-linolenic acid concentration, 283
Photosynthetic bacteria, lipid metabolism, 250
Phrenosine, 296
Phthienoic acids, structure, formation, 246

Physiologic states, characterized by Glc intolerance and elevated plasma FFA concns., 353
Phytanic acid, α- and β-oxidation, in breakdown by mammalian tissues, 271
—, in triglycerides, in serum in Refsum's disease, separation from typical triglycerides, 148
Phytosphingosine, catabolism to pentadecanoic acid, and ethanolamine, 185
Picramnia lendiniana, seed, FA composition, 266
Pig, brain, ganglioside metabolism, 214, 218–221
Pinocytosis, role in intestinal fat absorption, 8
Pituitary gland, anterior, and lipid mobilization, 23–29
—, effect on FFA mobilization in fasting animals, 23–29
Placenta, gangliosides, 206
Plant seeds, oil-rich, conversion of FA to sucrose in germination, role of glyoxylate cycle, 272
Plant tissues, FA metabolism, 265–290
Plasma, chylomicrons, *see* Chylomicrons, plasma
—, clearing-factor lipase, activity, post-heparin, in fed and fasting states, 99
—, —, —, in liver diseases, 101
—, —, —, in pancreatic diseases, 101
—, effects of fasting on FFA and Glc concn., 22
— lecithin:cholesterol acyltransferase, formation of lysolecithin in plasma, 189
— β-lipoproteins, *see* Low-density lipoproteins
— triglycerides, *see* Triglycerides, plasma
Plasmalogens, biosynthesis in nervous tissues, 310
—, content, in myelin lipid, 314
—, definition, 163
—, ether bond cleavage, enzymes, 183
—, fatty aldehydes incorporation, 171
—, formation by desaturation of glyceryl ether phospholipids, 171

Plasmalogens, formation (continued)
—, — of ether bond, 170
—, in nervous tissues, 308
—, structure, 170
—, vinylic ether side-chain, rate of incorporation of precursors, 170
Pneumococci, lipid metabolism, 251
Poly-β-hydroxybutyrate granules, in bacteria, 256, 257
Polymyxins, and bacterial surface lipids, 258
Polyphosphoinositides, myelin, phosphate group exchange, 312, 321
Polysaccharides, lipo-, bacteria, see Bacterial lipopolysaccharides
Polysulphatides, occurrence, 297
Post-heparin plasma, clearing-factor lipase, esterase and phospholipase activity, 89
Postprandial hyperlipaemia, see Alimentary lipaemic response
Potassium transport, cellular, and diphosphoinositide, 165
Potato, FA metabolism, 280, 282, 286
Pregnancy, Glc intolerance and elevated plasma FFA, 354
—, transitory lipaemia, and clearing-factor lipase activity, 91, 99
Pre-β-lipoproteins, plasma, see Very-low-density lipoproteins
Probenecid, inhibition of organic acid transport mechanism in proximal convoluted tubules of kidney, 40
Prolactin, in lipid mobilization, 27
Pronethalol, and FFA mobilization, 29
Propanolol, and FFA mobilization, 29
Propionic acid, formation in plant tissues, 275
—, metabolism, in bacteria, differences with metabolism in plant systems, 274, 275
—, —, in higher plants, β-hydroxypropionate pathway, function, 275
—, —, succinic acid pathway, 273
—, oxidation, germinating peanut cotyledons, mitochondria, β-hydroxylpropionate pathway, 273, 274
Prostaglandins, effects on cyclic AMP concentrations in different cell types, 97

Prostaglandin E, suppression of lipolytic effects of fast-acting hormones in vitro, 31, 32
Protamine sulphate, inhibition of clearing-factor lipase, 90
Protein, chylomicron —, chemical composition, 13
—, of lymph chylomicrons, 68
—, malnutrition, fatty liver in, 81
—, synthesis, and chylomicron formation, 13, 14
Proteolipids, in nervous system, 296, 315
Protozoa, lipid metabolism, 240, 241
Pseudomonadales, lipid metabolism, 230, 244, 246, 248, 260
Pseudomonas aeruginosa, lipid metabolism, 246, 252, 257, 258
Psychosine, acylation to cerebrosides, 208, 299
—, synthesis from sphingosine, 298
Puromycin, administration, effect on plasma triglyceride concn., and hepatic triglyceride content, 80, 81
—, effect on clearing-factor lipase activity, 95
—, — on FFA uptake and esterification in liver, and on secretion of lipoprotein, 39
—, — on lecithin biosynthesis, in rat-liver slices, 147
—, — on lipid GPX synthesis, in rat-liver homogenates and slices, 147
—, — on lipolysis stimulation in adipose tissue, in vitro, by growth hormone + glucocorticoid, 33
—, — on triglyceride synthesis, in rat-liver homogenates, 146, 147
—, inhibition of triglyceride release from intestine, 68
Pyruvate carboxylase, activation by acetyl-CoA, 345–348
— —, activity, increase, 349
— —, — in livers of fasted rats, 348
— —, and change-over from glycolysis to gluconeogenesis, 343
— —, in diabetes, 349
— —, in gluconeogenesis, 343
— —, inhibition by malonyl-CoA, 345
— —, intramitochondrial localization, 348
—, decarboxylation, suppression by long-chain acylcarnitines, 336

Pyruvate (*continued*)
— dehydrogenase, allosteric inhibition by acetyl-CoA, 345
— —, inactivation by phosphorylation, reactivation by Mg^{2+}-dependent phosphatase, 345
— —, inhibition by acetyl-CoA, 346
— kinase, inhibition by FFA, NADH, ATP and GTP, 345

Rancidity, and microbial lipases, 259
Rat(s), brain, ganglioside metabolism, 205, 206, 214, 215, 218, 220, 221
—, fasted, ketone oxidation, 351
—, kidney, ganglioside metabolism, 212
—, spleen, ganglioside metabolism, 209, 210
Refsum's disease, phytanic acid containing triglycerides in serum, separation, 148
Renal cortex, acetoacetate oxidation, 351
Reserpine, and FFA mobilization, 29
Rhamnolipid, excretion by *P. aeruginosa*, 246, 256
—, in *P. aeruginosa*, composition, biosynthesis, 252
Rhizobiaceae, lipid metabolism, 248
Ricinoleic acid, metabolism, 265, 266, 271, 273–275, 285, 288
Ricinus communis, (castor bean), FA metabolism, 265, 266, 271–273, 276–278, 280, 282, 285, 286, 288
Rumen bacteria, lipid metabolism, 236, 244

Saccharomyces cerevisiae, lipid metabolism, 240
Safflower, FA metabolism, 280, 282, 287
Salicylate, injection, effect on FFA uptake and triglyceride output by liver, 80
Salmonella sp., lipid metabolism, 253, 254
Sarcina sp., lipid metabolism, 244
Satiety centre, in hypothalamus and obesity, control by Glc levels in blood, 149, 150
Seed lipids, FA composition, (table), 266
Sheep, brain, ganglioside metabolism, 214
Sialic acid, mammalian brain gangliosides, see *N*-Acetylneuraminic acid
Sialidase, see Neuraminidase

Sialosylglycosylceramides, see Di-, Mono-, and Trisialosylglycosylceramides
Sialosyltransferase(s), *see also* CMP–NAN: ganglioside sialosyltransferases
—, acceptors, 210, 211
—, definition, functions, 210, 211
—, in ganglioside biosynthesis, 210, 211, 217, 301
— C, and monosialosyl-lactosylceramide, 217
α-Smegmamycolic acid, *M. smegmatis*, structure, formation, 247, 248
Snake venoms, and phospholipid degradation, 176
Sodium transport, cellular, and diphosphoinositide, 165
Soybean lipoxidase, substrates, inhibitors, 267–269
Sphingolipids, catabolism, summary outline, (fig.), 187
—, phosphorus-containing, formation, 171–174
Sphingomyelin, analogue, with ethanolamine, in molluscs, 174
—, biosynthesis, 173
—, —, from ceramide derived from cerebrosides *in vivo*, 208
—, —, from *erythro*- and *threo*-sphingosine, 174, 311
—, catabolism, 179
—, content, in human neural tissues, 295
—, —, in myelin lipid, 314
—, formation of sphingosylphosphorylcholine from, 173, 174
—, hydrolysis to ceramide and phosphorylcholine by rat-liver mitochondria preparation, 180
Sphingosine, acylation by acyl-CoA to ceramide, 311
—, biosynthesis, dihydrosphingosine in, 171, 172, 298
—, —, from serine and palmitic aldehyde, 171, 172, 298
—, brain gangliosides, changes in ageing, 208
—, branched long-chain base analogues in protista, 172
—, catabolism to palmitic acid and ethanolamine, 185
—, *erythro*- and *threo*-, in sphingomyelin synthesis, 174, 311
—, structure, 171

Sphingosine, (*continued*)
—, synthesis of psychosine from, 298
Sphingosylphosphorylcholine, formation of sphingomyelin with fatty acyl-CoA, 173, 174
Spinach, FA metabolism, 278, 279, 281, 282, 284
Spleen, ganglioside metabolism, 206, 209, 210, 223
Staphylococcus aureus, lipid metabolism, 250, 259
Staphylococcus sp., lipid metabolism, 250
Starvation, *see* Fasting
Steroid(s), effects on hepatic gluconeogenic enzyme, activities, latent period, 344
— therapy, increase of pyruvate carboxylase and PEP carboxykinase activity, 349
Sterol(s), isolation from nervous tissue, 302
—, synthesis, in developing brain, rate-controlling steps, 303, 304
—, transformation by microorganisms, to corticosteroids and steroidal hormones, 260
Streptococcus sp., lipid metabolism, 250
Stress conditions, effect on triglyceride output by liver 77
Succinic acid, decrease of oxidation of FA, 337
— pathway, metabolism of propionic acid, 273
Succinyl-CoA, activation of short-chain FA, 122
—, inhibition of enzymes of FA β-oxidation, 337
Sucrose, formation, from FA in plant seeds, 272
—, high dietary intake, and ischaemic heart diease, 79
Sulphatides, biosynthesis, from cerebrosides and PAPS, catalysis, by microsomal system, 299
—, —, and myelination, in developing brain, 299, 300
—, content, in human neural tissue, 295
—, hydrolysis by arylsulphatase, 300
—, location of sulphate group, 297
—, metabolic stability, 300

Sulphatides, (*continued*)
—, in metachromatic leucodystrophy, 297, 300
—, methods for analysis, 298
—, phrenosine and kerasin type, 297
—, poly-, occurrence, 297
—, separation from cerebrosides, 297
— synthetase, activity, correlation with active myelination, 299, 300
Sulphoquinovosyldiglyceride, in chloroplasts in photosynthesizing tissues, 282, 283
Surfactant, lung, formation from plasma palmitate, 43
Sympathetic nervous system, acceleration of lipolysis in adipose tissue, role of norepinephrine, 29
— —, effect on lipolysis by altering of blood flow through adipose tissue, 30
— —, in lipid mobilization in uncontrolled diabetes, 29
— —, pharmacological agents affecting, effects on FFA mobilization, 29, 30
— —, role in FA mobilization, 29, 30
— —, — in FFA mobilization, in fasting animals, 24

Tarisic acid, content in *P. lendiniana* seed lipid, 266
Taurocholate, displacement of pH optimum of pancreatic lipase, 3
Tetracyclines, and bacterial surface lipids, 258
N-Tetraglycosylceramide, ganglioside formation, 211
Tetrahymena pyriformis, AEP, free and bound, 174, 175
— —, ether bond cleavage of glyceryl ethers, and glyceryl vinylic ethers, 183, 184
— —, phosphonate lipid, 174, 175
2,4,6,8-Tetramethyloctacosanoic acid, tubercle bacilli, formation, 245, 246
Tetrasialosyl-*N*-tetraglycosylceramide, (G$_{Q1}$), chemical structure, 203
Theophylline, effect on lipolysis stimulation in adipose tissue, *in vitro*, by growth hormone + glucocorticoid, 33
—, increase of action of lipolytic fast-acting hormones, 32
Thiokinase enzymes, FA activation, 122

Thiol group reagents, effect on palmityltransferase, 339
Thiophorase, and activation of C_4–C_6 FA by succinyl-CoA, 122
Thyroid hormones, effects on regulation of plasma triglyceride concns., and tissue clearing-factor lipases, 101
Thyroid-stimulating hormone, lipolytic effects, 27, 31
α-Toxin, of *C. perfringens*, phospholipase C, 259
Transport, active, and bacterial lipids, 257, 258, 261
— phenomena of membranes, and phosphatidic acid cycle, 160, 181
Tricarboxylic acid cycle, increase of activity, inhibition of oxidative decarboxylation of pyruvate and stimulation of its conversion to oxaloacetate, role of ATP, 347
—, intermediates, increased levels generated by long-chain FA oxidation, significance for gluconeogenesis, 346
Triglycerides, of adipose tissue, replenishment in fed animals, 55, 103
—, chylomicron, *see* Chylomicron triglycerides
—, content in human neural tissues, 295
—, digestion and absorption, 121
—, emulsification, in intestinal absorption of fats, 3, 121
—, endogenous, turnover rate, liver, 133
—, —, —, plasma, 133
—, entry into the blood, 69–81
—, ether analogues, biosynthesis, 138, 140
—, fatty acid(s), distribution, alteration by reshuffling by transesterification, 125
— —, —, differences with phospholipids, 124, 125
— —, plasma, fate in extrahepatic tissues, 102, 103
— —, —, quantitative significance for energy needs of extrahepatic tissues, 102
— —, —, removal by adipose tissue in fed and fasted animals, 86
— —, —, — from blood, by extrahepatic tissues, 82–84
— —, —, — by liver, by direct uptake of triglycerides, 84, 85

Triglyceride fatty acid(s), plasma, removal by liver, (*continued*)
— —, —, — —, after lipolysis in liver, 85
— —, —, significance for caloric needs in tissues in fasting, 53, 76, 77
— —, —, uptake by body musculature, effect of exercise, 86
— —, removal from blood, 81–102
— —, — —, absence in brain, 86, 87
— —, — —, role of clearing-factor lipase, 87–93
— —, transport, *see also* Chylomicrons, HDL, LDL *and* VLDL
— —, —, in the fasting state, 53, 75, 99
— —, —, in the fed state, influx of chylomicron triglycerides into circulation, 54, 78–80
— —, —, under conditions of enhanced FA mobilization, 53, 54, 76–78
— —, uptake by heart *in vivo*, 86
— —, — in mammary gland, from blood, 87
— —, — and utilization by perfused isolated heart and heart-lung preparations, 86
—, formation, from Glc, saturation of FA, 141
—, —, α-glycerophosphate pathway, 9
—, from intestine, entry into blood, rate, 70
— lipase, adipose tissue, enzymes for stepwise hydrolysis of triglyceride, 33, 34
— —, —, hormone-sensitive, activation by cyclic AMP, 34
— —, —, maximal rate of hydrolysis *in vivo* and *in vitro*, 34
—, in liver, in cytoplasmic droplets, storage pool, 73
—, —, effects of hormones on content of — in hypophysectomized and adrenalectomized–pancreatectomized rats, 26–29
—, —, entry into blood, 71–81
—, —, output of —, methods for measuring, 73, 74
—, lymph, distribution of FA between 1, 3 and 2 positions of glycerol, similarity to dietary triglycerides, 12

Triglycerides, (*continued*)
—, mobilizing lipase, *see* Lipase, triglyceride-mobilizing
—, number of different molecular species, 119, 120
—, oil-rich seeds, conversion of FA to sucrose during germination, role of glyoxylate cycle, 272
—, output by liver, *see also* Triglycerides, release from liver
—, —, after alcohol administration in animals, 81
—, —, caloric values, 76
—, —, under conditions of enhanced FA mobilization, 76–78
—, —, in diabetes, 77
—, —, effect of chloropropamide injection, 80
—, —, effect of nicotinic acid injection, 80
—, —, effect of salicylate injection, 80
—, —, effect of stress conditions, 77
—, —, in experimental obesity, 81
—, —, in fed state, 76, 78–80
—, —, in man, on high-carbohydrate diet, 79
—, —, in nephrosis, 81
—, phytanic acid containing, 148
—, plasma, *see also* Chylomicron triglycerides, VLDL triglycerides, *and* Hypertriglyceridaemia
—, —, concn., decrease on carbohydrate administration after starvation, causes, 99
—, —, —, — in carbohydrate feeding, to fasted animals, 76
—, —, —, — with heavy exercise in man, causes, 99
—, —, —, dependence on rate of influx from liver and removal by extrahepatic tissues, 53, 74, 97–102
—, —, —, in diabetes, 77, 100
—, —, —, in diseases due to hormone imbalance, 101, 102
—, —, —, effect of dietary intake of FA, 53, 70, 71
—, —, —, — of fructose as carbohydrate in diet, 79
—, —, —, — of heavy exercise, 77
—, —, —, in ischaemic heart disease, 100
—, —, —, in nephrosis, 81

Triglycerides, plasma, concn., (*continued*)
—, —, —, in pregnancy, 55, 99
—, —, —, in stress conditions, 77
—, —, —, tissue clearing-factor lipase activity as determinant, 97–102
—, —, and formation of milk triglycerides, 103
—, —, function in FA transport, 51–104
—, —, hydrolysis in extrahepatic tissues, recirculation of FFA to liver, 85
—, —, release from intestinal and liver cells, mechanism, 52, 69–81
—, release from intestine, in animals on fat-free diet, 69
—, —, inhibition by ethionine and puromycin administration, 68
—, —, similarity in sequence of events, with triglyceride release from liver, 68
—, release from liver, *see also* Triglycerides, output by liver
—, —, dependence on formation and release of HDL and LDL, 62
—, —, effect of CCl$_4$ administration, 66
—, —, — of plasma proteins in medium, 64
—, —, EM studies, 63, 64
—, —, inhibition, of — and of LDL formation, in liver, by orotic acid administration, 63
—, —, —, by puromycin, 63, 64
—, —, linkage to LDL protein formation in liver, 63, 65, 67, 68, 72
—, —, and lipoprotein–phospholipid formation, 67
—, —, microsomes, and slices, effect of plasma proteins in medium, 64
—, —, and plasma lipoprotein "apoprotein", 66, 67
—, —, role of HDL, 65
—, —, steps involved, 64
—, —, *in vitro* experiments with isolated perfused rat liver and rat-liver slices, 62–64
—, separation methods, 120
—, synthesis, activity of liver mitochondria and microsomes, 123, 130–132
—, —, adipose tissue, *see* Adipose tissue
—, —, control 140–148

Triglycerides, synthesis, (*continued*)
—, —, effect of ACTH, 142, 143
—, —, from excess dietary carbohydrate, 140, 141
—, —, and feed back control, of FA synthesis, by FA and fatty acyl-CoA esters, 144
—, —, —, role of rate of formation of citrate, 144
—, —, glycerol kinase in, 124, 132
—, —, glycerol-phosphate pathway, and asymmetric pattern of FA distribution, 139
—, —, —, location of enzymes, 11, 12
—, —, —, reactions, 124
—, —, hormonal control, 141–144
—, —, increase by insulin, 141
—, —, after ingestion of large amounts of carbohydrate and/or lipid, and *de novo* synthesis of enzymes, 145
—, —, intestinal mucosa, diglycerides from glycerol-phosphate pathway, and from monoglyceride path, location in cells, 139
—, —, —, glycerol phosphate pathway, 9, 137–140
—, —, —, monoglyceride acylation, species differences, 140
—, —, —, monoglyceride pathway, 11, 12 137–140
—, —, —, random acylation of monoglycerides with different acyl-CoA esters, 138, 139
—, —, —, specificity of acylating enzymes converting monoglycerides to triglycerides, 138
—, —, —, synthesis of individual molecular species, 138
—, —, isolated fat cells, effects of hormones, 142, 143
—, —, labelled glycerol for assessing of net —, 145–148, 151
—, —, in liver, 123–134
—, —, liver cells, cellular localization, role of microsomes and mitochondria, 130–132
—, —, —, microsomal, dependence on added albumin or serum lipoproteins, 131
—, —, mammalian tissues, 117–152

Triglycerides, synthesis, (*continued*)
—, —, measuring based on net glycerol utilization, 142
—, —, NADPH production in control, 143, 144
—, —, *de novo* synthesis of enzymes for, after ingestion of large amounts of carbohydrates and/or lipids, 145
—, —, phosphatidic acid phosphatase in, 124
—, —, rat liver, effect of puromycin, 146, 147
—, —, by rat-liver cell-free systems, effect of ATP and Mg^{2+}, 125–128
—, —, rate of, with triglycerides of different FA pattern, 130
— synthetase complex, diglyceride transacylase in, 12
— — —, monoglyceride transacylase in, 12
— —, and membrane portion of the endoplasmic reticulum, intestinal mucosa cells, 12
γ-Trimethylammonium-β-hydroxybutyrate, see Carnitine
Triphosphoinositide, distribution in nervous tissues, relation with myelination, 166
—, formation in brain, 312
—, in myelin, phosphate exchange, 321
—, phosphatidyl inositol formation from, 182
— phosphomonoesterase, brain, 182
—, structure, biosynthesis, 165
—, turnover of phosphate groups, 166
Trisialosyl-*N*-tetraglycosylceramide, (G_{T1a}, G_{T1b}), biosynthesis, 215, 217
—, chemical structures, 203
Triton WR 1339, inhibition of clearing-factor lipase activity, 90
—, prevention of triglyceride removal from plasma, 74, 75
Tubercle bacilli, lipid metabolism, 245, 246
Tuberculostearic acid, (10-methylstearic acid), biosynthesis in *M. phlei*, 245

UDP–*N*-acetylgalactosamine:monosialosyllactosylceramide *N*-acetylgalactosaminosyltransferase, chicken embryonic brain, substrates, reactions, 213

UDP–galactose:ceramide galactosyltransferase activity, chicken embryonic brain homogenate, synaptosome fraction, 209
UDP–galactose:glucosylceramide galactosyltransferase, chicken embryonic brain homogenate, 209
—, rat-spleen homogenate, 209
UDP–galactose:monosialosyl-N-triglycosylceramide galactosyltransferase, activity variation with age, 214
—, chicken brain, substrates, products, 214
UDP–glucose:ceramide glucosyltransferase, activity, chicken embryonic brain homogenates, synaptosome fraction, detergent requirement, heat lability, 209

cis-Vaccenic acid, and precursors, in bacteria, 237
cis-Vaccenic series of FA, in bacteria, (table), 239
Vasopressin, lipolytic effects, 31
Verononia anthelmenties, seed, FA composition, 266
Very-low-density lipoproteins, and complexes derived from —, by triglyceride FA removal from plasma, 61, 62
—, components, size, density, 52, 57, 58
—, composition in the fasting state or on fat-poor diets, 57, 58
—, electrophoresis, 58
—, HDL and LDL components, 57, 58
—, penetration into space of Disse through fenestrated endothelial lining of blood capillaries, in liver, 84

Very-low-density lipoproteins, (*continued*)
—, rate of removal of triglyceride FA from blood in man and animals by extrahepatic tissues, 82
—, resemblance and differences with chylomicrons, 52
—, separation from HDL and LDL, 58
—, sequestration at capillary endothelial surface and association with clearing-factor lipase, 53, 90
—, site of formation, 62
—, sub groups, 57, 58
—, triglyceride, concn., in plasma, in man, 58
—, —, —, in uncontrolled diabetes, 59
—, —, from liver, release into plasma, 52, 71–81
—, — load, transport from liver to extrahepatic tissues, 57
—, —, uptake of glycerides in liver without hydrolysis, 84
Vibrio cholera, neuraminidase, similarity to brain neuraminidase, 220
VLDL, *see* Very-low density lipoproteins

Waxes, *Brassica oleracea*, elongation-decarboxylation pathway for C_{29} straight-chain hydrocarbon, 289, 290
—, plant, composition, 289
—, *Senecio odoris* leaves, formation by epidermal layers of paraffins and very-long-chain FA, 290
—, tobacco leaf, branched-chain hydrocarbons, formation, 290
Wheat, FA metabolism, 280, 284

Yeasts, lipid metabolism, 240, 241

Zymosterol, ($\Delta^{8,24}$-Cholestadienol), in developing brain, 302–304